ANNALS OF THE NEW YORK ACADEMY OF SCIENCES
Volume 926

MECHANISMS OF CELL DEATH II

THE THIRD ANNUAL CONFERENCE OF THE INTERNATIONAL CELL DEATH SOCIETY

Edited by Zahra Zakeri, Richard A. Lockshin,
and Carlos Martinez-A.

The New York Academy of Sciences
New York, New York
2000

Library of Congress Cataloging-in-Publication Data

International Cell Death Society. Conference (3rd : 2000 : Escorial, Spain)
Mechanisms of cell death II : the Third Annual Conference of the International Cell Death Society / edited by Zahra Zakeri, Richard A. Lockshin, and Carlos Martinez-A.
p. cm. — (Annals of the New York Academy of Sciences ; v. 926)
Includes bibliographical references and index.
ISBN 1-57331-320-3 (alk. paper). — ISBN 1-57331-321-1 (alk. paper)
1. Cell death—Congresses. 2. Apoptosis—Congresses. I. Zakeri, Zahra, 1952– II. Lockshin, R. A. (Richard A.) III. Martinez-A., C. (Carlos) IV. Title. V. Series.
Q11 .N5 vol. 926
[QH671]
500 s—dc21
[571.9'39]

00-048984
CIP

K-M Research/CCP
Printed in the United States of America
ISBN 1-57331-320-3 (cloth)
ISBN 1-57331-321-1 (paper)
ISSN 0077-8923

ANNALS OF THE NEW YORK ACADEMY OF SCIENCES

Volume 926

December 2000

MECHANISMS OF CELL DEATH II

THE THIRD ANNUAL CONFERENCE OF THE INTERNATIONAL CELL DEATH SOCIETY[a]

Editors and Conference Organizers

ZAHRA ZAKERI, RICHARD A. LOCKSHIN, AND CARLOS MARTINEZ-A.

CONTENTS

[a]Papers in this volume were presented at the **Third International Conference on Mechanisms of Cell Death**, held May 6–10, 2000 in Escorial, Spain.

Part II. Signals

Part III. Intracellular Changes

Part IV. Proteolysis

Part V. Phagocytosis

Financial assistance was received from:

- BD-Biosci
- Biocolor
- Biomol
- European Commission
- Intergen
- Merck
- National Institutes of Health (1 R13 AG 182207-01)
- Nature
- New York Academy of Sciences
- Pharmacia
- Queens College of the City University of New York
- R&D Systems
- St. John's University

Preface: Mechanisms of Cell Death 2000

This book derives from a meeting of The International Cell Death Society held in May, 2000. The theme of the meeting was perhaps best expressed by a quotation used to open the meeting, "Truth does not become more true by virtue of the fact that the entire world agrees with it, nor less so even if the whole world disagrees with it" (Maimonides). Many of the speakers either challenged ideas that recently dominated the field, or elaborated and nuanced ideas that previously were considered to be straightforward. Many theories are rapidly evolving, and the work of the contributors to this volume reflects the most current and provocative ideas regarding these theories.

The biology of apoptosis has become a more complex issue, or rather, the importance of cell biology has been rediscovered. This topic is addressed by several authors. Both Damjanovski *et al.*, looking at organ culture in tadpole metamorphosis, and Chiarini and Linden, studying development of embryonic retina explants, emphasize the importance of cell-cell relationships and extracellular matrix. Bursch *et al.* argue that autophagic death is distinct from classical apoptosis, reflecting not a mutually exclusive dichotomy but the flexibility of cells in accommodating different pressures and capabilities. Goldstein, Kluck, and Green emphasize that, because of differences in timing of the separate stages of apoptosis, study of the kinetics of a single cell gives a very different, and presumably more accurate, interpretation of sequence than the usual studies based on measurements of populations of cells. In a similar vein, Joseph, Lewensohn, and Zhivotovsky report the complexity of early and simplistic analyses. In tumors, in which the inefficiency of apoptosis is a concern, caspase 8 is inversely correlated with radiosensitivity, but it cannot be shown to be rate-limiting. Quaglino *et al.* find many necrotic, as opposed to apoptotic, cells in older thymuses. Bouillet *et al.*, studying Bim, and De Laurenzi and Melino, studying proteins related to p53, both identify important roles for these molecules in embryogenesis, which may be unrelated to their functions in apoptosis. Bouillet *et al.* further stress the importance of compartmentalization of molecules in carrying out their functions.

Signaling for cell death likewise is an issue whose complexity is only now being appreciated. Brás *et al.* point out that different mechanisms regulate apoptosis at each step and age of differentiating thymocytes. Thus, cell death is one of the mechanisms balancing between autoimmune disease, good protection, and immunological failure. Abe *et al.* make a similar point, noting that TRAIL interacts with at least five receptors and acts in a selective manner. Likewise, Goldstein, Kluck, and Green, who find that variance in time to onset of apoptosis does not correlate with variance in progress through apoptosis, conclude that the decision to die depends on the physiological state of the cell when challenged. This argument is consistent with observations of HIV (discussed below). Pinti Cossarizza *et al.*, who report a novel quantitative-competitive RT-PCR to determine mRNA production, find a complex balance between host-driven pro-apoptotic and virus-driven anti-apoptotic processes. Jouvet, Kozma, and Mehmet identify branched-chain amino acids and α-keto derivatives as inducers of apoptosis, and Mussche, Leybaert, and D'Herde describe how cells can sense extracellular [Ca^{++}] and incorporate it into the decision to undergo apoptosis.

Finally, heat shock proteins, which were early assumed to be important in apoptosis but whose role was poorly understood, have come back into prominence. As Charette and Landry report, for HSP27, and Nylansted, Brand, and Jäättelä, for HSP70, these proteins are effective protectors against apoptosis. Expression of HSP70 appears to be specifically required to protect tumorigenic cells.

Mitochondria remain central in many interpretations of the initiation of apoptosis, but, like most major ideas, this year saw the development of several nuances that were previously unappreciated. HIV variants conserve several proteins leading to mitochondrial permeability transition. As Ferri *et al.* point out, HIV employs multiple strategies to regulate host cell apoptosis by targeting mitochondria. Virus strategies are more normally anti-apoptotic. Gougeon *et al.* argue that the situation is reversed in AIDS because the infection leads to an unbalanced immune activation, which results in loss of cells that would otherwise make protective cytokines. Benítez-Bribiesca *et al.* find that even in traumatic situations such as ischemia-reperfusion the earliest ultrastructural changes occur in mitochondria and parallel loss of bcl-2. Florek and Ucker, using fluors to mark both cytochrome c and the MMP, find that mitochondrial disintegration does not follow a strictly ordered process and is not concerted.

One of the fastest growing areas of concern is proteolysis, particularly as more and more caspase-independent deaths are noted. In addition to the role of metaloproteinases and stromelysins found in metamorphosis by Damjanovski *et al.* (and in prostate involution by Guenette *et al.*),[1] Torriglia *et al.*, observing differentiation of lens, find that one protein can play two separate roles: proteolytic cleavage of one inhibitor both activates a serine protease and converts the former inhibitor into a DNase.

Finally, the role of phagocytosis is being taken more seriously as common mechanisms are identified. Following the argument of Bursch *et al.*, that cells that cannot be phagocytosed generate autophagic mechanisms of self-destruction, and Liu and Hengartner's report[2] that the mechanisms of phagocytosis are highly conserved, Schlegel *et al.* report that PS asymmetry is sufficient to activate phagocytosis, and it occurs by two independent mechanisms, inactivation of translocase and activation of scramblase. This suggests that some cells may intentionally not expose PS. Once they do expose the PS, apoptotic bodies are opsonized to guarantee their uptake, and Mevorach points out that disease may result from failure of opsonization.

In brief, it has been an exciting period since the publication of our last volume. We attempted in the meeting to call attention to these several topics, and in this publication to bring to a wider audience the highlights of the meeting. Our contributors have been most cooperative in delivering manuscripts promptly, so that you as readers can benefit from a most timely presentation of the most provocative currents in this very important field. We have edited the work (sometimes to the annoyance of the contributors) to maintain a consistent and, we hope, lucid, editorial style. We hope that the several papers presented here will challenge you re-evaluate your own findings, analyze the relationship between the molecular and the cell biology, and to further enrich this demanding topic. In the words of another great Spanish scientist (Santiago Ramón y Cajal), "...some advice about what should be known...and about the carelessness and inclination toward bias that must be avoided, is far more useful than all the rules and warnings of theoretical logic."

Since the topics interweave quite heavily, the order of chapters is divided into five sections: The Biology of Apoptosis, Signals, Intercellular Changes, Proteolysis, and Phagocytosis. As is suggested by the discussion above, we encourage the reader to contemplate these offerings as individual contributions, in the context of each section, and in the correlative fashion described in this preface.

REFERENCES

1. GUENETTE, R.S., M. MOOIBROEK, K. WONG, P. WONG & M. TENNISWOOD. 1994. Cathepsin B, a cysteine protease implicated in metastatic progression, is also expressed during regression of the rat prostate and mammary glands. Eur. J. Biochem. **226:** 311–321.
2. LIU, Q.A. & M.O. HENGARTNER. 1998. Candidate adaptor protein CED-6 promotes the engulfment of apoptotic cells in *C. elegans.* Cell **93:** 961–972.

ACKNOWLEDGMENTS

This book derives from a meeting sponsored by The National Institutes of Health (1 R13 AG182207-01), The European Commission, Queens College of the City University of New York, St. John's University, The New York Academy of Sciences, and the following corporate sponsors: BD-Biosci, Biocolor, Biomol, Intergen, Merck, Nature, Pharmacia, and R&D Systems. It was prepared while the organizers were supported by the following other grants: NIH R15GM/AG57614 (R.A.L.). The Department of Immunology and Oncology of Universidad Autónoma, Campus de Cantoblanco, Madrid was founded and is supported by the Spanish National Research Council (CSIC) and by the Pharmacia Corporation. Finally, we wish to thank the following for their help and service during the meeting and in the preparation of this book (in alphabetical order): (Queens College of C.U.N.Y.) Claudette Davis, Nidhi Gadura, Terri Kagan, Ho Kyung Lee, Lin Lin, Demetrius Matassov; (St. John's University) Carolyn Facey, Farhan Khan, Javier Negrón, Stephanie Rice, Carlos Romero. We especially thank Catherine Mark for her indefatigable and patient efforts.

RICHARD A. LOCKSHIN
ZAHRA ZAKERI
CARLOS MARTINEZ-A.
Madrid, July 2000

Programmed Cell Death (PCD)

Apoptosis, Autophagic PCD, or Others?

W. BURSCH,[a,b] A. ELLINGER,[c] CH. GERNER,[a] U. FRÖHWEIN,[a] AND R. SCHULTE-HERMANN[a]

[a]*Institut für Krebsforschung der Universität Wien, Borschkegasse 8a, A-1090 Wien, Austria*

[c]*Histologisch-Embryologisches Institut der Universität Wien, Schwarzspanierstrasse 17, A-1090 Wien, Austria*

ABSTRACT: The occurrence of cell death as a physiological event in multicellular organisms has been known for more than 150 years; in 1972 the term apoptosis was introduced on morphological grounds. However, accumulating evidence suggests that programmed cell death (PCD) is not confined to apoptosis, but that cells use different pathways for active self-destruction as reflected by different morphology: condensation prominent, type I or apoptosis; autophagy prominent, type II; etc. Autophagic PCD appears to be a phylogenetically old phenomenon; it may occur in physiological and disease states. We have studied the relation between morphological and biochemical events during autophagic and apoptotic PCD in human mammary, lymphoblast, and colon cancer cells using electron microscopy and proteom analysis. We find that autophagic cell death (type II) PCD includes degradation of Golgi apparatus, polyribosomes, and endoplasmic reticulum, which precedes nuclear destruction. Intermediate and microfilaments are largely preserved; presumably the cytoskeleton is required for autophagocytosis. Apoptosis (type I) PCD is characterized by condensation of cytoplasm and preservation of organelles; cytoskeletal elements disintegrate in early stages. Either type of PCD involves synthesis of distinct proteins. Finally, both types of PCD share features some of a cell's stress response (e.g., translocation of hsp90). In conclusion our findings support the concept that autophagic cell death is a separate pathway of PCD distinctly different from "classical" apoptosis. However, autophagic and apoptotic PCD should not be considered as mutually exclusive phenomena. Rather, they appear to reflect a high degree of flexibility in a cell's response to changes of environmental conditions, both physiological or pathological.

KEYWORDS: MCF-7; HT29/HI1; Jurkat cells; tamoxifen; tyrphostin A25; CD95-antibody; autophagic cell death; apoptosis; cytoskeleton; monodansylcadaverine

[b]Address for correspondence: Dr. W. Bursch, Institut für Krebsforschung der Universität Wien, Borschkegasse 8a, A-1090 Wien, Austria. Voice: ++43-1-4277 651 39; fax: ++43-1-4277 9651.

wilfried.bursch@univie.ac.at

CONCEPTS OF CELL DEATH

The occurrence of cell death under a variety of physiological and pathological conditions in multicellular organisms has been documented many times during the past 150 years (ref. 1; for review see refs. 2,3). In 1871, Virchow[4] described the diversity of cell death as "necrosis" and "necrobiosis." Subsequently, cell death was reported to occur during the development of mammals and metamorphosis of invertebrates and lower vertebrates, in hormone-dependent tissues (prostate, mammary gland) after removal of trophic hormones by gonadectomy, in lymphocytes and leukemic cells after treatment with glucocorticoids and after various kinds of damage by chemicals, injury, radiation, hypoxia, etc. In developmental biology cell death early was recognized as a "programmed" event.[5–8] On the other hand, in toxicology and pathology cell death traditionally was considered as a passive, degenerative phenomenon occurring after severe damage of tissues. It was not until the early 1970s when Farber *et al.*—upon the basis of a characteristic morphology of cell death and its requirement for protein synthesis—suggested the occurrence of a "suicide" type of cell death in liver, intestine, and other organs after treatment with cytotoxic anticancer drugs.[9] The widespread occurrence and biological relevance of programmed cell death was finally recognized by Kerr, Wyllie and Currie, who in 1972 proposed a classification of cell death into two broad categories. According to this proposal the term "necrosis," which was usually used for all types of cell death, was re-defined and restricted to events caused by environmental perturbation, "which must be violent and lead to rapid incapacitation of major functions and to collapse of internal homeostasis."[10] The term "apoptosis" was coined to describe a category of cell death "which appears to play a complementary but opposite role to mitosis in the regulation of animal cell populations. Its morphological features suggest that it is an active, inherently programmed phenomenon which can be initiated or inhibited by a variety of environmental stimuli, both physiological and pathological."[10] This proposal initiated a controversy on cell death in the scientific community, but eventually helped to elucidate the hypothesis that apoptosis is as important as cell proliferation for normal development and regulation of cell homeostasis in organisms serving two major functions: (1) as an antagonist to mitosis for the controlled elimination of excessive cells that are no longer needed by the organism (e.g., during atrophy or involution of tissues); (2) as one of the organism's defense lines against disease, e.g., to eliminate (pre)neoplastic cells or damaged cells after cytotoxic injury. Induction of apoptosis in addition to inhibition of cell proliferation now is a major goal of strategies for cancer prevention (for review see refs. 11–14). In determining molecular targets for apoptosis induction it should be noted, however, that accumulating evidence suggests that programmed cell death (PCD) is not confined to apoptosis but comprises morphologically and biochemically distinct subtypes.[15–20]

TYPES OF PROGRAMMED CELL DEATH

Apoptosis originally was defined on the basis of a specific pattern of morphological changes in the dying cell[10]: condensation of cytoplasm, in solid tissues separation from neighboring cells, condensation of chromatin at the nuclear membrane to sharply delineated masses, and cell fragmentation into apoptotic bodies. In highly condensed dead cells or cell fragments, organelles are still intact as shown by electron microscopy; cellular membranes are well preserved and, consequently, cell contents are not liberated. *In vivo*, apoptotic bodies are rapidly phagocytosed and degraded by neighboring cells. An important feature of apoptosis is that lysosomes (autophagic vacuoles) do not play a distinct role in early stages but are involved later in the heterophagic degradation of apoptotic bodies.[10,21,22] However, Schweichel and Merker[15] and Clarke[16] noted that programmed cell death may exhibit morphological features different from apoptosis. These authors described three morphologically distinct types of cell death in the developing embryo: type I is most likely identical to apoptosis. Type II is characterized by a prominent formation of autophagic vacuoles ("autophagic cell death"). *In vivo*, cell residues undergoing type II cell death, like those of apoptosis (type I) are finally phagocytosed by neighboring cells.[15] Type III is described as occurring through disintegration of cells into fragments without involvement of the lysosomal system and without marked condensation.[16] We have studied morphological and biochemical features of autophagic (type II) and apoptotic (type I) PCD in human mammary carcinoma cells (MCF-7), colon cancer cells (HT29/HI1), Jurkat cells and in the liver in some detail.[13,19,20,22,23]

AUTOPHAGIC CELL DEATH IN HUMAN MAMMARY CARCINOMA CELLS: MORPHOLOGICAL AND FUNCTIONAL FEATURES

Previous studies on human mammary carcinoma MCF-7 cells revealed an inhibitory effect of antiestrogens on DNA synthesis and enhanced cell death.[24–26] We have confirmed and extended these results using tamoxifen, 4-hydroxy-tamoxifen and ICI 164384.[19] Tamoxifen at high doses (10^{-5} M) causes lysis (necrosis) of almost all cells within 24 hours, which cannot be prevented by estradiol. The cytotoxic action of tamoxifen may result from perturbations in membrane fluidity,[27] formation of reactive oxygen species,[28] or DNA damage by DNA adducts or chromosomal aberrations, which have been found to occur in kidney and liver.[29,30] On the other hand, lower concentrations of tamoxifen (10^{-6} M and below) induced a gradual appearance of cell death starting to occur 3 days after treatment. This type of cell death is considered to be a receptor-mediated, active cell suicide because of its inhibition by estradiol ("mitogen rescue") even six days after anti-estrogen treatment. By electron microscopy we noted early formation of autophagic vacuoles (AV) and a gradual loss of cytoplasmic organization after TAM treatment. Interestingly, structures required for protein synthesis such as polyribosomes, ER, and Golgi disappear, whereas a few clusters of intact mitochondria persist in close vicinity to AVs and the nuclear envelope. The morphological studies were confirmed and extended by histochemical studies with monodansylcadaverine (MDC), which has been described to selectively accumulate in autophagic vacuoles.[31] We used MDC to visualize AVs in MCF-7 cells and scored MDC positive dots per cell and MDC positive cells per population for quantitative assessment of the role of AVs in autophagic cell death. MDC-positive cells steeply increased between 2 and 4 days after TAM treatment reaching a maximal incidence of about 60%. These findings provide quantitative support to previous electron-microscopical data showing that the number of cells exhibiting a normal appearing nucleus in the presence of extensive autophagy exceeds the number of those cells exhibiting a pyknotic nucleus. Obviously, AV formation preceded the execution stage of cell death.[19] In our studies we also found that cell death as indicated by nuclear condensation further increased beyond day 7, but cells could not be cultured longer than day 10 because general degenerative changes occurred in TAM-treated as well as in control cultures.[19] Thus, the seemingly inconsistent number of MDC-positive cells and those cells exhibiting nuclear condensation 7 days after TAM most probably reflect differences between the kinetics of initiation and completion of cell death. The importance of autophagic vacuoles in the preparation of cells for death was supported by functional criteria. Both the formation of AVs as well as nuclear condensation was inhibited by 3-methyladenine (3-MA),[19] (Török and Bursch, unpublished observation). 3-MA has previously been characterized as a specific inhibitor of formation of autophagic vacuoles in liver.[32] Formation of AVs preceding the nuclear manifestation of cell death has been also observed in leukemic and glioblastoma cells; overexpression of Ras served as death signal.[33,34] As in our studies with MCF-7 cells, 3-MA inhibits both the formation autophagic vacuoles and the progression to the death of leukemic and glioblastoma cells.[33,34] In view of functional criteria for the differentiation between subtypes of PCD, it is of interest to mention preliminary data suggesting that 3-MA does not inhibit TGF-β1–induced apoptosis of hepatocytes (W. Parzefall, personal communication).

FATE OF CYTOSKELETON DURING AUTOPHAGIC AND APOPTOTIC CELL DEATH

In further studies we addressed the fate of the cytoskeleton during autophagic cell death as well as apoptosis. Depolymerization or cleavage of actin, cytokeratins, lamins and other cytoskeletal proteins has been found to be involved in the cell's preparatory as well as execution steps of apoptosis.[35–50] On the other hand, autophagocytosis is known to depend on the *intact* cytoskeleton (for review see ref. 51). Thus, sequestration of cytoplasmic structures, the initial step in formation of autophagosomes, depends on intermediate filaments such as cytokeratin and vimentin.[52,53] Furthermore, all steps including the degradation of cytoplasmic material in AVs are ATP-dependent.[54–57] Therefore, we wondered whether cleavage of the cytoskeleton as described for apoptosis would occur during autophagic cell death. First, cytokeratin and F-actin were visualized in individual MCF-7 cells by immuno- and histochemical means. In intact cells, cytokeratin and F-actin form fine-meshed networks. In dead cells as indicated by pyknotic or fragmented nuclei the following features were found:

1. Cytokeratin appears disassembled into dot-like structures.
2. Actin was still polymerized into filaments as demonstrated by its interaction with phalloidin; the filaments aggregated around the nucleus with star-like extensions to the cell membrane.[20]

Quantitative analysis of phalloidin-stained cell cultures revealed that about 85% of dead MCF-7 cells still contained F-actin.[20] Remarkably, most of the dead cells remain attached to the substrate until 8–10 days after treatment.[19] Secondly, Western blot analysis of cytokeratin and actin was performed for time course analysis. Controls exhibited the typical cytokeratin pattern of MCF-7 cells with CK 8 (52 kD), CK 18 (48 kD) and CK 19 (40 kD). An additional band appeared at 33 kD in both controls and TAM treated cells. TAM-treatment did not result in significant changes of this pattern until day 7, except some cleavage as indicated by the occurrence of a weak band running between CK18 and CK19. This band was detected not before day 3–4, i.e., when the number of MDC-positive cells already reached its maximum. Furthermore, a more pronounced fragmentation of the cytokeratin could be detected in MCF-7 cells detached from the substrate and therefore probably in a stage of secondary necrosis.[20] Taken together, these findings suggest that the cytokeratin is redistributed but not degraded to a significant extent and is probably functionally intact until late stages during the course of autophagic degradation and cell death. Furthermore, actin was still polymerized into F-actin in the vast majority of cells with pyknotic nuclei. Western blot analysis of total actin revealed no signs of proteolytic cleavage; this finding has been confirmed by two-dimensional electrophoresis (Gerner and Bursch, unpublished observation). Moreover, polymerization of G- to F-actin is an ATP-dependent process and therefore F-actin is a sensitive indicator of the metabolic state of a cell. These observations are supported by previous studies showing that at late stages of the death process, when most of the cytoplasm appeared amorphous, the few remaining AVs were associated with clusters of structurally and functionally intact mitochondria as shown by electron microscopy and rhodamine 123 staining.[19] The preservation of mitochondria and thus ATP synthesis throughout autophagic degradation was also supported by the observation that mitochondrial dehydrogenase-activity did not decrease with time in TAM-treated MCF-7 cultures (Török and Bursch, unpublished observation). Hypothetically, ATP synthesis is required during type II programmed cell death while extensive autophagic degradation of polyribosomes and Golgi apparatus occurs. Moreover, the protein cross-linking enzyme transglutaminase, which is activated in apoptotic hepatocytes,[58] apparently is not involved in tamoxifen-induced PCD of MCF-7 cells.

In conclusion, our observations on autophagic driven cell death match with current concepts on the role of the cytoskeleton for initiation and execution of autophagy. It should be emphasized that our studies also revealed some selection of cell components for autophagy such as elimination of ER and Golgi apparatus, but preservation of cytoskeleton and mitochondria (ATP). Likewise, selective autophagic elimination has been observed during regression of rat liver hypertrophy after removal of growth inducer: smooth endoplasmic reticulum after phenobarbital withdrawal but peroxisomes after withdrawal of peroxisome proliferators such as clofibrate; selective elimination of damaged mitochondria has been associated with Reye's syndrome (for review see ref. 51). On the other hand, rat liver ribosomes were found to be excluded from glucagon-induced autophagy. The underlying mechanisms for selection or exclusion of cell components for autophagy are poorly understood (for review, ref. 51).

We also analyzed the fate of cytoskeletal components during "classical" apoptosis. In colon cancer cells (HT29/HI1) the tyrosine kinase inhibitor tyrphostin A25 was found to induce apoptosis, as shown unequivocally by electron microscopy.[20] As shown by immunocytochemistry and Western blot analysis, intermediate filaments in HT29/HI1 cells are subjected to substantial degradation during apoptosis, apparently starting before detachment of apoptotic cells from the substrate and fragmentation into apoptotic bodies and/or secondary necrosis.[20] Similar observations have been made in studies on olomoucine-, roscovitin- or cycloheximide-induced apoptosis of human non-small cell lung cancer and neuroblastoma cells.[48–50] Thus, the first signs of aggregation of cytokeratin and vimentin were found to coincide with the exposure of phosphatidylserine on the outer cell membrane. Further downstream in the apoptotic process the intermediate filaments appear to be proteolytically cleaved and, subsequently, the apoptotic cells detach from the substrate. Likewise, Ku *et al.*[43] reported specific proteolytic cleavage of human type I keratin 18 into 28 and 20 kD as well as of keratin 19 into 29 and 23 kD fragments during anisomycin-induced apoptosis of HT29 cells. Furthermore, we found that F-actin was depolymerized to G-actin in about 95% of the HT29/HI1 cells exhibiting an apoptotic nucleus but were not yet detached from the substrate, i.e., an early stage of apoptosis in HT29/HI1 cells. Western blot analysis revealed no signs of proteolytic cleavage of total actin. Likewise, little proteolysis of actin but accumulation of nonpolymerized actin has been observed during early stages of apoptosis in rat embryo fibroblasts,[59] in non-small cell lung cancer cells[49] and renal tubular cells.[42] Most recently, a causal link between depolymerization of F-actin and eventual cell death was established by Korichneva and Hämmerling:[60] in murine EL4 T lymphoma cells, disorganization of the actin cytoskeleton by cytochalasin B renders the cells more sensitive to anhydroretinol-induced cell death; conversely, stabilization of the actin filaments by jasplakinolide promotes cell resistance to apoptosis.

AUTOPHAGIC AND APOPTOTIC PCD: SCREENING FOR DIFFERENCES AND COMMONALITIES

More recently, we initiated a study to screen for differences between autophagic and apoptotic type of PCD at the proteome level. For this purpose, subcellular fractions (cytosol, crude membranes, nuclear matrix) were isolated from normal and dying cells and analyzed by high-resolution two-dimensional (2D) gel electrophoresis; using this method we were able to separate more than 2000 proteins from the cell fractions. Labeling of proteins by ^{35}S-methionine and subsequent autoradiography of 2D gels allowed the differentiation of newly synthesized proteins, while silver staining of these gels revealed the whole protein pool of the respective fraction. Proteins were identified based upon their immunoreactivity or by mass spectrometry of tryptic digests. Jurkat cells treated with CD95 antibody were used as an additional cell death model because of the rapid and relatively synchronous induction of apoptosis.[61] As shown in TABLE 1, our results on the formation of autophagic vacuoles and its dependence on functional cytoskeletal elements in TAM-treated MCF-7 cells was confirmed and extended; we also detected not-yet-identified proteins specifically synthesized during both types of PCD. Furthermore, these analyses revealed profound differences in the fate of nuclear proteins, since lamin B and scaffold attachment factor, etc. that appear not to be cleaved during autophagic PCD in MCF-7 cells are cleaved but during apoptosis of Jurkat cells (TABLE 1). The underlying causes are not yet understood. It should be noted that MCF-7 cells lack functionally active caspase-3 and TNF-α, and staurosporine-induced death of MCF-7 cells was

TABLE 1. **Differences and commonalities in autophagic and apoptotic cell death**

	Autophagic (MCF-7)	Apoptotic (Jurkat)
Autophagic vacuoles	+	–
Stress response		
hsp 90 translocation	+	+
Protein synthesis		
annexinVI	+	–
immunophilin p59	+	+
Protein modification		
PCNA, calreticulin	+	–
endoplasmin, hsp27	–	+
Nuclear protein breakdown		
(SAF-A[2], SATB1[2], p40, lamin B[2,3])	–	+
Cytoskeleton breakdown		
fodrin,[1,2] myosin light chain, actin,[2] intermediate filaments (cytokeratin, vimentin)	–	+

NOTE: Cell death induction: MCF-7: tamoxifen; Jurkat: anti-CD956 antibody. (1) calpain target, (2) caspase 3 target, (3) caspase 6 target; SAF-A, scaffold attachment factor A; SATB1, special AT-rich sequence binding protein 1; p40, laminin-binding receptor-precursor p40.

not associated with DNA fragmentation, cell shrinkage, and blebbing.[62] Therefore, "classical" apoptosis might not ensue in MCF-7 cells on antiestrogen treatment. However, glioblastoma cells undergo autophagic death induced by overexpression of Ras; the autophagy did not result from the lack of caspases.[33] Likewise, preliminary results indicate the formation of autophagic vacuoles preceding TAM-induced cell death in MCF-7 cell clones containing active caspase 3 (kindly provided by Drs. R.U. Jänicke and A.G. Porter; Bursch and Geyerhofer, unpublished observation). Even in the case of "classical" apoptosis, alternative cell death pathways exist, such as those involving caspase 7 or even without activation any of the known caspases (ref. 61; for review, refs. 45,63). Recent studies in our laboratory on UV-induced apoptosis in U 937 cells suggest that morphological nuclear condensation/fragmentation may precede caspase-3 cleavage (Gerner, Fröhwein, unpublished observation). Taken together, the induction of autophagic death of MCF-7 cells by TAM is unlikely to be due to the lack of caspase 3. Furthermore, as recently pointed out by Zakeri *et al.*[18] type I cell death (apoptosis) and type II may not be strictly separated or mutually exclusive phenomena. Likewise, in our MCF-7 cell culture studies, a subfraction of dying cells shows autophagic cell death with an apoptotic nuclear morphology.[19] Therefore, in a given cell type several pathways leading to active death may coexist. This notion is supported by our findings showing that autophagic and apoptotic PCD seem to share commonalities as exemplified by the cell's "stress response" (translocation of heat shock proteins; TABLE 1).

TABLE 2. Occurrence of autophagic/lysosomal cell death (examples)

Organism	Physiological state	Disease
Fungi imperfecti *Dictyostelium discoideum*	starvation: sorocarp formation (stalk cells)[65]	
Insects *Manduca sexta*	metamorphosis: labial gland, intersegmental muscle[18,66]	
Mammalia		
rats, mice	embryogenesis regression of webs,[67] central epithelial cells of intestine (cavity formation)[15]	
	postnatal/adults mammary gland post weaning[18]	mammary tumors post ablation of trophic hormones[68]
	prostate post castration (mixed type: apoptosis-autophagic PCD)[18]	brain: postnatal degeneration of dopaminergic neurons (mice, weaver gene mut.)[69]
human	endometrium: menstrual cycle[70]	brain: Alzheimer, Parkinson[71,72]

NOTE: Autophagic/lysosomal death and apoptosis may occur in the same tissue.

Finally, so far no specific biochemical pattern has been attributed to type III cell death. Schwartz *et al.* have identified a type of cell death in insect tissues characterized by strong expression of the polyubiquitin gene and of the multicatalytic proteinase (proteasome).[17,64] Ubiquitin binds to cellular proteins to label them for proteolytic degradation by the proteasome protease. So far, enhanced expression of ubiquitin was not observed in models of apoptosis.[17] Likewise, in our studies MCF-7 cells treated with tamoxifen showed no induction of ubiquitin expression above controls.

CONCLUSION

In conclusion, distinct biochemical and molecular events can be attributed to the apoptotic and autophagic morphology of cell death as exemplified by our *in vitro* studies with human mammary, lymphoblast, and colon carcinoma cell lines. This notion is supported by *in vivo* observations, indicating that autophagic cell death is a biological phenomenon of general importance in physiological and disease states (TABLE 2). However, autophagic and apoptotic PCD should not be considered as mutually exclusive phenomena. Rather, they appear to reflect a high degree of flexibility in a cell's response to changes of environmental conditions, both physiological or pathological.

ACKNOWLEDGMENTS

The excellent technical assistance of B. Bublava, A. Dutter, M. Eisenbauer, J. Stockinger, and E. Scherzer is gratefully acknowledged. The studies of the present paper were supported by grants from the Anton-Dreher-Gedächtnisschenkung für medizinische Forschung, the Austrian Herzfelder'sche Familienstiftung, the Austrian Ministry of Traffic and Science, and by Hafslund-Nycomed Austria.

REFERENCES

1. VOGT, C. 1842. Untersuchungen bei der Entwicklungsgeschichte der Geburtshelferkroete (*Alytes obstericans*). Jent and Gassmann. Solothurn.
2. MAJNO, G. & I. JORIS. 1995. Apoptosis, oncosis, and necrosis. Am. J. Pathol **146**(1): 3–15.
3. CLARKE, P. & ST. CLARKE. 1995. Historic apoptosis. Nature **378:** 230.
4. VIRCHOW, R. 1871. Die Cellularpathologie in ihrer Begründung auf physiologische Gewebelehre. 4. Auflage. Hirschwald. Berlin.
5. GLÜCKSMANN, A. 1951. Cell death in normal vertebrate ontogeny. Biol. Rev. Cambridge Phil. Soc. **26:** 59-86.
6. SAUNDERS, J.W., JR. 1966. Death in embryonic systems. Science **154:** 604–612.
7. LOCKSHIN, R.A. & C.M. WILLIAMS. 1964. Programmed cell death II. Endocrine potentiation of the breakdown of the intersegmental muscles of silkmoths. J. Insect Physiol. **10:** 643–649.
8. LOCKSHIN, R.A. & C.M. WILLIAMS. 1965. Programmed cell death I. Cytology of muscles of the Pernyi silkmoth. J. Insect Physiol. **11:** 123–133.
9. FARBER, E., R.S.VERBIN & M. LIEBERMAN. 1971. Cell suicide and cell death. *In* A Symposium on Mechanisms of Toxicology. N. Aldridge, Ed.: 163–173. Macmillan. New York.
10. KERR, J.F.R., A.H. WYLLIE & A.R. CURRIE. 1972. Apoptosis: a basic biological phenomenon with wide-ranging implications in tissue kinetics. Br. J. Cancer **26:** 239–257.
11. HICKMAN, J. & C.C. BOYLE. 1997. Toxin-induced increase in survival factor receptors: modulation of the threshold for apoptosis. Cancer Res. **57:** 2404–2409.
12. ROBERTS, R.A., D.W. NEBERT, J.A. HICKMAN, J.H. RICHBURG & T.L. GOLDSWORTHY. 1997. Perturbation of the mitosis/apoptosis balance: a fundamental mechanism in toxicology. Fund. Appl. Toxicol. **38:** 107–115.
13. SCHULTE-HERMANN, R., W. BURSCH & B. GRASL-KRAUPP. 1995. Active cell death (apoptosis) in liver biology and disease. *In* Progress in Liver Disease. J.L. Boyer & R.K. Ockner, Eds. **13:** 1–35. WB Saunders Co. Philadelphia.
14. WYLLIE, A.H. 1997. Apoptosis and Carcinogenesis. Eur. J. Cell Biol. **73:** 189–197.
15. SCHWEICHEL, J.U. & H.J. MERKER. 1973. The morphology of various types of cell death in prenatal tissues. Teratology **7:** 253–266.
16. CLARKE, P.G.H. 1990. Developmental cell death: morphological diversity and multiple mechanisms. Anat. Embryol. **181:** 195–213.
17. SCHWARTZ, L.M., S.W. SCHMITH, M.E.E. JONES & B. OSBORNE. 1993. Do all programmed cell deaths occur via apoptosis? Proc. Natl. Acad. Sci. USA **90:** 980–984.
18. ZAKERI, Z., W. BURSCH, M. TENNISWOOD & R.A. LOCKSHIN. 1995. Cell death: programmed apoptosis, necrosis, or other? Cell Death Diff. **2:** 83–92.
19. BURSCH, W., H. KIENZL, A. ELLINGER, L. TÖRÖK, R. WALKER, M. SIKORSKA, S. PANDEY & R. SCHULTE-HERMANN. 1996. Active cell death induced by antiestrogens tamoxifen and ICI 164384 in human mammary carcinoma cells (MCF-7) in culture: the role of autophagy. Carcinogenesis **17:** 1595–1607.
20. BURSCH, W., K. HOCHEGGER, L. TÖRÖK, B. MARIAN, A. ELLINGER & R. SCHULTE HERMANN. 2000. Autophagic and apoptotic types of programmed cell death exhibit different fates of cytoskeletal filaments. J. Cell Sci. **113** (Pt. 7): 1189–1198.

21. Wyllie, A.H., J.F.R. Kerr & A.R. Currie. 1980. Cell death: the significance of apoptosis. Internat. Rev. Cytol. **68:** 251–300.
22. Bursch, W., H.S. Taper, B. Lauer & R. Schulte-Hermann. 1985. Quantitative histological and histochemical studies on the occurrence and stages of apoptosis (controlled cell death) during regression of rat liver hyperplasia. Virchows Archiv Abt. B. Zellpathol. **50:** 153–166.
23. Bursch, W., A. Ellinger, L. Török, W. Parzefall, S. Coulibaly, K. Hochegger, M. Schörkhuber, G. Partik, B. Marian & R. Schulte-Hermann. 1997. In vitro studies on subtypes and regulation of active cell death. Toxicol. in Vitro **11:** 579–588.
24. Bardon, S., F. Vignon, P. Montcourier & H. Rochefort, H. 1987. Steroid receptor-mediated cytotoxicity of an antiestrogen and antiprogestin in breast cancer cells. Cancer Res **47:** 1441–1448.
25. Jordan, V.C. & C.S. Murphy. 1990. Endocrine pharmacology of antiestrogens as antitumor agents. Endocr. Rev. **11:** 1152–1159.
26. Wärri, A., R.L. Huovinen, A.M. Laine, P.M. Martikainen & P.L. Härkönen. 1993. Apoptosis in toremifene-induced growth inhibition of human breast cancer cells in vivo and in vitro. J. Nat. Cancer Inst. **85:** 1412–1418.
27. Wiseman, H. 1994. Tamoxifen: new membrane-mediated mechanisms of action and therapeutic advances. Trends Pharmacol. Sci. **15:** 83–89.
28. Tuner, M.J., C.E. Fields & D.B. Everman. 1991. Evidence for superoxide formation during hepatic metabolism of tamoxifen. Biochem. Pharmacol. **41:** 1701–1705.
29. Han, X.I. & J.G. Liehr. 1992. Induction of covalent DNA adducts in rodents by tamoxifen. Cancer Res. **52:** 1360–1363.
30. Sargent, L.M., Y. Dragan, N. Bahnhub, J. Wiley, C. Sattler, P. Schroeder, G. Sattler, V. Jordan & H. Pitot. 1994. Tamoxifen induces hepatic aneuploidy and mitotic spindle disruption after a single in vivo administration to female Sprague-Dawley rats. Cancer Res. **4:** 3357–3360.
31. Biederbick, A., H.F. Kern & H.P. Elsässer. 1995. Monodansylcadaverine (MDC) is a specific in vivo marker for autophagic vacuoles. Eur. J. Cell Biol. **66:** 3–14.
32. Seglen, P.O. & P.B. Jordan. 1982. 3-Methyladenine specific inhibitor of autophagic/lysosomal protein degradation in isolated rat hepatocytes. Proc. Natl. Acad. Sci. USA **79:** 1889–1892.
33. Chi, S., C. Kitanaka, K. Noguchi, T. Mochizuki, Y. Nagashima, M. Shirouzu, H. Fujita, M. Yoshida, W. Chen, A. Asai, M. Himeno, S. Yokoyama & Y. Kuchino. 1999. Oncogenic Ras triggers cell suicide through the activation of a caspase-independent cell death program in human cancer cells. Oncogene **18:** 2281–2290.
34. Jia, L., R.R. Dourmashkin., P.D. Allen, A.B. Gray, A.C. Newland & S.M. Kelsey. 1997. Inhibition of autophagy abrogates tumour necrosis factor alpha induced apoptosis in human T-lymphoblastic leukaemic cells. Br. J. Haematol. **98**(3)**:** 673–859.
35. Bonfoco, E., M. Leist, B. Zhivotovsky, S. Orrenius, S.A. Lipton & P. Nicotera. 1996. Cytoskeletal breakdown and apoptosis elicited by NO donors in cerebellar granule cells require NMDA receptor activation. J. Neurochem. **67:** 2484–2493.
36. Brancolini, C., D. Lazarevic, J. Rodriguez & C. Schneider. 1997. Dismantling cell-cell contacts during apoptosis is coupled to a caspase-dependent proteolytic cleavage of β-catenin. J. Cell Biol. **139:** 759–771.
37. Brown, S.B., K. Bailley & J. Savill. 1997. Actin is cleaved during constitutive apoptosis. Biochem. J. **323:** 233–237.
38. Caulin, C., G.S. Salvesen & R.G. Oshima. 1997. Caspase cleavage of keratin 18 and reorganization of intermediate filaments during epithelial cell apoptosis. J. Cell Biol. **138:** 1379–1394.
39. Chen, Z., M. Naito, T. Mashima & T. Tsuruo. 1996. Activation of actin-cleavable interleukin 1beta-converting enzyme (ICE) familiy protease CPP-32 during chemotherapeutic agent-induced apoptosis in ovarian carcinoma cells. Cancer Res. **56:** 5224–5229.

40. KAYALAR, C., T. ÖRD, M.P. TESTA, L. ZHONG & D.E. BREDESEN. 1996. Cleavage of actin by interleukin 1β-converting enzyme to reverse DNase inhibition. Proc. Natl. Acad. Sci. USA **93:** 2234–2238.

41. KOTHAKOTA, S., T. AZUMA, C. REINHARD, A. KLIPPEL, J. TANG, K. CHU, T. MCGARRY, M.W. KIRSCHNER, K. KOTHS, J. KWIATKOWSKI & L.T. WILLIAMS. 1997. Caspase-3-generated fragment of gelsolin: effector of morphological change in apoptosis. Science **278:** 294–298.

42. KRUIDERING, M., B.VAN DE WATER, Y. ZHAN, J.J. BAELDE, E. DEHEER, G.J. MULDER, J.L. STEVENS & F. NAGELKERKE. 1998. Cisplatin effects on F-actin and matrix proteins precede renal tubular cell detachment and apoptosis in vitro. Cell Death Diff. **5:** 601–614.

43. KU, N.O., J. LIAO & M.B. OMARY. 1997. Apoptosis generates stable fragments of human type I keratins. J. Biol. Chem. **272:** 33197–33203.

44. LEVEE, M.G., M.I. DABROWSKA, J.L. LELLI, JR. & D.B. HINSHAW. 1996 Actin polymerization and depolymerization during apoptosis in HL-60 cells. Am. J. Physiol. **271:** 1981–1992.

45. PORTER, A.G. & R.U. JÄNICKE. 1999. Emerging roles of caspases-3 in apoptosis. Cell Death Diff. **6:** 99–104.

46. SCHMEISER, K. & J.A. GRAND. 1999. The fate of E- and P-cadherin during the early stages of apoptosis. Cell Death Diff. **6:** 377–386.

47. TAKAHASHI, A., E.S. ALNEMRI, Y.A. LAZEBNIK, T. FERNANDES-ALNEMRI, G. LITWACK, R.D. MOIR, G.G. POIRER, S.H. KAUFMANN & W.C. EARNSHAW. 1996. Cleavage of lamin A by Mch2a but not CPP32: multiple interleukin 1β-converting enzyme-related proteases with distinct susbtrate recognition properties are active in apoptosis. Proc. Natl. Acad. Sci. USA **93:** 8395–8400.

48. TINNEMANS, M.M.F.J., M.H.J.H. LENDERS, G.P.M. TEN VELDE, F.C.S. RAMAEKERS & B. SCHUTTE. 1995. Alterations in cytoskeletal and nuclear matrix-associated proteins during apoptosis. Eur. J. Cell Biol. **68:** 35–46.

49. VAN ENGELAND, M., F.C. RAMAEKERS, B. SCHUTTE & C.P. REUTELINGSPERGER. 1996. A novel assay to measure loss of membrane asymmetry during apoptosis of adherent cells in culture. Cytometry **24:** 131–139.

50. VAN ENGELAND, M., H.J.H. KUIJPERS, F.C. RAMAEKERS, C.P. REUTELINGSPERGER & B. SCHUTTE. 1997. Plasma membrane alterations and cytoskeletal changes in apoptosis. Exp. Cell Res. **235:** 421–430.

51. BLOOMMART, E.F.C., J.J.F.P. LUIKEN & A.J. MEIJER. 1997. Autophagic proteolysis, control and specificity. Histochem. J. **29:** 365–385.

52. APLIN, A., T. JASIANOWSKI, D.L. TUTTLE, S. LENK & W.A. DUNN. 1992. Cytoskeletal elements are required for the formation and maturation of autophagic vacuoles. J. Cell Physiol. **153:** 458–466.

53. BLANKSON, H., I. HOLEN & P.O. SEGLEN. 1995. Disruption of the cytokeratin cytoskeleton and inhibition of hepatic autophagy by okadaic acid. Exp. Cell Res. **218:** 522–530.

54. LUIKEN, J.J.F.P., J.M.F.G. AERTS & A.J. MEIJER. 1996. The role of intralysosomal pH in the control of autophagic proteolytic flux in rat hepatocytes. Eur. J. Biochem. **235:** 564–573.

55. PLOMP, P.J.A., P.B. GORDON, A.J. MEIJER, H. HOYVIK & P.O. SEGLEN. 1989. Energy dependence of different steps in the autophagic-lysosomal pathway. J. Biol. Chem. **254:** 6699–6704.

56. PLOMP, P.J.A., E.J. WOLVETANG, A.K. GROEN, A.J. MEIJER, P.B. GORDON & P.O. SEGLEN. 1988. Energy dependence of autophagic protein degradation in isolated rat hepatocytes. Eur. J. Biochem. **164:** 197–203.

57. SCHELLENS, J.P.M. & A.J. MEIJER. 1991. Energy depletion and autophagy. Cytochemical and biochemical studies in isolated rat hepatocytes. Histochem. J. **23:** 460–466.

58. BURSCH, W., L. FESUS & R. SCHULTE-HERMANN. 1992. Apoptosis ("programmed" cell death) and its relevance in liver injury and carconogenesis. *In* Tissue Specific Toxicology. W. Dekant & H.G. Neumann, Eds.: 95–117. Academic Press. London.

59. GUÉNAL, I., Y. RISLER & B. MIGNOTTE. 1997. Down-regulation of actin genes precedes microfilament network disruption and actin cleavage during p53-mediated apoptosis. J. Cell Sci. **110:** 489–495.

60. KORICHNEVA, I. & U. HÄMMERLING. 1999. F-actin as a functional target for retro-retinoids: a potential role in anhydroretinal-triggered cell death. J. Cell Sci. **112:** 2521–2528.

61. BERNDT, C., B. MÖPPS, S. ANGERMÜLLER, P. GIERSCHIK & P. KRAMMER. 1998. CXCR4 and CD4 mediate a rapid CD95-independent cell death in $CD4^+$ T cells. Proc. Natl. Acad. Sci. USA **95:** 12556–12561.

62. JÄNICKE, R.U., M.L. SPRENGART, M.R. WATI & A.G. PORTER. 1998. Caspase-3 is required for DNA fragmentation and morphological changes associated with apoptosis. J. Biol. Chem. **273:** 9357–9360.

63. KITANAKA, C. & Y. KUCHINO. 1999. Caspase-independent programmed cell death with necrotic morphology. Cell Death. Diff. **6:** 508–515.

64. JONES, M.E.E., M.F. HAIRE, P.M. KLOETZEL, D.L. MYKLES & L.M. SCHWARTZ. 1995. Changes in structure and function of the multicatalytic proteinase (proteasome) during programmed cell death in the intersegmental muscles of the hawmoth, *Munduca sexta*. Develop. Biol. **169:** 436–447.

65. CORNILLON, S., C. FOA, J. DAVOUST, N. BUONAVISTA, J.D. GROSS & P. GOLSTEIN. 1994. Programmed cell death in Dictyostelium. J. Cell Sci. **107:** 2691–2704.

66. JOCHOVA, J., D. QUAGLINO, Z. ZAKERI, K. WOO, M. SIKORSKA, V. WEAVER & R.A. LOCKSHIN. 1997. Protein synthesis, DNA degradation, and morphological changes during programmed cell death in labial glands of *Manduca sexta*. Dev. Genet. **21**(4): 249–257.

67. BEAULATON, J. & R.A. LOCKSHIN. 1977. Ultrastructural study of the normal degeneration of the intersegmental muscles of *Antherea polyphemus* and *Manduca sexta* (Insecta, Lepidoptera) with particular reference of cellular autophagy. J. Morphol. **154:** 39–57.

68. GULLINO, P.M. 1980.The regression process in hormone-dependent mammary carcinomas. *In* Hormones and Cancer. S. Iacobelli, R.B. King, R. Lindner & M.E. Lippman, Eds.: 271–279. Raven Press Ltd. New York.

69. MIGHELI, A., R. PIVA, J. WEI, A. ATTANASIO, S. CASOLINO, M.E. HODES, S.R. DLOUHY, S.A. BAYER & B. GHETTI. 1997. Diverse cell death pathways result from a single missense mutation in weaver mouse. Am. J. Pathol. **151**(6): 1629-38

70. VERMA, V. 1983. Ultrastructural changes in human endometrium at different phases of the menstrual cycle and their functional significance. Gynecol. Obstet. Invest. **15:** 193–212.

71. CATALDO, A.M., J.L. BARNETT, S.A. BERMAN, J. LI, S. QUARLESS, S. BURSZTAJN, C. LIPPA & R.A. NIXON. 1995. Gene expression and cellular content of cathepsin D in Alzheimer's Disease brain: evidence for early up-regulation of the endosomal-lysosomal system. Neuron **14:** 671–680.

72. ANGLADE, P., S. VYAS, F. JAVOY-AGID, M.T. HERRERO, P.P. MICHEL, J. MARQUEZ, A. MOUATT-PRIGENT, M. RUBERG, E.C. HIRSCH & Y. AGID. 1997. Apoptosis and autophagy in nigral neurons of patients with Parkinson's disease. Histol. Histopathol. **12:** 25–31.

Apoptosis as a Scaffold for Building up the B Cell Repertoire

ALEXANDRA BRÁS, ANTONIO RUIZ-VELA, DAVID GARCÍA-DOMINGO, AND CARLOS MARTÍNEZ-A.[a]

Department of Immunology & Oncology, Centro Nacional de Biotecnología/CSIC, Universidad Autónoma, Campus de Cantoblanco, E-28049 Madrid, Spain

ABSTRACT: Control of cell number is determined by a balance between cell proliferation and cell death, both of which are highly regulated processes, with numerous checks and balances. Cells control their own death through activation of an internally coded suicide program that, when activated, initiates a characteristic form of cell death called apoptosis. This type of regulation allows elimination of cells that have been produced in excess, that have developed improperly, or that have sustained genetic damage. Apoptosis is, therefore, the most common physiological form of cell death and occurs during embryonic development, tissue remodeling, immune regulation, cell activation and tumor regression.

KEYWORDS: B cells; B cell receptor; Apoptosis; Fas; FasL

B CELL TOLERANCE AND SELECTION FOR LYMPHOCYTE MAINTENANCE

B lymphocyte development can be divided into two major phases. In the first, which is antigen-independent, a programmed sequence of immunoglobulin V(D)J gene rearrangements yields a lymphocyte population, each of which makes and expresses on the cell surface its own characteristic immunoglobulin molecule, the B cell receptor (BCR).[b] In the second phase, encounter with foreign antigen leads to the selective expansion of those B cells that produce a cognate antibody: they differentiate into either antibody-secreting plasma cells or memory B cells. It is the BCR that mediates the response of the B lymphocyte to foreign antigen (reviewed in ref. 1). When BCRs on emerging bone marrow B cells recognize a nearby self tissue, differentiation is blocked, and they either undergo cell death or proceed with light chain gene rearrangements.[2] This developmental arrest is reversible; a new B cell

[a]Address for correspondence: Dr. Carlos Martínez-A., Department of Immunology & Oncology, Centro Nacional de Biotecnología/CSIC, Universidad Autónoma, Campus de Cantoblanco, E-28049 Madrid, Spain. Voice: (+34) 91/ 585-4559; fax: (+34) 91/ 372-0493.
cmartineza@cnb.uam.es

[b]**Abbreviations:** BCR, B cell receptor; ERK, extracellular signal-regulated kinases; DIO-1, death inducer-obliterator-1; FasL, Fas ligand; FLIP, FLICE inhibitory protein; GC, germinal center; HEL, hen egg lysozyme; IL-4R, interleukin-4 receptor; MAPK, mitogen-activating protein kinase; PI3K, phosphatidylinositol-3-kinase; Pten, phosphatase tensine homologue; SAPK/JNK, stress-activated protein kinase/c-jun N-terminal kinase; SHIP, Src homology 2 domain-containing 5' inositol phosphatase; Th1, type 1 T helper cell; RAG, recombination activating gene; Xid, X-linked immunodeficiency.

antigen receptor can be created through secondary Ig gene rearrangements that can alter BCR specificity, a process termed receptor editing, increasing the possibility of cell survival. This mechanism rescues autoreactive cells, allowing them to alter their specificities and to mature. Unlike clonal selection, receptor editing thus provides a novel mechanism of immune tolerance that imposes selection at the receptor level.[3,4]

Recent studies are bringing into focus the relative roles of apoptosis and receptor editing in central tolerance. Compared to mature B cells, immature B cells have classically been considered more sensitive to apoptosis and tolerance mechanisms. Highly purified immature B cells either undergo rapid apoptosis after BCR crosslinking, or edit their receptors after BCR ligation,[5] whereas mature B cells are more resistant to death induced by anti-BCR antibodies.[6] Bcl-2 expression promotes receptor editing in autoreactive immature B cells, suggesting that apoptosis limits the time window in which these B cells can edit their receptors.[7] Transitional B cells, which represent an intermediate between immature and recirculating mature B cells, are highly sensitive to antigen-induced apoptosis *in vivo*. These studies indicate, therefore, that receptor editing may be the preferred tolerance mechanism of newly formed B cells, but that as these cells mature they lose their ability to edit their receptors and undergo apoptosis upon self-antigen binding.

Newly produced B cells are exported from bone marrow to spleen, and only a subset of these new emigrants enter the long-lived pool. Tolerance contributes to the elimination of that proportion of newly formed spleen B cells that are autoreactive, a process called negative selection. In contrast, positive selection mediated by basal signaling *via* BCR, usually by foreign antigens, may enhance B cell survival.[8] B cells encountering antigen in the peripheral immune system can be functionally "anergized" or can undergo cell death.[9] Recent data indicate that the tolerance state of a B cell can dictate the signaling pathways receptive to BCR activation. In fact, anergic B cells activate BCR signaling pathways differently than do naïve B cells.[10] The impaired BCR signaling of anergic B cells renders them sensitive to at least two death pathways. The first requires competition for survival signals with other, non-anergic B cells, and is associated with the BCR-induced migration of the cells to the zone rich in T cells (periarteriolar lymphoid sheath). In the absence of B cell competition, anergic B cells find refuge in the B cell follicles, an environment that extends their survival.[11,12] This interpretation is controversial, however, since normal antigen-reactive cells undergo a similar migration response to the zone rich in T cells and, like anergic cells, they require helper signals from T cells for rescue. In the second pathway, B cells can be killed by cognate T cells through the Fas pathway, and anergic cells seem particularly sensitive to this pathway.[13–17] Fas/FasL interactions are also important in B cell homeostasis in the absence of self-reactivity,[18,19] as we will discuss later. Even in the absence of these pathways, anergic cells appear to have a reduced lifespan compared to non-anergic cells.[11,12]

In response to antigenic stimulation, B lymphocytes proliferate in germinal centers (GC) and differentiate into antibody-secreting cells. Most cells do not differentiate into plasma cells, however, but revert to the resting state to become memory B lymphocytes. Many of these cells ultimately take up residence within lymphoid follicles, where they survive for years; if they are subsequently activated, they will undergo further cycles of replication to produce still more memory and effector cells. In general, the progeny at each stage continue to express the same Ig genes as their parents; however, two specialized types of genetic processes occur at high frequency whenever memory B cells proliferate in the periphery. These processes, known as class switching and somatic hypermutation, further diversify the Ig genes permanently to alter the characteristics of their progeny. Somatic hyperpointmutation of antibody variable region genes was the first mechanism of receptor modification to be discovered.[20] As this process may create self-reactive BCRs, these cells are subjected to a second window of immune tolerance in which ongoing GC reactions can be disrupted either by BCR crosslinking or by follicular dendritic cell interaction,[21–23] leading to the rapid death of antigen-reactive cells. The expression of RAG genes and secondary Ig gene rearrangements in GC has recently been described.[24,25] Since many of these cells that express RAG undergo apoptosis,[26] one might think that receptor editing in this context promotes V-gene diversification, which may occur in cells that are escaping GC selection. V(D)J recombination, in supposedly mature B cells, occurs shortly after immunization, that is, before GC development.[27] Receptor diversification by receptor editing may thus also be initiated before somatic hyperpointmutation.

Receptor editing in the GC is, however, a process very different from the tolerance-induced receptor editing in bone marrow. Self-reactive B cells in the spleen are eliminated rather than being edited,[28] even in the GC during an immune response.[21–23] In contrast, receptor editing in GC B cells appears to result from diminished BCR signaling, due to low or poor affinity for foreign antigen.[25]

In conclusion, B cells that fail to pass a developmental checkpoint, as either immature or mature B cells, can be rescued by creating a new BCR through receptor editing. Tolerance-mediated receptor editing occurs in self-reactive immature B cells, while peripheral receptor editing probably occurs in low-affinity B cells competing for antigen and for survival signals within the GC response.

TOLERANCE DURING B CELL DIFFERENTIATION AND ACTIVATION

Apoptosis in Immature B Cells

The ability of a B cell to recognize an antigen is first manifested at the immature development stage, at which the mature BCR form is first expressed. It has long been known that extensive crosslinking by the BCR in immature B cells leads to clonal deletion, since these cells are sensitive to tolerance induction and negative selection.[2,29–32] During this developmental stage, the antigens encountered in bone marrow, neonatal spleen or fetal liver are likely to be self-antigens, and reactive clones are therefore appropriately eliminated. As self-antigens available in these anatomic compartments may not delete potentially self-reactive B cells, mature peripheral B cells may also be negatively selected; however, this occurs through a complex regulation of the activation process,[33] as will be described below. The molecular mechanisms that account for the sensitivity of immature B cells to tolerance are largely unclear.

The immature B cell line WEHI-231 has phenotypic characteristics resembling those responsible for clonal deletion or anergy, since BCR crosslinking initiates a program leading to growth arrest and cells death by apoptosis; it is therefore commonly used as a model for self-tolerance. Self-antigens that crosslink antigen receptors extensively cause elimination of immature B cells in bone marrow[29] and of mature B cells in the periphery.[6,8] Antigens that trigger less extensive crosslinking do not induce deletion of the B cell clones that recognize them, but do produce unresponsiveness[34] by rendering the B cells anergic or preventing their access to lymphoid follicles.[35]

For WEHI-231 cells, it has been reported that BCR engagement leads sequentially to a drop in c-Myc,[36–38] induction of the cdk inhibitor $p27^{Kip}$,[39,40] and finally, apoptosis. This $p27^{Kip}$ induction has been shown to lead to growth arrest,[40] although it has also recently been shown that an increase in $p27^{Kip}$ expression promotes apoptosis in WEHI-231 cells.[39] The $p53/p21^{cip}$ pathway has also been implicated in WEHI-231 apoptosis mediated by BCR crosslinking, independent of the Myc/$p27^{Kip}$ pathway. Anti-IgM treatment also results in p53-mediated $p21^{cip}$ induction, and inhibition of this expression partially protects cells from apoptosis.[41] We recently showed that BCR-crosslinking in WEHI-231 induces caspase activation mediated by caspase-7 prior to induction of apoptosis.[42] This caspase activation was independent of caspase-8 and of cytochrome c release. In addition, we demonstrated that calpain specifically activates caspase-7 in B cells.[43] After BCR crosslinking, calpain is activated, a process associated with a decrease in the levels of calpastatin, an endogenous calpain inhibitor.[43]

As originally reported by Honjo's laboratory, and confirmed by others, CD40 signaling blocks BCR-induced deletion in immature B cells.[44–51] One of the mechanisms through which CD40 signaling prevents BCR-induced deletion of immature B cells is by blocking caspase activity.[42] CD40 ligation upregulates calpastatin levels, and therefore interferes with the caspase activation pathway in immature B cells.[43]

Studies with Bcl-2 transgenic mice also show that Bcl-2 expression may impair negative selection of B cells;[2] neverthless, the mechanism involved in this process is unknown. Bcl-2 overexpression counteracts BCR-triggered caspase activation in immature B cells,[42] but Bcl-2 is not able to prevent BCR crosslinking–induced growth arrest. This suggests that the pathways leading to growth arrest and apoptosis are distinct, an argument that is supported by the data showing that BCR triggering of growth arrest precedes apoptosis. Growth arrest in WEHI-231 cells occurs in the G1 phase, immediately before entry into the S phase. Anti-IgM treatment in these cells induces Rb dephosphorylation, which is detected prior to apoptosis induction and is consistent with the kinetics of cell cycle arrest.[52] CD40 signaling counteracts Rb-induced dephosphorylation, concurring with data showing that only CD40 signaling is able to prevent growth arrest induced by BCR crosslinking.[42] In contrast, Bcl-2 does not affect BCR-mediated Rb dephosphorylation,[42] in accordance with the fact that Bcl-2 does not interfere with the growth arrest.

Bcl-x_L also plays a role in resistance to BCR-induced apoptosis. This is supported by the observation that Bcl-x_L-overexpressing WEHI-231 cells are protected against sIgM-induced apoptosis and that CD40, but not sIgM, mediates Bcl-x_L expression.[48,50,51,53] CD40-mediated induction of Bcl-x_L has not been found in all such experiments, however, and it is unclear whether differences in results are due to the use of distinct WEHI-231 sublines, or to other factors.[49] These results suggest that inducible expression of Bcl-x_L may counteract sIg-mediated apoptotic signals in some, but perhaps not all, situations. In the same way as Bcl-2, Bcl-x_L prevents apoptosis, but not growth arrest induced by BCR,[49] indicating that signals other than Bcl-x_L and Bcl-2 are necessary to promote cell cycle progression.

Bcl-2 and Bcl-x_L are thought to block apoptosis by preventing caspase activation;[54–58] however, the mechanisms are still not clear. In fact, Bcl-2 prevents cytochrome *c* release, which activates the mitochondrial pathway of apoptosis; this pathway is not activated by BCR crosslinking. Caspases are essential effector molecules in apoptosis, but specific cofactor molecules are required for caspase activation and apoptosis.[59,60] One of these adaptors, Apaf-1 (apoptotic protease–activating factor-1), is necessary for caspase activation during cytochrome *c*-dependent apoptosis.[61,62] Apaf-1 binds to caspase-9 when liberated from mitochondria and, in the presence of cytochrome *c* and dATP, initiates the caspase activation cascade.[63] Bcl-2 is able to regulate the subcellular localization of Apaf-1.[64] This may be one mechanism by which Bcl-2 prevents apoptosis by BCR crosslinking.

All together, the data are compatible with the simplest model described in FIGURE 1, which depicts activation of the apoptotic pathway in two ways, (i) caspase-7 activation by calpain, which is blocked by calpain inhibitors, and (ii) caspase-9 activation by Apaf-1 translocation, which is blocked by Bcl-2.

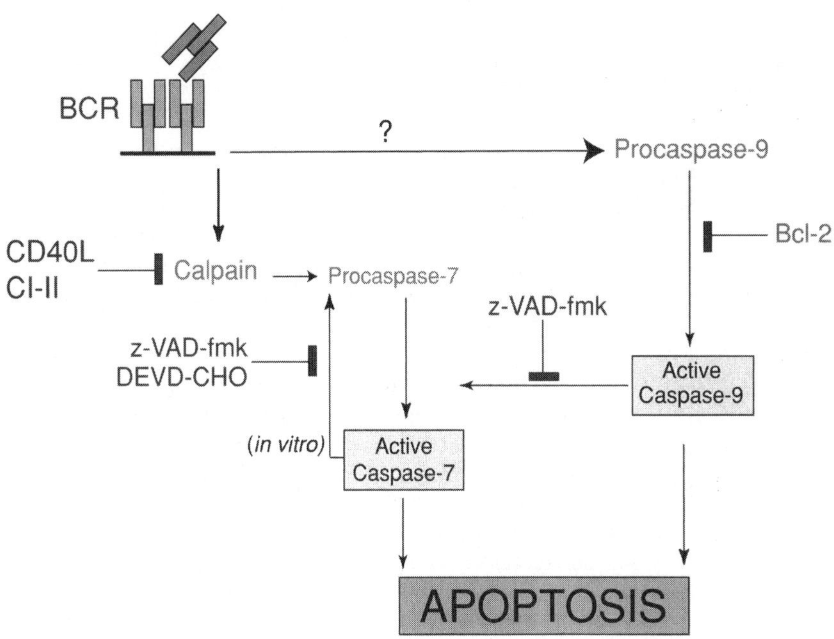

FIGURE 1. Schematic representation of a B cell clonal deletion model. BCR crosslinking activates the apoptotic pathway by proteolyzing calpain, which activates caspase-7. This activation is blocked by the calpain inhibitor CI-II and CD40L, but not by Bcl-2 overexpression. In this model, both calpain/caspase-7 and caspase-9/Bcl-2 systems are critical for caspase activation in B cell clonal deletion.

Role of BCR in Mature B Cell Apoptosis

As mentioned, the BCR has a central role in signaling the fate of a B cell, even before it has encountered antigen. Progression through the preB cell stage of differentiation, during which antigen receptors are being assembled, depends on signals from the BCR or its precursor, the preBCR. Furthermore, the persistence of mature B lymphocytes depends on the continued presence of BCR,[33] probably because it provides survival signals to B cells.[17,65–67] In fact, work from Rajewsky's group[68] using Cre-mediated inducible gene targeting techniques showed that mature B cells lacking a BCR are not maintained in the peripheral immune system. The cell surface phenotype of B cells that lose BCR expression resembles that of "transitional" B cells that have yet to be selected into the long-lived peripheral pool[5] and of "anergic" B cells that have downregulated BCR expression due to chronic exposure to self-antigen.[2] As in the case of transitional and anergic B cells, B cells that lose BCR expression upregulate Fas.

The fact that loss of BCR on mature B cells leads to increased Fas expression suggests that lack of BCR on mature B lymphocytes induces selective modulation of the expression of certain cell surface molecules and that these molecules in turn influence the fate of these cells. The immune system therefore appears to have developed strategies, in addition to cell-autonomous apoptosis, to ensure elimination of nonfunctional B lymphocytes. The upregulation of Fas expression on B cells devoid of BCR may thus render these cells susceptible to T cell-mediated killing by FasL, in analogy to the putative elimination of autoreactive B cells through a Fas-FasL-mediated mechanism.[16]

Role of Fas in B Cell Apoptosis

Several lines of evidence suggest that Fas participates in regulating the elimination of activated, tolerant, autoreactive B cells. Fas-deficient *lpr/lpr* mice and Fas knockout mice express high titers of autoantibodies and have hyper-IgM secretion.[69–71] In addition, the recovery of tolerant B cells after adoptive transfer is improved if the B cells are Fas-deficient or if the co-transferred T cells are FasL-deficient.[15,16] Finally, transgenic mice that overexpress IL-4, a cytokine that prevents B cell elimination through Fas, produce an excess of autoantibodies.[14]

Mature, recirculating B cells exhibit little Fas, and anti-IgM-mediated apoptosis is not caused by a Fas/FasL interaction.[72] After activation, however, Fas expression is induced and B cells acquire susceptibility to cytotoxicity produced by type 1 T helper (Th1) effector cells that express Fas ligand.[17,73,74] Goodnow and colleagues have thus shown that in HEL/Ig^{HEL} double transgenic mice (that is, mice expressing soluble hen egg lysozyme (HEL) as a circulating self-antigen crossed with mice expressing HEL-specific receptor on most B cells), the elimination of autoreactive B cells occurs in a Fas-dependent manner that requires $CD4^+$ T cells.[16] Regulation of Fas-mediated apoptosis may therefore play a role in determining the fate of B cells that produce autoreactive antibodies and that have been discarded from the immunocompetent pool. In the presence of antigen, such autoreactive B cells may develop into antibody-forming cells, resulting in serological autoreactivity. This may occur in normal individuals as a means of enlarging the available repertoire when necessary, by recruiting B cells with antigen receptors that may crossreact with self as well as with microbial pathogens.[75,76] In fact, crossreactivity between responses to microbial antigens and autoantigens has been described,[77–80] explaining the fact that certain infections may induce autoimmune diseases.

CD40/Fas-Dependent Regulation of B Cell Tolerance

As described above, expansion or elimination of a B cell clone is determined by its susceptibility to cell death following antigenic recognition, and by a complex interaction between the tumor necrosis factor receptor family members CD40 and Fas, expressed on the B cell surface, and their respective physiological ligands (CD40L and FasL), expressed on the surface of $CD4^+$ T cells.[81–83] The regulation of B cell growth by signals transduced through BCR, CD40 and Fas therefore contributes to the maintenance of peripheral tolerance. B cell stimulation as a result of CD40 engagement by CD40L is accompanied by induction of Fas expression and susceptibility to Fas-mediated apoptosis.[17,84,85] Further, B cells co-cultured with CD40L and anti-IgM are resistant to Fas-mediated apoptosis, an active, dominant, protective process.[17,67]

The finding that CD40L is required to induce Fas in order to eliminate self-tolerant B cells may help explain the autoantibody disorders often found in human CD40L-deficient X-linked hyper-IgM (X-HIM) patients.[86–89] These patients often present autoimmune hemolytic anemia and thrombocytopenia, showing the same spectrum of autoantibodies that occur in Fas-deficient individuals[90,91] and in the genetically undefined human disorder Evans syndrome.[92]

The negative regulatory role of the CD40L/CD40 pathway, that is, sensitizing B cells to Fas-induced cell death, therefore stands in contrast to its ability to promote survival of germinal center B cells and to protect primary immature B cells and B lymphomas against BCR-mediated apoptosis.[44–50] It thus appears that elimination of tolerant B cells by the CD40L/CD40 pathway is Fas-FasL dependent. In fact, despite functional CD40L, Fas-deficient tolerant B cells are not eliminated.

BCR Crosslinking and IL-4 Protect B Cells from Fas-Induced Apoptosis

B cells sensitized to Fas-mediated apoptosis can be protected by signals from BCR.[17,65–67] This protection is achieved by inhibition of caspase activation.[67] It has also recently been described that induction of FLIP through the BCR mediates the BCR crosslinking-induced caspase inhibition.[93] This provides a novel protective mechanism for antigen-specific B cells in terms of bystander killing during a potent T cell immune response, in which unstimulated Fas-positive B cells would otherwise be killed. Later, CD40/CD40L interaction would promote B cell growth and differentiation, followed perhaps by increased Fas expression to limit clonal expansion. One could therefore consider that this BCR protection from Fas killing may regulate the fate of B cells within T cell regions of lymphoid tissue. In fact, it has been shown that autoantibody-producing B cells in Fas-deficient *lpr/lpr* mice accumulate in T cell-rich regions of the inner periarteriolar lymphoid sheath, since Fas-mediated apoptosis is blocked in these animals.[94] A model that accounts for the role of BCR crosslinking in preventing Fas-FasL-induced cell death is depicted in FIGURE 2.

Another agent capable of modulating susceptibility to Fas killing is the Th2 lymphokine IL-4.[13,14] Accordingly, sera from IL-4-overexpressing transgenic mice

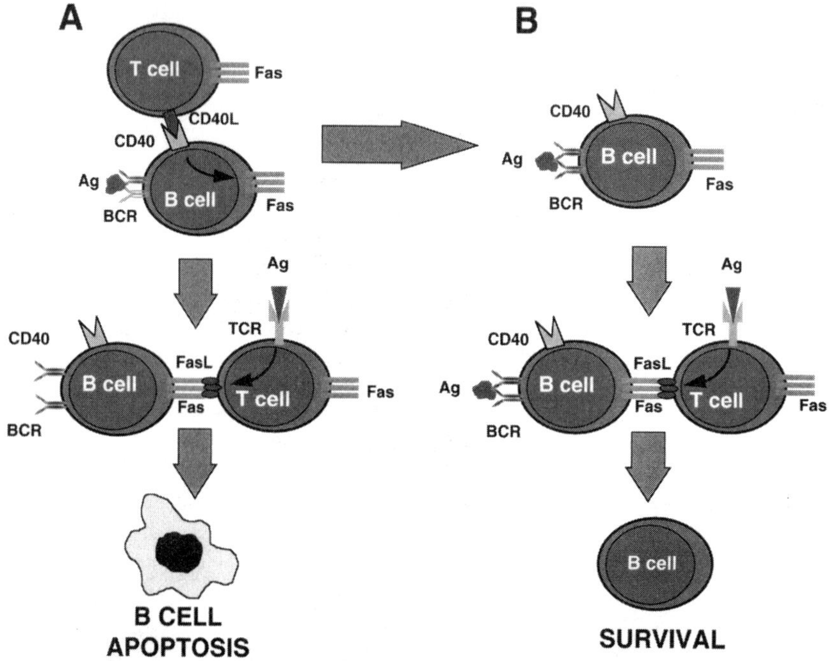

FIGURE 2. T-B cell interactions in the regulation of B cell responses. **(A)** In immature T cells and Th1 populations, TCR activation triggers the production and membrane expression of FasL. In addition, B-T cell interaction via CD40/CD40L leads to an upregulation of Fas in mature B cells and susceptibility to subsequent T-cell FasL-mediated apoptosis. **(B)** B cells can be protected from this latter pathway via prior B cell recognition of antigen, which does not block Fas expression, but rather renders resistance to Fas-mediated apoptosis.

show autoreactive antibodies,[14] suggesting a role for this cytokine in regulating the fate of autoreactive B cells.

OTHER PATHWAYS IMPLICATED IN B CELL HOMEOSTASIS

PI3 Kinase Pathway

A number of survival signals interfere with the apoptotic machinery, preventing cell death. Several pathways are known to transduce these survival signals, among them the mitogen-activated protein kinase (MAPK) cascade. A number of studies suggest that a balance between signal transduction pathways controls apoptosis. Effector kinases with a pro-apoptotic role include SAPK/JNK and p38,[95,96] whereas effector kinases with an antiapoptotic role include the serine/threonine kinases, extracellular signal-regulated kinases (ERK) and the phosphatidylinositol 3-kinase (PI3K) pathway.[96,97] PI3K is activated after BCR crosslinking and generates phosphatidylinositol 3,4,5-triphosphate (PI3,4,5-P_3), which triggers a wide range of signaling molecules through recruitment to the plasma membrane.[98–101] PI3,4,5-P_3-mediated activation of the enzymes Bruton's tyrosine kinase (Btk) and phospholipase C (PLC) γ results in the generation of inositol phosphates (inositol 1,4,5-triphosphates, IP_3), leading to intracellular calcium release and extracellular calcium influx.[102–106] PI3,4,5-P_3 also regulates activation of additional signaling pathways, including those involving Akt[100,107,108] and the MAP kinases,[99] as well as of transcription factors such as NF-κB.[109–111] Loss of PI3,4,5-P_3 generation, by genetic ablation of the PI3K regulatory subunit p85α in mice, results in B cell immunodeficiency.[112,113] These knockout animals have impaired B cell development at the pro-B cell stage, with reduced numbers of peripheral mature B cells and decreased serum Ig levels. The few B cells that develop show diminished proliferative responses to BCR, anti-CD40, and LPS stimulation. The phenotype in p85–/– mice is quite similar to that of mice lacking the tyrosine kinase Btk or Xid (X-linked immunodeficiency).[112,113] Signaling cascades that mediate PI3,4,5-P_3 production are thus critical not only for active BCR-mediated responses, but also for B cell development and survival.

The role of PI3K in autoimmune and lymphoproliferative disorders became clear from studies using Pten herterozygous knockout and $p65^{PI3K}$ transgenic mice.[114,115] Pten acts as a negative regulator of the PI3K/Akt pathway, and $p65^{PI3K}$ is an active PI3K form derived from a thymic lymphoma. Both Pten +/– knockout and $p65^{PI3K}$ transgenic mice have increased PI3K activity and develop autoimmune disease.

PI3,4,5-P_3 are attenuated following BCR coaggregation with the low affinity Fc receptor, FcγRIIB1.[116–118] This inhibition occurs *via* recruitment of the Src homology 2 domain-containing 5′ inositol phosphatase (SHIP), which hydrolyzes the 5′ phosphate from PI3,4,5-P_3 to produce PI3,4-P_2.[119–121] This hydrolysis disrupts BCR-mediated activation of Btk and PLCγ and consequently, the calcium mobilization response.[121,122] B cells from SHIP-deficient mice have increased resistance to BCR-mediated cell death.[123] In these mice, SHIP regulates the B cell development rate in bone marrow and spleen, since B cell precursors progress more rapidly through the immature and transitional developmental stages.[123]

DIO-1, a New Gene Implicated in B Cell Apoptosis

A new gene, DIO-1 (death inducer-obliterator-1; now termed DATF-1, death activation transcription factor) has been identified by differential display PCR in preB cells undergoing apoptosis.[124] When overexpressed, DIO-1 induces apoptosis in several cell lines, and inhibition of caspase activation prevents DIO-1–induced apoptosis. Bcl-2 has also been shown to block DIO-1–induced apoptosis.[124] During BCR crosslinking in immature WEHI-231 cells, DIO-1 is upregulated,[124] suggesting a role for this gene in BCR-induced B cells apoptosis.

CONCLUDING REMARKS

The proliferation and survival of a B cell population must be extremely well regulated to avoid undesired responses that could lead to malignancy or autoimmunity, as well as potentially deleterious non-responses that could entail immunodeficiency.

TABLE 1. Effect of specific receptors on the regulation of B cell apoptosis depends on the developmental cell stage and triggering signal

	Immature	Mature
sIgM	+	–
CD40	–	+
IL-4R	–	–
Apoptotic signal	sIgM	Fas

NOTE: +, induces or increases susceptibility to apoptosis; –, prevents or decreases susceptibility to apoptosis.

This rigid control of B cell growth is exerted by $CD4^+$ T lymphocytes through cognate and noncognate interactions that result in the clonal proliferation of B cells bearing functional antigen receptor to non-self, or the clonal deletion by apoptosis of B cells bearing BCR directed against self. Specific receptor signals modulate B cell apoptosis depending on the B cell developmental stage (TABLE 1). Fas and CD40 receptors thus mediate signaling events involved in eliminating self-reactive and bystander B cells in the periphery. There is, however, crosstalk between BCR-mediated signals and the Fas pathway, since BCR can prevent Fas killing, and thus has the functional capacity to determine B cell fate. On the other hand, antigen engagement and CD40 protect B cells from activation-induced apoptosis during the initial phases of an immune response. Later, Fas upregulation by CD40/CD40L interactions would create targets for FasL-mediated death by activated T cells, controlling lymphocyte population size. In conclusion, engagement of BCR, CD40, and IL-4R surface receptors modulate programmed B cell death, depending on the developmental stage and apoptotic trigger, although the influence of each of these receptors on B cell fate is not always the same.

ACKNOWLEDGMENTS

This work was supported by grant from the Comunidad Autónoma de Madrid, the Spanish CICyT/FEDER, and the Dirección General de Investigación y Ciencia. The Department of Immunology and Oncology was founded and is supported by the Spanish Research Council (CSIC) and by the Pharmacia Corporation.

REFERENCES

1. NEUBERGER, M.S. 1997. Antigen receptor signaling gives lymphocytes a long life. Cell **90:** 971–973.
2. HARTLEY, S.B., M.P. COOKE, D.A. FULCHER, A.W. HARRIS, S. CORY, A. BASTEN & C.C. GOODNOW. 1993. Elimination of self-reactive B lymphocytes proceeds in two stages: arrested development and cell death. Cell **72:** 325–335.
3. CHEN, C., E.L. PRAK & M. WEIGERT. 1997. Editing disease-associated autoantibodies. Immunity **6:** 97–105.
4. PELANDA, R., S. SCHWERS, E. SONODA, R.M. TORRES, D. NEMAZEE & K. RAJEWSKY. 1997. Receptor editing in a transgenic mouse model: site, efficiency and role in B cell tolerance and antibody diversification. Immunity **7:** 765–775.

5. CARSETTI, R., G. KÖHLER & M.C. LAMERS. 1995. Transitional B cells are the target of negative selection in the B cell compartment. J. Exp. Med. **181:** 2129–2140.
6. NORVELL, A., L. MANDIK & J.G. MONROE. 1995. Engagement of the antigen-receptor on immature murine B lymphocytes results in death by apoptosis. J. Immunol. **154:** 4404–4413.
7. LANG, J., B. ARNOLD, G. HAMMERLING, A.W. HARRIS, S. KORSMEYER, D. RUSSELL, A. STRASSER & D. NEMAZEE. 1997. Enforced Bcl-2 expression inhibits antigen-mediated clonal elimination of peripheral B cells in an antigen dose-dependent manner and promotes receptor diting in autoreactive, immature B cells. J. Exp. Med. **186:** 1513–1522.
8. NOSSAL, G.J.V. 1994. Negative selection of lymphocytes. Cell **76:** 229–239.
9. KLINMAN, N.R. 1997. The "clonal selection hypothesis" and current concepts of B cell tolerance. Immunity **5:** 189–195.
10. HEALY, J.I., R.E. DOLMETSCH, L.A. TIMMERMAN, J.G. CYSTER, M.L. THOMAS, G.R. CRABTREE, R.S. LEWIS & C.C. GOODNOW. 1997. Different nuclear signals are activated by the B cell receptor during positive versus negative signaling. Immunity **6:** 419–428.
11. COOK, M.C., A. BASTEN & B. FAZEKAS DE ST. GROTH. 1997. Outer periarteriolar lymphoid sheath arrest and subsequent differentiation of both naive and tolerant immunoglobulin transgenic B cells is determined by B cell receptor occupancy. J. Exp. Med. **186:** 631–643.
12. CYSTER, J.G., J.I. HEALY, K. KISHIHARA, T.M. MAK, M.L. THOMAS & C.C. GOODNOW. 1996. Regulation of B lymphocyte negative and positive selection by tyrosine phosphatase CD45. Nature **381:** 325–328.
13. FOOTE, L., A. MARSHAK-ROTHSTEIN & T.L. ROTHSTEIN. 1998. Tolerant B lymphocytes acquire resistance to Fas-mediated apoptosis after treatment with interleukin 4 but not after treatment with specific antigen unless a surface immunoglobulin threshold is exceeded. J. Exp. Med. **187:** 847–853.
14. FOOTE, L.C., R.G. HOWARD, A. MARSHAK-ROTHSTEIN & T.L. ROTHSTEIN. 1996. IL-4 induces Fas resistance in B cells. J. Immunol. **157:** 2749–2753.
15. RATHMELL, J.C., S.E. TOWNSEND, J.C. XU, R.A. FLAVELL & C.C. GOODNOW. 1996. Expansion or elimination of B cells *in vivo*: dual roles for CD40- and Fas (CD95)-ligands modulated by the B cell antigen receptor. Cell **87:** 319–329.
16. RATHMELL, J.C., M.P. COOKE, W.Y. HO, J. GREIN, S.E. TOWNSEND, M.M. DAVIS & C.C. GOODNOW. 1995. CD95 (Fas)-dependent elimination of self-reactive B cells upon interaction with $CD4^+$ T cells. Nature **376:** 181–184.
17. ROTHSTEIN, T.L., J.K. WANG, D.J. PANKA, L.C. FOOTE, Z. WANG, B. STANGER, H. CUI, S.T. JU & A. MARSHAK-ROTHSTEIN. 1995. Protection against Fas-dependent Th1-mediated apoptosis by antigen receptor engagement in B cells. Nature **374:** 163–165.
18. REAP, E.A., K. ROOF, K. MAYNOR, M. BORRERO, J. BOOKER & P.L. COHEN. 1997. Radiation and stress-induced apoptosis: a role for Fas/Fas ligand interactions. Proc. Natl. Acad. Sci. USA **94:** 5750–5755.
19. ROTHSTEIN, T.L. 1996. Signals and susceptibility to programmed death in B cells. Curr. Opin. Immunol. **8:** 362–371.
20. RAJEWSKY, K. 1996. Clonal selection and learning in the antibody system. Nature **381:** 751–758.
21. SHOKAT, K. & C.C. GOODNOW. 1995. Antigen-induced B cell death and elimination during germinal-centre immune responses. Nature **375:** 334–338.
22. PULENDRAN, B., G. KANNOURAKIS, S. NOURI, K.G. SMITH & G.J.V. NOSSAL. 1995. Soluble antigen can cause enhanced apoptosis of germinal-centre B cells. Nature **375:** 331–334.
23. LINTON, P.J., A. RUDIE & N.R. KLINNAM. 1991. Tolerance susceptibility of newly generating memory cells. J. Immunol. **146:** 4099–4104.
24. PAPAVASILIOU, F., R. CASELLAS, H. SUH, X.F. QIN, E. BESMER, R. PELANDA, D. NEMAZEE, K. RAJEWSKY & M.C. NUSSENZWEIG. 1997. V(D)J recombination in mature B cells: a mechanism for altering antibody responses. Science **278:** 298–301.

25. HAN, S., S.R. DILLON, B. ZHENG, M. SHIMODA, M.S. SCHLISSEL & G. KELSOE. 1997. V(D)J recombinase activity in a subset of germinal center B lymphocytes. Science **278:** 301–305.
26. HIKIDA, M., M. MORI, T. KAWABATA, T. TAKAI & H. OHMORI. 1997. Characterization of B cells expressing recombination activating genes in germinal centers of immunized mouse lymph nodes. J. Immunol. **158:** 2509–2512.
27. MEFFRE, E., F. PAPAVASILIOU, P. COHEN, O. BOUTEILLER, D. BELL, H. KARASUYAMA, C. SCHIFF, J. BANCHEREAU, Y.-J. LIU & M.C. NUSSENZWEIG. 1998. Antigen receptor engagement turns off the V(D)J recombination machinery in human tonsil B cells. J. Exp. Med. **188:** 765–772.
28. RUSSELL, D.M., Z. DEMBIC, G. MORAHAN, J.F. MILLER, K. BURKI & D. NEMAZEE. 1991. Peripheral deletion of self-reactive B cells. Nature **354:** 308–311.
29. NEMAZEE, D.A. & K. BURKI. 1989. Clonal deletion of B lymphocytes in a transgenic mouse bearing anti-MHC class I antibody genes. Nature **337:** 562–566.
30. BROMBACHER, F., G. KOHLER & H. EIBEL. 1991. B cell tolerance in mice transgenic for anti-CD8 immunoglobulin mu chain. J. Exp. Med. **174:** 1335–1346.
31. HARTLEY, S.B., J. CROSBIE, R. BRINK, A.B. KANTOR, A. BASTEN & C.C. GOODNOW. 1991. Elimination from peripheral lymphoid tissues of self-reactive B lymphocytes recognizing membrane-bound antigens. Nature **353:** 765–769.
32. OKAMOTO, M., M. MURAKAMI, A. SHIMIZU, S. OZAKI, T. TSUBATA, S. KUMAGAI & T. HONJO. 1992. A transgenic model of autoimmune hemolytic anemia. J. Exp. Med. **175:** 71–79.
33. LAM, K.P. & K. RAJEWSKY. 1998. Rapid elimination of mature autoreactive B cells demonstrated by Cre-induced change in B cell receptor specificity *in vivo.* Proc. Natl. Acad. Sci. USA **95:** 13271–13175.
34. GOODNOW, C.C., J. CROSBIE, S. ADELSTEIN, T.B. LAVOIE, S.J. SMITH-GILL, R.A. BRINK, H. PRITCHARD-BRISCOE, J.S. WOTHERSPOON, R.H. LOBLAY & K. RAPHAEL. 1988. Altered immunoglobulin expression and functional silencing of self-reactive B lymphocytes in transgenic mice. Nature **334:** 676–682.
35. CYSTER, J.G., S.B. HARTLEY & C.C. GOODNOW. 1994. Competition for follicular niches excludes self-reactive cells from the recirculating B-cell repertoire. Nature **371:** 389–395.
36. FISCHER, G., S.C. KENT, L. JOSEPH, D.R. GREEN & D.W. SCOTT. 1994. Lymphoma models for B cell activation and tolerance. X. Anti-μ-mediated growth arrest and apoptosis of murine B cell lymphomas is prevented by the stabilization of myc. J. Exp. Med. **179:** 221–228.
37. WU, M., M. ARSURA, R.E. BELLAS, M.J. FITZGERALD, H. LEE, S.L. SCHAUER, D.H. SHERR & G.E. SONENSHEIN. 1996. Inhibition of c-myc expression induces apoptosis of WEHI 231 murine B cells. Mol. Cell. Biol. **16:** 5015–5025.
38. WU, M., W. YANG, R.E. BELLAS, S.L. SCHAUER, M.J. FITZGERALD, H. LEE & G.E. SONENSHEIN. 1997. c-myc promotes survival of WEHI 231 B lymphoma cells from apoptosis. Curr. Topics Microbiol. Immunol. **224:** 91–101.
39. WU, M., R.E. BELLAS, J. SHEN, W. YANG & G.E. SONENSHEIN. 1999. Increased $p27^{Kip}$ cyclin-dependent inhibitor gene expression following anti-IgM treatment promotes apoptosis if WEHI 231 B cells. J. Immunol. **163:** 6530–6535.
40. EZHEVSKY, S.A., H. TOYOSHIMA, T. HUNTER & D.W. SCOTT. 1996. Role of cyclin A and p27 in anti-IgM-induced G_1 growth arrest of murine B-cell lymphomas. Mol. Biol. Cell **7:** 553–564.
41. WU, M., R.E. BELLAS, J. SHEN & G.E. SONENSHEIN. 1998. Roles of the tumor suppressor p53 and the cyclin-dependent kinase inhibitor p21WAF1/CIP1 in receptor-mediated apoptosis of WEHI 231 B lymphoma cells. J. Exp. Med. **187:** 1671–1679.
42. BRÁS, A., A. RUIZ-VELA, G. GONZALEZ DE BUITRAGO & C. MARTÍNEZ-A. 1999. Caspase activation by BCR crosslinking in immature B cells: differential effects on growth arrest and apoptosis. FASEB J. **13:** 931–944.
43. RUIZ-VELA, A., G. GONZALEZ DE BUITRAGO & A.C. MARTINEZ. 1999. Implication of calpain in caspase activation during B cell clonal deletion. EMBO J. **18:** 4988–4998.

44. LIU, Y.J., D.E. JOSHUA, G.T. WILLIAMS, C.A. SMITH, J. GORDON & I.C. MACLENNAN. 1989. Mechanism of antigen-driven selection in germinal centres. Nature **342:** 929–931.
45. VALENTINE, M.A. & K.A. LICCIARDI. 1992. Rescue from anti-IgM induced programmed cell death by the B cell surface proteins CD20 and CD40. Eur. J. Immunol. **22:** 3141–3148.
46. TSUBATA, T., J. WU & T. HONJO. 1993. B-cell apoptosis induced by antigen receptor crosslinking is blocked by a T-cell signal through CD40. Nature **364:** 645–648.
47. SANTOS-ARGUMEDO, L., J. GORDON, A.W. HEATH & M. HOWARD. 1994. Antibodies to murine CD40 protect normal and malignant B cells from induced growth arrest. Cell Immunol. **156:** 272–285.
48. CHOI, M.S. K., L.H. BOISE, A.R. GOTTSCHALK, J. QUINTANS, C.B. THOMPSON & G.G.B. KLAUS. 1995. The role of Bcl-x_L in CD40-mediated rescue from anti-μ-induced apoptosis in WEHI-231 B lymphoma cells. Eur. J. Immunol. **25:** 1352–1357.
49. MERINO, R., D.A.M. GRILLOT, P.L. SIMONIAN, S. MUTHUKKUMAR, W.C. FANSLOW, S. BONDADA & G. NUÑEZ. 1995. Modulation of anti-IgM-induced B cells apoptosis by Bcl-x_L and CD40 in WEHI-231 cells. Dissociation from cell cycle arrest and dependence on the avidity of the antibody-IgM receptor interaction. J. Immunol. **155:** 3830–3838.
50. WANG, Z., J.G. KARRAS, R.G. HOWARD & T.L. ROTHSTEIN. 1995. Induction of bcl-x by CD40 engagement rescues sIg-induced apoptosis in murine B cells. J. Immunol. **155:** 3722–3725.
51. ISHIDA, T., N. KOBAYASHI, T. TOJO, S. ISHIDA, T. YAMAMOTO & J.-I. INOUE. 1995. CD40 signaling-mediated induction of Bcl-x_L, Cdk4, and Cdk6. Implication of their cooperation in selective B cell growth. J. Immunol. **155:** 5527–5535.
52. JOSEPH, L.F., S.A. EZHESKY & D.W. SCOTT. 1995. Lymphoma models for B-cell activation and tolerance: anti-immunoglobulin M treatment induces growth arrest by preventing the formation of an active kinase complex which phosphorylates retinoblastoma gene product in G_1. Cell Growth Diff. **6:** 51–57.
53. FANG, W., J J. RIVARD, J.A. GANSER, T.W. LEBIEN, K.A. NATH, D.L. MUELLER & T.W. BEHRENS. 1995. Bcl-x_L rescues WEHI 231 B lymphocytes from oxidant-mediated death following diverse apoptotic stimuli. J. Immunol. **155:** 66–75.
54. RAO, L. & E. WHITE. 1997. Bcl-2 and the ICE family of apoptotic regulators: making connection. Curr. Opin. Gen. Dev. **7:** 52–58.
55. CHINNAIYAN, A.M., K. ORTH, K. O'ROURKE, H. DUAN, G.G. POIRIER & V.M. DIXIT. 1996. Molecular ordering of the cell death pathway. Bcl-2 and Bcl-x_L function upstream of the CED-3-like apoptotic proteases. J. Biol. Chem. **271:** 4573–4576.
56. CHINNAIYAN, A.M., K. O'ROURKE, B.R. LANE & V.M. DIXIT. 1997. Interaction of CED-4 with CED-3 and CED-9: a molecular framework for cell death. Science **275:** 1122–1126.
57. HU, Y., M.A. BENEDICT, D. WU, N. INOHARA & G. NUNEZ. 1998. Bcl-X_L interacts with Apaf-1 and inhibits Apaf-1-dependent caspase-9 activation. Proc. Natl. Acad. Sci. USA **95:** 4386–4391.
58. PAN, G., K. O'ROURKE & V.M. DIXIT. 1998. Caspase-9, Bcl-X_L, and Apaf-1 form a ternary complex. J. Biol. Chem. **273:** 5841–5845.
59. CECCONI, F., B.G. ALVAREZ, B.I. MEYER, K.A. ROTH & P. GRUSS. 1998. Apafl (CED-4 homolog) regulates programmed cell death in mammalian development. Cell **94:** 727–737.
60. YOSHIDA, H., Y.Y. KONG, R. YOSHIDA, A.J. ELIA, A. HAKEM, R. HAKEM, J.M. PENNINGER & T.W. MAK. 1998. Apafl is required for mitochondrial pathways of apoptosis and brain development. Cell **94:** 739–750.
61. LI, P., D. NIJHAWAN, I. BUDIHARDJO, S.M. SRINIVASULA, M. AHMAD, E.S. ALNEMRI & X. WANG. 1997. Cytochrome c and dATP-dependent formation of Apaf-1/caspase-9 complex initiates an apoptotic protease cascade. Cell **91:** 479–489.
62. ZOU, H., W.J. HENZEL, X. LIU, A. LUTSCHG & X. WANG. 1997. Apaf-1, a human protein homologous to *C. elegans* CED-4, participates in cytochrome c-dependent activation of caspase-3. Cell **90:** 405–413.

63. SRINIVASULA, S.M., M. AHMAD, A.T. FERNANDES & E.S. ALNEMRI. 1998. Autoactivation of procaspase-9 by Apaf-1-mediated oligomerization. Mol. Cell **1:** 949–957.
64. RUIZ-VELA, A., B.B. WOLF, J.P. ALBAR, D.R. GREEN & C. MARTÍNEZ-A. 2000. Bcl-2 controls subcellular Apaf-1 localization in lymphocytes. Submitted manuscript.
65. LAGRESLE, C., P. MONDIERE, C. BELLA, P.H. KRAMMER & T. DEFRANCE. 1996. Concurrent engagement of CD40 and the antigen receptor protects naive and memory human B cells from APO-1/Fas-mediated apoptosis. J. Exp. Med. **183:** 1377–1388.
66. WANG, J., I. TANIUCHI, Y. MAEKAWA, M. HOWARD, M.D. COOPER & T. WATANABE. 1996. Expression and function of Fas antigen on activated murine B cells. Eur. J. Immunol. **26:** 92–96.
67. BRÁS, A., C. MARTINEZ-A. & E. BAIXERAS. 1997. B cell receptor cross-linking prevents Fas-induced cell death by inactivating the IL-1β-converting enzyme proteases and regulating Bcl-2/Bcl-x expression. J. Immunol. **159:** 3168–3177.
68. LAM, P.-K., R. KÜHN & K. RAJEWSKY. 1997. *In vivo* ablation of surface immunoglobulin on mature B cells by inducible gene targeting results in rapid cell death. Cell **90:** 1073–1083.
69. IZUI, S., V.E. KELLEY, K. MASUDA, H. YOSHIDA, J.B. ROTHS & E.D. MURPHY. 1984. Induction of various autoantibodies by mutant gene lpr in several strains of mice. J. Immunol. **133:** 227–233.
70. COHEN, P.L. & R.A. EISENBERG. 1991. Lpr and gld: single gene models of systemic autoimmunity and lymphoproliferative disease. Annu. Rev. Immunol. **9:** 243–69.
71. ADACHI, M., S. SUEMATSU, T. SUDA, D. WATANABE, H. FUKUYAMA, J. OGASAWARA, T. TANAKA, N. YOSHIDA & S. NAGATA. 1996. Enhanced and accelerated lymphoproliferation in Fas-null mice. Proc. Natl. Acad. Sci. USA **93:** 2131–2136.
72. RACILA, E., R. HSUEH, R. MARCHES, T.F. TUCKER, P.H. KRAMMER, R.H. SCHEUERMANN & J.W. UHR. 1996. Tumor dormancy and cell signaling: anti-μ-induced apoptosis in human B-lymphoma cells is not caused by an APO-1-APO-1 ligand interaction. Proc. Natl. Acad. Sci. USA **93:** 2165–2168.
73. DANIEL, P.T. & P.H. KRAMMER. 1994. Activation induces sensitivity toward APO-1 (CD95)-mediated apoptosis in human B cells. J. Immunol. **152:** 5624–5632.
74. JU, S.T., H. CUI, D.J. PANKA, R. ETTINGER & A. MARSHAK-ROTHSTEIN. 1994. Participation of target Fas protein in apoptosis pathway induced by $CD4^+$ Th1 and $CD8^+$ cytotoxic T cells. Proc. Natl. Acad. Sci. USA **91:** 4185–4189.
75. CARROLL, P., D. STAFFORD, R.S. SCHWARTZ & B.D. STOLLAR. 1985. Murine monoclonal anti-DNA autoantibodies bind to endogenous bacteria. J. Immunol. **135:** 1086–1890.
76. DUGGAN, D.B., C. MACKWORTH-YOUNG, A. KARI-LEFVERT, J. ANDRE-SCHWARTZ, D. MUDD, K.P. MCADAM & R.S. SCHWARTZ. 1988. Polyspecificity of human monoclonal antibodies reactive with *Mycobacterium leprae*, mitochondria, ssDNA, cytoskeletal proteins, and the acetylcholine receptor. Clin. Immunol. Immunopathol. **49:** 327–340.
77. ROSE, N.R. 1998. The role of infection in the pathogenesis of autoimmune disease. Semin. Immunol. **10:** 5–13.
78. GASTON, J.S. 1998. Heat shock proteins as potential targets in the therapy of inflammatory arthritis. Biotherapy **10:** 197–203.
79. ESAGUY, N., A.P. ÁGUAS, J.D. VAN EMBDEN & M.T. SILVA. 1991. Mycobacteria and human autoimmune disease: direct evidence of cross-reactivity between human lactoferrin and the 65-kilodalton protein of tubercle and leprosy bacilli. Infect. Immun. **59:** 1117–1125.
80. ÁGUAS, A., N. ESAGUY, C.E. SUNKEL & M.T. SILVA. 1990. Cross-reactivity and sequence homology between the 65-kilodalton mycobacterial heat shock protein and human lactoferrin, transferrin, and DR beta subsets of major histocompatibility complex class II molecules. Infect. Immun. **58:** 1461–1470.
81. KEHRY, M.R. 1996. CD40-mediated signaling in B cells. Balancing cell survival, growth, and death. J. Immunol. **156:** 2345–2348.
82. SCHATTNER, E.J., J. MASCARENHAS, J. BISHOP, D.H. YOO, A. CHADBURN, M.K. CROW & S.M. FRIEDMAN. 1996. $CD4^+$ T-cell induction of Fas-mediated apoptosis in Burkitt's lymphoma B cells. Blood **88:** 1375–1382.

83. NAGATA, S. 1997. Apoptosis by death factor. Cell **88:** 355–365.
84. GARRONE, P., E.M. NEIDHARDT, E. GARCIA, L. GALIBERT, K.C. VAN & J. BANCHEREAU. 1995. Fas ligation induces apoptosis of CD40-activated human B lymphocytes. J. Exp. Med. **182:** 1265–1273.
85. SCHATTNER, E.J., K.B. ELKON, D.H. YOO, J. TUMANG, P.H. KRAMMER, M.K. CROW & S.M. FRIEDMAN. 1995. CD40 ligation induces Apo-1/Fas expression on human B lymphocytes and facilitates apoptosis through the Apo-1/Fas pathway. J. Exp. Med. **182:** 1557–1565.
86. ARUFFO, A., M. FARRINGTON, D. HOLLENBAUGH, X. LI, A. MILATOVICH, S. NONOYAMA, J. BAJORATH, L.S. GROSMAIRE, R. STENKAMP, M. NEUBAUER, *et al.* 1993. The CD40 ligand, gp39, is defective in activated T cells from patients with X-linked hyper-IgM syndrome. Cell **72:** 291–300.
87. KORTHAUER, U., D. GRAF, H.W. MAGES, F. BRIERE, M. PADAYACHEE, S. MALCOLM, A.G. UGAZIO, L.D. NOTARANGELO, R.J. LEVINSKY & R.A. KROCZEK. 1993. Defective expression of T-cell CD40 ligand causes X-linked immunodeficiency with hyper-IgM. Nature **361:** 539–541.
88. RAMESH, N., R. FULEIHAN, V. RAMESH, S. LEDERMAN, M.J. YELLIN, S. SHARMA, L. CHESS, F.S. ROSEN & R.S. GEHA. 1993. Deletions in the ligand for CD40 in X-linked immunoglobulin deficiency with normal or elevated IgM (HIGMX-1). Int. Immunol. **5:** 769–573.
89. DISANTO, J.P., J.Y. BONNEFOY, J.F. GAUCHAT, A. FISCHER & G. DE SAINT BASILE. 1993. CD40 ligand mutations in x-linked immunodeficiency with hyper-IgM. Nature **361:** 541–543.
90. FISHER, G.H., F.J. ROSENBERG, S.E. STRAUS, J.K. DALE, L.A. MIDDLETON, A.Y. LIN, W. STROBER, M.J. LENARD & J.M. PUCK. 1995. Dominant interfering Fas gene mutations impair apoptosis in a human autoimmune lymphoproliferative syndrome. Cell **81:** 935–946.
91. RIEUX-LAUCAT, F., F. LE DEIST, C. HIVROZ, I.A.G. ROBERTS, K.M. DEBATIN, A. FISCHER & J.P. VILLARTAY. 1995. Mutations in Fas Associated with human lymphoproliferative syndrome and autoimmunity. Science **268:** 1347–1349.
92. PUI, C.H., J. WILIMAS & W. WANG. 1980. Evans syndrome in childhood. J. Pediatr. **97:** 754–758.
93. WANG, J., A.A. LOBITO, F. SHEN, F. HORNUNG, A. WINOTO & M.J. LENARDO. 2000. Inhibition of Fas-mediated apoptosis by the B cell antigen receptor through c-FLIP. Eur. J. Immunol. **30:** 155–163.
94. JACOBSON, B.A., D.J. PANKA, K.A. NGUYEN, J. ERIKSON, A.K. ABBAS & R.A. MARSHAK. 1995. Anatomy of autoantibody production: dominant localization of antibody-producing cells to T cell zones in Fas-deficient mice. Immunity **3:** 509–519.
95. KUMMER, J.L., P.K. RAO & K.A. HEIDENREICH. 1997. Apoptosis induced by withdrawal of trophic factors is mediated by p38 mitogen-activated protein kinase. J. Biol. Chem. **272:** 20490–20494.
96. XIA, Z., M. DICKENS, J. RAINGEAUD, R.J. DAVIS & M.E. GREENBERG. 1995. Opposing effects of Erk and JNK-p38 MAP kinases on apoptosis. Science **270:**1326–1330.
97. KENNEDY, S.G., A.J. WAGNER, S.D. CONZEN, J. JORDAN, A. BELLACOSA, P.N. TSICHLIS & N. HAY. 1997. The PI 3-kinase/Akt signaling pathway delivers an anti-apoptotic signal. Genes Dev. **11:** 701–713.
98. DEFRANCO, A. 1997. The complexity of signaling pathways activated by the BCR. Curr. Opin. Immunol. **9:** 296–308.
99. CAMPBELL, K.S. 1999. Signal transduction from the B cell antigen-receptor. Curr. Opin. Immunol. **11:** 256–264.
100. GOLD, M.R., M.P. SCHEID, L. SANTOS, M. DANG-LAWSON, R.A. ROTH, L. MATSUUCHI, V. DURONIO & D.L. KREBS. 1999. The B cell antigen receptor activates the Akt (protein kinase B)/glycogen synthase kinase-3 signaling pathway via phosphatidylinositol 3-kinase. J. Immunol. **163:** 1894–1905.
101. GOLD, M.R. & R. AEBERSOLD. 1994. Both phosphatidylinositol 3-kinase and phosphatidylinositol 4-kinase products are increased by antigen receptor signaling in B cells. J. Immunol. **152:** 42–50.

102. BUHL, A.M., C.M. PLEIMAN, R.C. RICKERT & J.C. CAMBIER. 1997. Qualitative regulation of B cell antigen receptor signaling by CD19: selective requirement for PI3-kinase activation, inositol-1,4,5-trisphosphate production and Ca^{2+} mobilization. J. Exp. Med. **186:** 1897–1910.

103. BUHL, A.M. & J.C. CAMBIER. 1999. Phosphorylation of CD19 Y484 and Y515, and linked activation of phosphatidylinositol 3-kinase, are required for B cell antigen receptor-mediated activation of Bruton's tyrosine kinase. J. Immunol. **162:** 4438–4446.

104. BOLLAND, S., R.N. PEARSE, T. KUROSAKI & J.V. RAVETCH. 1998. SHIP modulates immune receptor responses by regulating membrane association of Btk. Immunity **8:** 509–516.

105. FALASCA, M., S.K. LOGAN, V.P. LEHTO, G. BACCANTE, M.A. LEMMON & J. SCHLESSINGER. 1998. Activation of phospholipase C gamma by PI 3-kinase-induced PH domain-mediated membrane targeting. EMBO J. **17:** 414–422.

106. FLUCKIGER, A.C., Z. LI, R.M. KATO, M.I. WAHL, H.D. OCHS, R. LONGNECKER, J.P. KINET, O.N. WITTE, A.M. SCHARENBERG & D.J. RAWLINGS. 1998. Btk/Tec kinases regulate sustained increases in intracellular Ca^{2+} following B-cell receptor activation. EMBO J. **17:** 1973–1985.

107. AMAN, M.J., T.D. LAMKIN, H. OKADA, T. KUROSAKI & K.S. RAVICHANDRAN. 1998. The inositol phosphatase SHIP inhibits Akt/PKB activation in B cells. J. Biol. Chem. **273:** 3922–33928.

108. JACOB, A., D. COONEY, S. TRIDANDAPANI, T. KELLEY & K.M. COGGESHALL. 1999. FcgammaRIIb modulation of surface immunoglobulin-induced Akt activation in murine B cells. J. Biol. Chem. **274:** 13704–13710.

109. ROMASHKOVA, J.A. & S.S. MAKAROV. 1999. NF-kappaB is a target of AKT in anti-apoptotic PDGF signalling. Nature **401:** 86–90.

110. OZES, O.N., L.D. MAYO, J.A. GUSTIN, S.R. PFEFFER, L.M. PFEFFER & D.B. DONNER. 1999. NF-kappaB activation by tumour necrosis factor requires the Akt serine-threonine kinase. Nature **401:** 82–85.

111. KANE, L.P., V.S. SHAPIRO, D. STOKOE & A. WEISS. 1999. Induction of NF-kappaB by the Akt/PKB kinase. Curr. Biol. **9:** 601–604.

112. SUZUKI, H., Y. TERAUCHI, M. FUJIWARA, S. AIZAWA, Y. YAZAKI, T. KADOWAKI & S. KOYASU. 1999. Xid-Like immunodeficiency in mice with disruption of the p85α subunit of phosphoinositide 3-kinase. Science **283:** 390–392.

113. FRUMAN, D.A., S.B. SNAPPER, C.M. YBALLE, L. DAVIDSON, J.Y. YU, F.W. ALT & L.C. CANTLEY. 1999. Impaired B cell development and proliferation in absence of phosphoinositide 3-kinase p85α. Science **283:** 393–397.

114. DI CRISTOFANO, A., P. KOTSI, Y.F. PENG, C. CORDON-CARDO, K.B. ELKON & P.P. PANDOLFI. 1999. Impaired Fas response and autoimmunity in $Pten^{+/-}$ mice. Science **285:** 2122–2125.

115. R.-BORLADO, L., C. REDONDO, B. ALVAREZ, C. JIMENEZ, L.M. CRIADO, J. FLORES, M.A.R. MARCOS, C. MARTÍNEZ-A., D. BALOMENOS & A.C. CARRERA. 2000. Increased phosphoinositide 3-kinase induces a lymphoproliferative disorder and contributes to tumor generation *in vivo*. FASEB J. **14:** 895–903.

116. ONO, M., S. BOLLAND, P. TEMPST & J.V. RAVETCH. 1996. Role of the inositol phosphatase SHIP in negative regulation of the immune system by the receptor Fc(gamma)RIIB. Nature **383:** 263–266.

117. FONG, D.C., O. MALBEC, M. AROCK, J.C. CAMBIER, W.H. FRIDMAN & M. DAERON. 1996. Selective in vivo recruitment of the phosphatidylinositol phosphatase SHIP by phosphorylated Fc gammaRIIB during negative regulation of IgE-dependent mouse mast cell activation. Immunol. Lett. **54:** 83–91.

118. CHACKO, G.W., S. TRIDANDAPANI, J.E. DAMEN, L. LIU, G. KRYSTAL & K.M. COGGESHALL. 1996. Negative signaling in B lymphocytes induces tyrosine phosphorylation of the 145-kDa inositol polyphosphate 5-phosphatase, SHIP. J. Immunol. **157:** 2234–2238.

119. GUPTA, N., A.M. SCHARENBERG, D.A. FRUMAN, L.C. CANTLEY, J.P. KINET & E.O. LONG. 1999. The SH2 domain-containing inositol 5′-phosphatase (SHIP) recruits the p85 subunit of phosphoinositide 3-kinase during FcgammaRIIb1-mediated inhibition of B cell receptor signaling. J. Biol. Chem. **274:** 7489–7494.
120. DAMEN, J.E., L. LIU, P. ROSTEN, R.K. HUMPHRIES, A.B. JEFFERSON, P.W. MAJERUS & G. KRYSTAL. 1996. The 145-kDa protein induced to associate with Shc by multiple cytokines is an inositol tetraphosphate and phosphatidylinositol 3,4,5-triphosphate 5-phosphatase. Proc. Natl. Acad. Sci. USA **93:** 1689–1693.
121. SCHARENBERG, A.M., O. EL-HILLAL, D.A. FRUMAN, L.O. BEITZ, Z. LI, S. LIN, I. GOUT, L. C. CANTLEY, D.J. RAWLINGS & J.P. KINET. 1998. Phosphatidylinositol-3,4,5-trisphosphate (PtdIns-3,4,5-P3)/Tec kinase-dependent calcium signaling pathway: a target for SHIP-mediated inhibitory signals. EMBO J. **17:** 1961–1972.
122. ONO, M., H. OKADA, S. BOLLAND, S. YANAGI, T. KUROSAKI & J.V. RAVETCH. 1997. Deletion of SHIP or SHP-1 reveals two distinct pathways for inhibitory signaling. Cell **90:** 293–301.
123. BRAUWEILER, A., I. TAMIR, J. DAL PORTO, R.J. BENSCHOP, C.D. HELGASON, R.K. HUMPHRIES, J.H. FREED & J.C. CAMBIER. 2000. Differential regulation of B cell development, activation, and death by the src homology 2 domain-containing 5′ inositol phosphatase (SHIP). J. Exp. Med. **191:** 1545–1554.
124. GARCIA-DOMINGO, D., E. LEONARDO, A. GRANDIEN, P. MARTINEZ, J.P. ALBAR, J.C. IZPISUA-BELMONTE & A.C. MARTINEZ. 1999. DIO-1 is a gene involved in onset of apoptosis in vitro, whose misexpression disrupts limb development. Proc. Natl. Acad. Sci. USA **96:** 7992–7997.

HIV, Cytokines and Programmed Cell Death

A Subtle Interplay

MARIE-LISE GOUGEON,[a] ERIC LEDRU, HONAMI NAORA, MARIALUISA BOCCHINO, AND HERVÉ LECOEUR

Unité d'Oncologie Virale and CNRS URA 1930, Département SIDA et Rétrovirus, Institut Pasteur, 75724 Paris Cedex 15, France

ABSTRACT: HIV infection is marked by the progressive destruction of the CD4 T lymphocyte subset, an essential component of the immune system and a vital source of cytokines required for differentiation of natural killer (NK) and γδ T cells, for maturation of B lymphocytes into plasmocytes, and for differentiation of CD8⁺ T cells into virus-specific cytotoxic T lymphocytes. CD4 T lymphocytes are also a source of chemokines which control migration of lymphocytes to the site of infection and which also inhibit HIV entry into CD4-expressing targets. Continuous production of viral proteins leads to an unbalanced immune activation and to the triggering of apoptotic programs, turning mononuclear cells, including CD4 T cells, CD8 T cells and APC, into effectors of apoptosis, leading to fratricidal destruction of healthy uninfected cells expressing the death receptors. Inappropriate PCD is also responsible for the disappearance of T helper cells primed for type-1 cytokine synthesis, thus contributing to the lack of survival factors which could prevent spontaneous lymphocyte apoptosis. Under potent anti-retroviral therapies, a significant decrease in spontaneous, TCR- and CD95-induced lymphocyte apoptosis is observed, concomitant with a partial quantitative and qualitative restoration of the immune system in treated patients. However, owing to the suppressive effect of anti-retroviral drugs on physiological apoptosis, these therapies are associated with alteration of TNF-α–regulated T cell homeostasis, leading to an accumulation in the blood of T cells primed for TNF-α synthesis, and contributing to the development of a new syndrome associated with these treatments, the lipodystrophy syndrome.

KEYWORDS: **T cells; HAART; TNF-α; HIV; Lipodystrophy syndrome**

INTRODUCTION

The hallmark of HIV infection is the progressive destruction of the CD4 T lymphocyte subset. CD4 T lymphocytes represent an essential component of the immune system. This subset is a vital source of cytokines required for differentiation of natural killer (NK) and γδ T cells, which possess a natural antiviral cytotoxic activity and may help in eliminating virus-infected cells,[1] for maturation of B lymphocytes into plasmocytes, which can produce neutralizing antibodies, and for

[a]Address for correspondence: Dr. Marie-Lise Gougeon, Unité d'Oncologie Virale and CNRS URA 1930, Département SIDA et Rétrovirus, Institut Pasteur, 28 Rue du Dr. Roux, 75724 Paris Cedex 15, France. Voice: 33 1 45 68 8907; fax: 33 1 45 68 8909.
mlgougeo@pasteur.fr

differentiation of $CD8^+$ T cells into virus-specific cytotoxic T lymphocytes.[2] In addition CD4 T lymphocytes are a source of chemokines, which control migration of lymphocytes to the site of infection and which also inhibit HIV entry into CD4-expressing targets.[3] Therefore, the disappearance of CD4 T lymphocytes is responsible for the lack of control of HIV replication and the development of severe immune deficiency.

Mechanisms responsible for the disappearance of CD4 T cells in HIV-infected persons are still a matter of debate. It is generally accepted that CD4 T cell destruction is directly mediated by HIV replication as a consequence of viral gene expression. However, the observation by several groups[4–6] that a significant fraction of peripheral lymphocytes from asymptomatic HIV-infected patients are prematurely primed for activation-induced programmed cell death has raised the hypothesis that indirect virus-driven mechanisms are involved in the priming for apoptosis of uninfected cells, leading to their destruction upon antigenic activation.[7–9] Thus, HIV can influence the cell death machinery and trigger apoptosis in not only infected cells but also in uninfected cells.

HIV-infection is characterized by an early impairment of cytokine synthesis involved in cell-mediated immunity.[10] Cytokines such as IL-2 play a crucial role in promoting both the survival and proliferation of lymphocytes, and cytokine deprivation renders activated lymphocytes highly susceptible to apoptosis.[11,12] Therefore it is of interest to discuss the interrelationship between cytokines and apoptosis during HIV infection.

CD4 T CELL HOMEOSTASIS IN HIV INFECTION

HIV-Induced Destruction and Impaired CD4 T Cell Renewal

Despite years of investigation, the dynamic basis for CD4 T cell depletion in HIV infection remains controversial. Since 1995, a series of studies have increasingly challenged the paradigm that HIV infection induces high CD4 T cell production and turnover as the result of virus-induced T cell destruction. First, it was proposed that a large (10^9) number of CD4 T cells is infected and destroyed every day, and an equal number of T cells is produced to compensate for the loss.[13,14] The outcome of this process of massive production/destruction of cells is exhaustion of the T cell regenerative process resulting in the progressive depletion in CD4 T cells. However, this hypothesis is not compatible with the very low frequency of productively HIV-infected T cells in lymphoid tissues (1/100 to 1/1000),[15] and with the estimate of the turnover of CD4 T cells. The turnover of T cells in the course of HIV infection, as estimated by analyzing T cell telomere length, has been found to be considerably increased in the CD8 T cell subset of HIV-infected subjects, but much less in the CD4 T subset.[16] More recent analyses of the proportion of lymphocytes engaged in the cell cycle, either *ex vivo* by the detection of Ki67,[17–18] or *in vivo* following injection of [^{2}H] glucose,[19,20] have shown that CD4 T cell turnover is two- to three-fold higher in HIV-infected patients compared with HIV-negative controls. These studies also found that production of new CD4 T cells in HIV-infected subjects is not significantly different from that in healthy controls. In addition to inducing apoptosis, virus replication inhibits CD4 T cell production in primary lymphoid organs.

Following suppression of HIV by highly active anti-retroviral therapy (HAART), a dramatic increase in CD4 T cell production is observed.[20] This is correlated with a relative thymic activity and with the appearance in the blood of naïve T cells expressing some products of the T cell receptor rearrangement (TREC) characterizing recent thymic emigrants.[21] Therefore, the current understanding of CD4 T cell homeostasis in the course of HIV infection is that the progressive depletion of CD4 T lymphocytes is the consequence of both their destruction by several virus-dependent mechanisms and the lack of compensation by newly produced CD4 T cells because of a possible blockade of the CD4 T cell renewal machinery at the level of bone marrow and of thymus.[22] Under HAART, a rapid increase in the number of CD4 T cells is detected in patients, which reflects both the survival and the regeneration of the pool of CD4 T cells.

Molecular Pathways of Apoptosis Triggered by HIV

Peripheral blood T cells from asymptomatic HIV-infected persons are prematurely prone to activation-induced cell death, triggered *ex vivo* either by TCR ligation by specific antibodies or superantigens, or by ligation of the death receptors CD95, TNF-RI and TNF-RII, or following activation by mitogens or PMA$^+$ ionomycin.[4, 23–27] In addition, spontaneous apoptosis can be detected in patients' T cells following a short-term incubation in medium alone.[4,23,25] This premature cell death concerns not only CD4 T cells, but also CD8 T cells, B cells and antigen-presenting cells (macrophages and dendritic cells), and it can detected *in vivo* in lymph nodes and in tonsillar tissue of HIV-infected patients.[28–30] The central paradox of HIV pathogenesis is that the viral burden seems too low to deplete the CD4 population by direct killing. The observation that an important fraction of T cells is prematurely primed for apoptosis in HIV-infected subjects prompted the hypothesis that some indirect mechanisms are responsible for inappropriate cell death and significantly contribute to CD4 T cell depletion[7,8] and also to CD8 T cell destruction in AIDS.[31] This hypothesis was further supported by the observation that apoptotic T cells in lymph node sections of HIV-infected children and SIV-infected macaques are predominantly uninfected bystander cells, whereas infected cells are not found to be apoptotic.[32] This has been confirmed by studies highlighting the substantial number of apoptotic cells in lymph nodes of HIV-infected adults.[28,29]

The relevance of PCD to AIDS pathogenesis has been suggested by several studies. For example, the proportion of CD4 and CD8 T lymphocytes undergoing apoptosis spontaneously or after ligation of the TCR or the CD95 receptor increases with disease evolution, evaluated by the *in vivo* reduction of the CD4 T cell number.[26, 27,33] In addition, comparative studies in pathogenic models of lentiviral infection, including macaques infected with SIV[23,34] or cats infected with FIV[35,36] versus nonpathogenic models including SIV-infected african green monkeys[34] or chimpanzees infected with HIV or SIVcpz,[23,34,37] revealed that increased lymphocyte apoptosis is observed only in pathogenic lentiviral infections. This suggests that apoptosis can significantly contribute to AIDS pathogenesis and that it could be the mechanism responsible for the clearance of activated but healthy T cells and consequently contribute to impoverishment of the pool of effectors (Th and CTL) and antigen-presenting cells.

Apoptosis plays a crucial role in the homeostatic control of cell numbers following primary viral infection, ensuring the clearance of primed lymphocytes in order to terminate the immune response and to avoid autoimmune reactions.[38,39] Nevertheless, this normal process of elimination of activated cells might be detrimental for the immune system in the case of a chronic infection such as that induced by HIV. Indeed, a general state of immune activation is observed in the asymptomatic phase of HIV infection both in lymphoid tissues and peripheral blood lymphocytes and persists throughout the entire course of HIV infection. This is reflected by the expression of activation markers such as HLA-DR, CD45R0 and CD38 in CD4 and CD8 T cells in the lymph nodes[28,40] and in the peripheral blood.[41] In addition to the unceasing expression of HIV antigens, exogenous factors, such as opportunistic pathogens, stimulate the production of proinflammatory cytokines, including TNF-α, IL1β and IL-6, which drive cellular activation and viral replication.[42] This unbalanced immune activation might be the primary mechanism responsible for the premature cell death in AIDS. Indeed, apoptotic cells in patients' lymphoid tissues and in blood exhibit an activated phenotype[28,33] and there is a statistically significant correlation between the intensity of spontaneous or TCR-triggered apoptosis in both CD4 and CD8 subsets and their *in vivo* activation state.[33] In addition, the lack of immune activation in the nonpathogenic HIV-1 infection in chimpanzees is associated with a very low level of T cell apoptosis[23,37] and we have recently reported that the lower pathogenicity of HIV-2-infection compared to HIV-1 in Africa is associated with a lower level of immune activation and less T cell apoptosis.[43]

At the molecular level, the unbalanced immune activation in HIV-1-infection is associated with the *in vivo* down-regulation of Bcl-2 expression and up-regulation of CD95/CD95L expression in patients' lymphocytes and monocytes. A significant correlation has been found between the reduced level of Bcl-2 expression in T cells, particularly in the CD8 subset, and the propensity of lymphocytes to undergo apoptosis.[44] Phenotypic analysis of the low Bcl-2 CD8 T cells from patients suggested that they were cytotoxic since they lacked CD28 expression and expressed the cytotoxic TIA-1 granules.[44] Interestingly, a similar subset, characterized as $CD8^+$ $CD45R0^+$ $TIA\text{-}1^+$ Bcl-2 low, has also been found in patients' lymph nodes.[40] The lack of survival factors might be responsible for the priming for apoptosis of this subset, as detailed below, and may contribute to the loss of anti-viral cytotoxic activity throughout HIV infection.

The *in vivo* involvement of the CD95 pathway in T cell apoptosis during HIV infection has been suggested by concomitant increased expression of CD95 on both CD4 and CD8 T cell subsets from patients, and increased susceptibility to CD95-induced apoptosis, which is positively correlated with disease progression.[27,37,44–46] Elevated levels of soluble CD95 have been detected in plasma of patients and found to a be predictive marker for progression to AIDS.[48] High levels of CD95L transcripts[49] and CD95L cell surface molecule[27] have been also detected in both CD4 and CD8 T cells from HIV-infected persons. Also, a significant increase in macrophage-associated CD95L has been detected in lymphoid tissue from HIV^+ subjects, which is correlated with the degree of tissue apoptosis.[50] The up-regulation of CD95L on CD4 T lymphocytes and macrophages, which can be induced *in vitro* by HIV infection or through the direct effect of gp120, Tat or Nef[51–53] makes these viral proteins possible effectors in killing CD95 expressing cells.[52,54] Interestingly, experiments performed on lymphocytes from HIV-infected chimpanzees showed that their resistance to CD4 T cell depletion is associated with the lack of susceptibility of their T lymphocytes to CD95-induced apoptosis.[37] Thus, the CD95 system is dysregulated in HIV infection.

In addition to CD95L, other members of the TNF family are involved in HIV-induced apoptotic cell death. TRAIL (TNF-related-apoptosis-inducing-ligand) was identified as an apoptosis-inducing factor on T cells from HIV-infected patients.[55] Apoptosis in CD8 T cells may involve the TNF/TNF-R system since *in vitro* binding of HIV gp120 on the chemokine receptor CXCR4, co-receptor for HIV, induces both the up-regulation of membrane TNF on macrophages and TNF-RII on peripheral CD8 T cells, leading to macrophage-dependent apoptosis of CD8 T cells.[56] The binding of HIV gp120 on CXCR4 also induces a rapid cell death in normal CD4 T lymphocytes, independent of CD95, which is inhibited by SDF-1.[57] The induction of apoptosis through CXCR4 by gp120 or SDF-1 was also reported on human neuronal cells in the absence of CD4 molecules.[58]

HIV-DEPENDENT DYSREGULATION OF CYTOKINE SYNTHESIS IS RELATED TO THE PRIMING FOR APOPTOSIS OF T HELPER CELLS

HIV-Infection Is Associated with Alteration in the Frequency of Th1 Cells Consequent to Their Priming for Activation-Induced Apoptosis

Upon activation, naïve T cells differentiate into memory/effector cells, which are able to proliferate following subsequent exposure to an antigen. According to the cytokine environment in which they develop, and to the nature of the co-stimulatory signals delivered by the antigen-presenting cells, naïve T cells can differentiate into effector cells with polarized patterns of cytokine production. Th1 cells secrete IL-2, IFN-γ and TNF-α, which are important mediators of cellular immunity and are involved in macrophage activation, whereas Th2 cells secrete IL-4, IL-5, IL-6, IL-10 and IL-13, which provide help to B cells, and antagonize the macrophage-activating action of Th1 cytokines.[59] HIV-infection induces an early impairment of cytokine synthesis. A progressive change in the balance of type-1 vs. type-2 cytokines produced by patients' PBMC in response to *in vitro* stimulation with recall or opportunistic antigens has been reported, associated with disease progression.[60] Thus, it was suggested that progression of HIV-infection is accompanied by a decline in levels of IL-2, IL-12 and IFN-γ and increased production of IL-4, IL-5, IL-6 and IL-10, and that this shift was predictive for the reduction in CD4 cell counts, time to AIDS diagnosis and time to death.[60] We have used single cell analysis to enumerate Th1/Th2 subsets derived from peripheral T cells stimulated *in vitro* and determine the frequency, phenotype and priming for apoptosis of cells that are potentially capable of producing a given cytokine.[61] We found that HIV-infection is associated with a differential alteration in representation of Th1 subsets, rather than commitment of T cells to secrete Th2 cytokines (FIG. 1). Indeed, a significant decrease in the frequency of T cells expressing IL-2 or TNF-α was observed, whereas those producing IFN-γ remained preserved.[62] The disappearance of IL-2–producing T cells in HIV-infected persons was found to be a good indicator of disease progression and correlated with the progressive shrinkage of the naïve $CD45RA^+$ $CD4^+$ T cell compartment, the main source of IL-2.[62] Analysis of the priming for activation-induced cell death of Th cell subsets revealed that the intrinsic capacity of lymphocytes to produce a given cytokine can influence their survival. Indeed, T lymphocytes committed to IFN-γ or TNF-α production are more sensitive to activation-induced

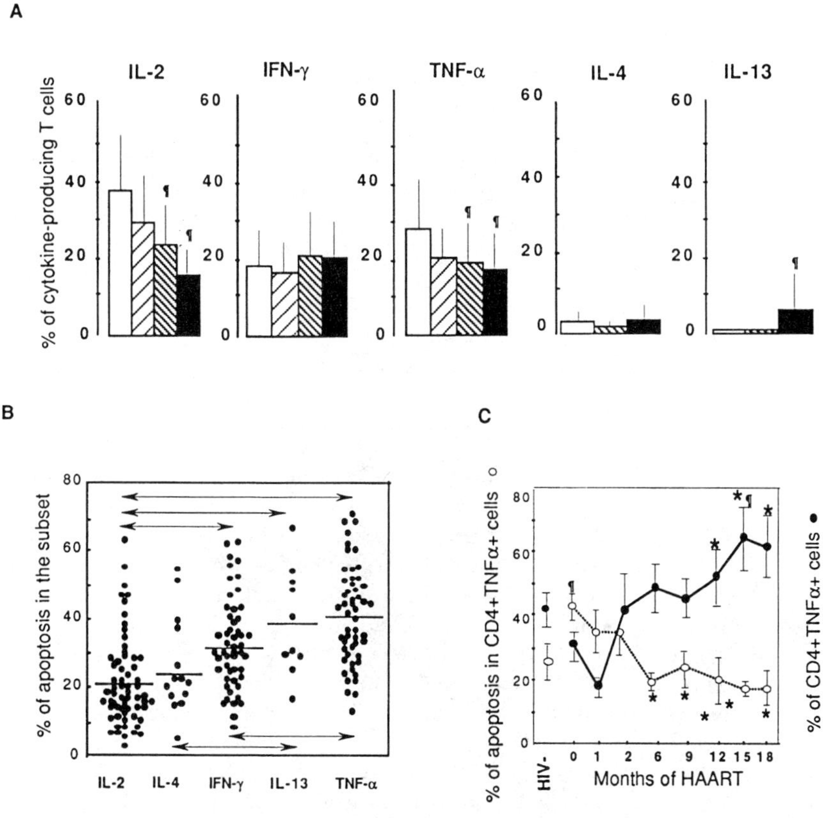

FIGURE 1. Concomitant detection of cytokine production and priming for apoptosis in peripheral CD3$^+$ T cells. PBMCs were stimulated 16 h with PMA, PHA and Ionomycin, and Brefeldin A was added during the last 12 h of culture to inhibit protein secretion. Susceptibility to apoptosis of cytokine-producing T cells was determined by concomitant nuclear 7-AAD staining. **A.** Determination of the percentage of T cells committed to type 1 or type 2 cytokine synthesis during HIV-disease evolution in non-PI treated patients ([□] HIV-negative controls; [▨] HIV-infected patients with *ex vivo* CD4 percentage > 29%; [▧], < 28% >13% [■] < 13 %). **B.** Differential susceptibility to apoptosis in the different cytokine-producing CD3 T cell subsets from non-PI treated HIV-infected patients. **C.** Follow-up of the priming of CD4$^+$ T cells for TNF-α synthesis and of their susceptibility to activation-induced apoptosis in a cohort of 15 HIV-infected patients submitted to HAART (including two nucleosidic inhibitors of the reverse transcriptase, and the protease inhibitor indinavir). [*] $p < 0.05$ vs. baseline level (Wilcoxon matched pairs test); [¶] $p < 0.05$ vs. HIV-negative controls (Mann-Whitney U test); [↔] $p > 0.05$ between the two T cell populations (Mann-Whitney U test).

apoptosis than lymphocytes committed to IL-2-production.[62] This gradient of susceptibility to apoptosis (IL-2 < IFN-γ < TNF-α) was detected in both CD4 and CD8 T cell subsets, in both control donors and HIV-infected patients, in whom the susceptibility to apoptosis of IL-2 and IFN-γ producers was increased compared with controls. This differential intrinsic apoptosis susceptibility of Th1 effectors was found to be tightly regulated by Bcl-2 expression. In HIV-infected persons, the progressive decrease in the proportion of IL-2-producing T cells was found to be correlated with their susceptibility to apoptosis and disease progression. On the other hand, IFN-γ synthesis was found preserved in patients despite an increased rate of apoptosis in the IFN-γ–producing subset. Indeed, the increased cell death of IFN-γ producers is probably counterbalanced by the increased percentage of $CD8^+$ T cells, the main source of IFN-γ producers.[62] These observations indicate the important relationship between apoptosis in peripheral T cells and the functional collapse of the immune system in AIDS.

Anti-Apoptotic Effect of Th1 Cytokines, Including IL-2 and IL-15, on Lymphocytes from HIV-Infected Patients: Consequences for Anti-Viral Immunity

IL-2 has been shown to exhibit pro- or anti-apoptotic properties according to the cellular activation status and to the cellular microenvironment. Preincubation of mouse T cells with IL-2 can increase their susceptibility to apoptosis after TCR stimulation. In contrast, IL-2 can rescue cells, including lymphocytes from HIV^+ subjects, from spontaneous apoptosis induced by growth factor withdrawal.[23] This rescue is controlled by molecules of the Bcl-2 family.[63–65] Administration of IL-2 to HIV-infected individuals has been shown to significantly increase CD4 T cell numbers and preserve anti-viral immune function.[66,67] IL-15 exhibits many activities in common with IL-2, including stimulating T cell proliferation, induction of cytolytic effector cells and NK cell activity, which are impaired in HIV infection[68–71] and recent attention has turned towards IL-15 as a possible alternative immunotherapy. The overlapping biological activities of IL-15 and IL-2 are attributable to the utilization by these cytokines of two common receptor chains β and γ.[72] In recent studies we compared biological effects of IL-2 and IL-15 from HIV-infected patients, and found that IL-15 had a more potent survival effect on NK cells as well as on naïve CD4 T cells, which are preferentially lost in HIV-infected individuals.[69,70] In addition, IL-15 was found to act as a potent survival factor in the prevention of spontaneous apoptosis (FIG. 2). The greater enhancement of lymphocyte survival promoted by IL-15 as compared with IL-2 when used at an equivalent concentration was associated with higher up-regulation of bcl-2 expression. Furthermore, IL-15 was more potent than IL-2 in not only stimulating proliferation of lymphocytes, but also increasing their apoptosis-susceptibility, which appeared to be associated with the greater ability of IL-15 to activate lymphocytes. Therefore, lymphocyte loss occurring by growth factor deprivation in HIV infection may be potentially prevented by IL-15, although its benefits for survival need to be closely assessed against its ability to augment lymphocyte activation.[65]

In addition to IL-2 and IL-15, various cytokines have been reported to regulate apoptosis of T cells from HIV-infected patients. IL-10 appears to promote antigen-stimulated lymphocyte death.[73] We[65] and others[74] have found IL-12 to partially

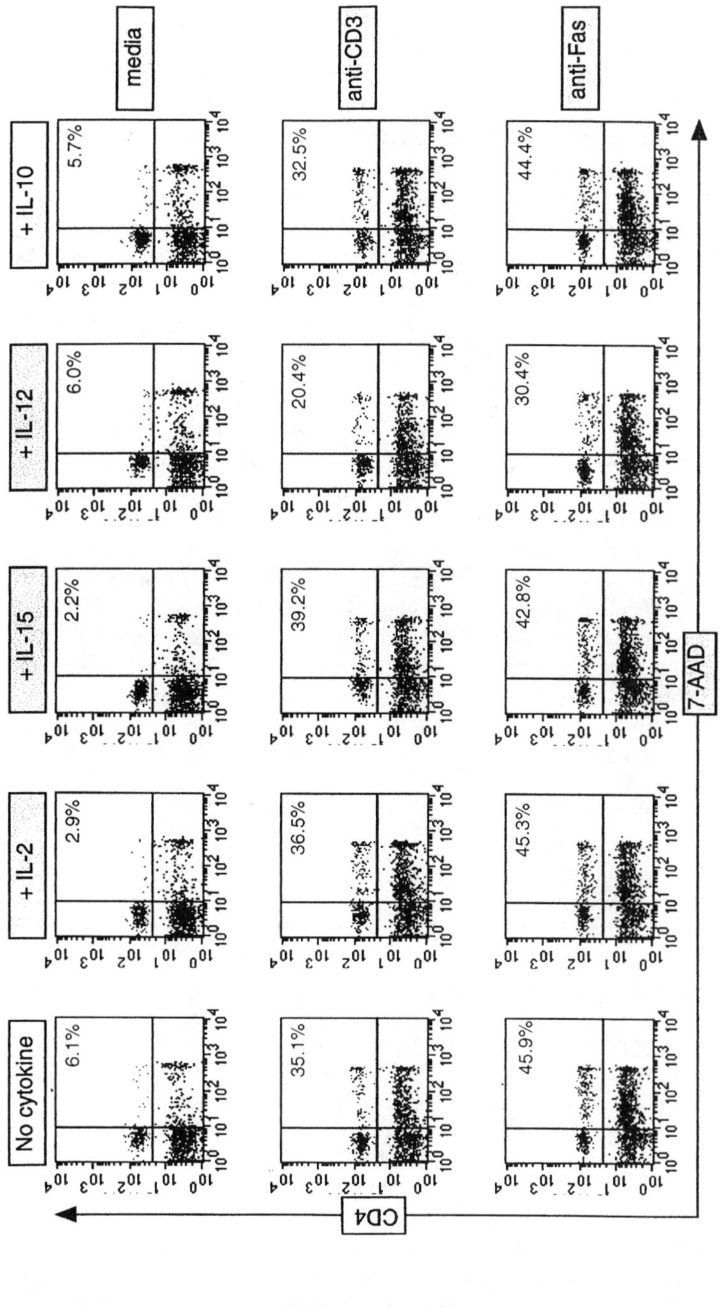

FIGURE 2. PBMC from a representative HIV-infected subject were cultured overnight in the presence of medium, anti-CD3 or anti-CD95 mAbs, in the presence of indicated cytokines. The proportion of apoptotic cells within CD4 T cells has been determined following dual staining with anti-CD4 mAbs and 7-AAD.

inhibit anti-CD95-induced death, while having little effect on spontaneous apoptosis. Reduced apoptosis has also been observed when IL-2 or IL-15 were added to anti-CD95-stimulated lymphocyte cultures (FIG. 2). However such inhibition was not consistently observed for all patients,[65] and was only marked in cultures that exhibited an appreciable reduction of spontaneous apoptosis in the presence of these cytokines, suggesting that the decrease in anti-CD95-induced apoptosis may be attributable to prevention of spontaneous apoptosis by these cytokines.[65] Interestingly, IL-15 was found *in vivo* to be more effective than IL-2 in protecting against lethal CD95-induced hepatic failure in mice.[75]

Antiviral therapy for HIV-infected patients has greatly improved in the recent years, and administration of highly active antiviral therapy (HAART) has been successful in decreasing plasma viremia to undetectable levels and has dramatically changed the clinical course in a substantial proportion of patients. However, as discussed below, the use of these combinations for prolonged periods of time is not without drawbacks such as risks of side effects and viral resistance. Furthermore, the persistence of a pool of latently infected cells from which virus could be isolated in essentially all treated-infected patients and the decline of HIV-specific helper CD4 and cytotoxic CD8 T cells (CTL) responses during HAART have stimulated the search for therapy regimens that tip the balance in favor of the host against the virus. Several trials have evaluated the effect of IL-2 administration, intravenously or subcutaneously, in HIV-infected patients. IL-2 was found to induce substantial expansion of the CD4 T cell compartment and significant improvement of T cell proliferative response against recall antigens and CTL against HIV proteins.[66,76,77] In a recent study, we also showed that, in contrast to HAART, prolonged IL-2 therapy for 4 years progressively decreased the viral reservoir, suggesting that IL-2 has a role in the diminution of the frequency of the number of CD4 T cells that harbor replication-competent virus, while maintaining virus-specific immunological CD4 T cell responses.[67] Thus immunotherapy with IL-2 may be an interesting alternative to the use of more complex HAART regimes and may delay the use of potent drugs in patients.

IMPACT OF ANTI-VIRAL THERAPIES ON PROGRAMMED CELL DEATH, T CELL RENEWAL AND HOMEOSTASIS OF CYTOKINE-PRODUCING CELLS

T Cell Apoptosis is Down-Regulated under HAART

The clinical responses to potent anti-retroviral therapies, a combination of HIV reverse transcriptase (RTI) and protease inhibitors (PI)) that reduce viral load to undetectable levels and concomitantly increase CD4 T cell counts, have considerably challenged the question of the mechanisms controlling the homeostasis and reconstitution of the immune system in treated-HIV-infected persons. Recent studies have pointed out the complexity of the mechanisms involved. The increase in CD4 T cells after HAART results from a combination of both the release of sequestered cells from lymphoid compartments to peripheral blood[78] and the increased production rate of circulating T lymphocytes. The latter appears to be due to peripheral T cell expansion,[18–20] suppression of apoptosis[47,79] and production of recent thymic emigrants,[21] suggesting that the mechanisms of renewal of CD4 T cells are still operational. A substantial drop in spontaneous, activation-induced and CD95-triggered apoptosis is observed in both CD4 and CD8 T cells from patients soon after initiation of HAART. This occurs before the decrease in immune activation, and the resistance to CD95-induced apoptosis precedes the down-regulation of CD95 expression.[47,79] Thus, suppression of the plasma viral load is associated with the regulation of apoptosis, which reaches normal values detected in T cells from healthy donors. Suppression of activation-induced apoptosis under HAART is likely to be a multifactorial process. Indeed, the mechanisms involved in apoptosis susceptibility of patients' lymphocytes during the chronic phase of HIV-infection, i.e., the *in vivo* immune activation state of these lymphocytes and the proapoptotic effect of some HIV proteins, are down-regulated under HAART, restoring the global survival of patients' lymphocytes. In addition, we cannot exclude a direct immunomodulatory effect of HIV PIs on T cells. Indeed PIs were shown to suppress physiological apoptosis by inhibition of cellular proteases[79] and exert a direct impact on cytokine production.[80]

HAART Induces an Alteration of TNF-α–Regulated T Cell Homeostasis Linked to the Suppression of Physiological Apoptosis: Consequences for the Development of the Lipodystrophy Syndrome

Long-term HAART has been associated with a unique and unexpected syndrome, called lipodystrophy, consisting of alterations in lipid metabolism and body fat redistribution. Lipodystrophy is associated with hyperlipidemia, insulin resistance, peripheral loss of fatty tissue and visceral abdominal fat accumulation.[81] The pathogenesis of this syndrome, which may lead to premature coronary artery disease recently reported in PI-treated HIV-infected patients, remains unknown. Cytokines, such as TNF-α, induce a number of alterations in lipid metabolism which can produce hyperlipidemia. Indeed, TNF-α inhibits the intake of free fatty acids (FFA) by adipocytes via the inhibition of lipoprotein lipase, leading to fat wasting, but can also increase lipogenesis via the stimulation of the hepatic triglyceride synthetase, leading to hyperlipidemia.[82] In addition, TNF-α promotes insulin resistance via both increasing circulating FFA and directly acting on the insulin receptor. In a recent study, we investigated the influence of HAART on TNF-α synthesis and whether alterations in lipid metabolism were related to modifications in the frequency of TNF-α T cell producers.[83] We had previously demonstrated that apoptosis plays an essential role in the negative regulation of TNF-α synthesis by T cells. Indeed, even in healthy donors, TNF-α producers are highly sensitive to apoptosis; this is in part related to the decreased expression of Bcl-2 molecule.[62] During the natural progression of HIV infection, susceptibility to apoptosis of TNF-α producers progressively increased and this was correlated with their decreased representation in the blood of patients.[62] Under HAART, apoptosis in TNF-α–T cell producers is highly suppressed, leading to their progressive accumulation in the blood.[83] This was found to be related to the cosynthesis by TNF-α–T cell producers of IL-2,[83] a survival factor. As discussed above, the direct effect of PI on cell proteases has also to be considered as a mechanism contributing to the accumulation of TNF-α–T cell producers. Interestingly, this T cell polarization to TNF-α synthesis seems to favor the development of the lipodystrophy syndrome by contributing to alteration in lipid metabolism.

Indeed, significant positive correlations were found between the absolute number of TNF-α–CD8 T-cell precursors and lipid parameters usually altered in lipodystrophy including cholesterol, triglycerides and the atherogenic ratio apolipoprotein B/apoA1.[83] In addition, recent studies have reported unusual clinical inflammatory syndromes after initiation of HAART, such as immune recovery vitritis (IRV) in CMV retinitis patients or focal mycobacterial lymphadenitis, and our observations suggest that, through the dysregulation of TNF-α–regulated T-cell homeostasis, HAART creates a proinflammatory environment, which might contribute to the development of these "immune restoration inflammatory diseases."

CONCLUSIONS

The chronicity of the lentiviral infection is responsible for the induction of premature cell death of lymphocytes from HIV-infected persons. Continuous production of viral proteins leads to an unbalanced immune activation and to the triggering of apoptotic programs, turning mononuclear cells, including CD4 T cells, CD8 T cells and APC, into effectors of apoptosis, leading to fratricidal destruction of healthy uninfected cells expressing the death receptors. Inappropriate PCD is also responsible for the disappearance of T helper cells primed for type-1 cytokine synthesis, thus contributing to the lack of survival factors which could prevent spontaneous lymphocyte apoptosis. Under potent anti-retroviral therapies, a significant decrease in spontaneous, TCR- and CD95-induced lymphocyte apoptosis is observed, concomitant with a partial quantitative and qualitative restoration of the immune system in treated patients. However, owing to the suppressive effect of antiretroviral drugs on physiological apoptosis, these therapies are associated with alteration of TNF-α–regulated T cell homeostasis, leading to an accumulation in the blood of T cells primed for TNF-α synthesis, and contributing to the development of a new syndrome associated with these treatments, the lipodystrophy syndrome.

ACKNOWLEDGMENTS

This work was supported by grants from the Agence Nationale de Recherche sur le SIDA (ANRS), the Fondation pour la Recherche Médicale (FRM) (Sidaction), the Centre National de la Recherche Scientifique (CNRS), the Pasteur Institute and the EU (contracts ERB-IC15-CT97-0901 and BMH4-CT97-2055). E.L. and H.N. were supported by FRM (Sidaction) and M.B. by EU.

REFERENCES

1. Malkovsky, M., J.-J. Fournié, P. Fisch, F. Poccia, M. Wallace & M.-L. Gougeon. 2000. Alternative cytotoxic effector mechanisms in infections with immunodeficiency viruses: γδ T lymphocytes and NK cells. *In* AIDS 2000. B. Hann & J. Heeney, Eds. In press

2. Koup, R.A., J.T. Safrit, Y. Cao, C.A. Andrews, G. McLeod, W. Borkowsky, C. Farthing & D.D. Ho. 1994. Temporal association of cellular immune responses with the initial control of viremia in primary human immunodeficiency virus type 1 syndrome. J. Virol. **68:** 4650.
3. Cocchi, R. 1995. Identification of RANTES, MIP-1α, and MIP-1β as the major HIV-suppressive factors produced by $CD8^+$ T cells. Science **270:**1811.
4. Gougeon, M.-L., R. Olivier, S. Garcia, D. Guétard, T. Dragic, C. Dauguet & L. Montagnier. 1991. Evidence for an engagement process towards apoptosis in lymphocytes of HIV infected patients. C. R. Acad. Sci. **312:** 529–537.
5. Groux, H., G. Torpier, D. Monté, Y. Mouton, A. Capron & J.C. Ameisen. 1992. Activation-induced death by apoptosis in lymphocytes from human immunodeficiency virus-infected asymptomatic individuals. J. Exp. Med. **175:** 331–340.
6. Meyaard, L., S.A. Otto, R.R. Jonker, M.J. Mijnster, R. Keet & F. Miedema. 1992. Programmed death of T cells in HIV-1 infection. Science **257:** 217–219.
7. Ameisen, J.C. & A. Capron. 1991. Cell dysfunction and depletion in AIDS: the program cell death hypothesis. Immunol. Today **12:** 102–105.
8. Gougeon, M.-L. & L. Montagnier. 1993. Apoptosis in AIDS. Science **260:** 1269–1270.
9. Finkel, T.H., G. Tudor-Williams, N.K. Banda, *et al.* 1995. Apoptosis occurs predominantly in bystander cells and not in productive cells of HIV- and SIV-infected lymph nodes. Nature Med. **1:** 129–.
10. Clerici, M. & G.M. Shearer. 1993. A Th1->Th2 switch is a critical step in the etiology of HIV infection. Immunol. Today **14:** 107–111.
11. Boise, L.H. & C.B. Thompson. 1996. Hierarchical control of lymphocyte survival. Science **274:** 67–68.
12. Lenardo, M.J. 1997. The molecular regulation of lymphocyte apoptosis. Semin. Immunol. **9:** 1–5.
13. Ho, D.D., A.U. Neumann, A.S. Perelson, W. Chen, J.M. Leonard & M. Markowitz. 1995. Rapid turnover of plasma virions and CD4 lymphocytes in HIV-1 infection. Nature **373:** 123–127.
14. Mohri, H., S. Bonhoeffer, S. Monard, A.S. Perelson & D.H. Ho. 1998. Rapid turnover of T lymphocytes in SIV-infected rhesus macaques. Science **279:** 1223–1227.
15. Haase, A.T. *et al.* 1996. Quantitative image analysis of HIV-1 infection in lymphoid tissues. Science **274:** 985–989.
16. Wolthers, K.C. *et al.* T cell telomere length in HIV-1 infection: no evidence for increased $CD4^+$ T cell turnover. Science **274:** 1543–1545.
17. Sachsenberg, N. *et al.* 1998. Turnover of CD4 and CD8 T lymphocytes in HIV-1 infection as measured by Ki67 antigen. J. Exp. Med. **187:** 1295–1303.
18. Fleury, S. *et al.* 1998. Limited $CD4^+$ T cell renewal in early HIV-1 infection: effect of highly active antiviral therapy. Nature Med. **4:** 794–801.
19. Macallan, D.C. *et al.* 1998. Measurements of cell proliferation by labeling of DNA with stable isotope-labeled glucose: studies in vitro, in animals and in humans. Proc. Natl. Acad. Sci. USA **95:** 708–713.
20. Hellerstein, M., M.B. Hanley, D. Cesar, S. Siler, C. Papageorgopoulos, E. Wieder, D. Schmidt, R. Hoh, R. Neese, D. Macallan, S. Deeks & J.M. McCune. 1999. Directly measured kinetics of circulating T lymphocytes in normal and HIV-infected humans. Nature Med. **5:** 83–89.
21. Douek, D.C., R.D. McFarland, P.H. Keiser, *et al.* 1998. Changes in thymic function with age and during the treatment of HIV infection. Nature **396:** 690–695.
22. Heeney, J.L. 1995. AIDS: a disease of impaired Th-cell renewal? Immunol. Today. **16:** 515.
23. Gougeon, M.-L., S. Garcia, J. Heeney, R. Tschopp, H. Lecoeur, D. Guetard, V. Rame, C. Dauguet & L. Montagnier. 1993. Programmed cell death of T lymphocytes in AIDS related HIV and SIV infections. AIDS Res. Hum. Retrov. **9:** 553–563.

24. GROUX, H., G. TORPIER, D. MONTÉ, Y. MOUTON, A. CAPRON & J.C. AMEISEN. 1992. Activation-induced death by apoptosis in lymphocytes from human immunodeficiency virus-infected asymptomatic individuals. J. Exp. Med. **175:** 331–340.

25. MEYAARD, L., S.A. OTTO, R.R. JONKER, M.J. MIJNSTER, R. KEET & F. MIEDEMA. 1992. Programmed death of T cells in HIV-1 infection. Science **257:** 217–219.

26. BÖHLER, T., C. BÄUMLER, I. HERR, *et al.* 1997. Activation of the CD95 system increases with disease progression in HIV1-infected children and adolescents. Pediatr. Infect. Dis. J. **16:** 754–759.

27. SLOAND, E.M., N.S. YOUNG, P. KUMAR, F.F. WEICHOLD, T. SATO & J.P. MACIEJEWSKI. 1997. Role of Fas ligand and receptor in the mechanism of T-cell depletion in AIDS: effect on $CD4^+$ lymphocyte depletion and HIV replication. Blood **89:** 1357–1363.

28. MURO-CACHO, C.A., G. PANTALEO & A. FAUCI. 1995. Analysis of apoptosis in lymph nodes of HIV-infected persons. Intensity of apoptosis correlates with the general state of activation of the lymphoid tissue and not with the stage of disease or viral burden. J. Immunol. **154:** 5555–5566.

29. AMENDOLA, A., M.-L. GOUGEON, F. POCCIA, A. BONDURAND, L. FESUS & M. PIACENTINI. 1996. Induction of tissue transglutaminase in HIV pathogenesis. Evidence for a high rate of apoptosis of CD4 T lymphocytes and accessory cells in lymphoid tissues. Proc. Natl. Acad. Sci. USA **93:** 11057–11062.

30. ROSOK, J.E., G. STENT, R. BJERKNES, P. VOLTERSVIK, J. OLOFSSON & B. ASJO. 1998. Correlates of apoptosis of $CD4^+$ and $CD8^+$ T cells in tonsillar tissue in HIV type 1 infection. AIDS Res. Hum. Retroviruses **14:** 1635–1643.

31. GOUGEON, M.-L. 1995. Does apoptosis contribute to CD4 T cell depletion in HIV infection? Cell Death Diff. **2:** 1–.

32. FINKEL, T.H., G. TUDOR-WILLIAMS, N.K. BANDA, *et al.* 1995. Apoptosis occurs predominantly in bystander cells and not in productive cells of HIV- and SIV-infected lymph nodes. Nature Med. **1:** 129–.

33. GOUGEON, M.-L., H. LECOEUR, A. DULIOUST, M.G. ENOUF, *et al.* 1996. Programmed cell death in peripheral lymphocytes from HIV-infected persons: the increased susceptibility to apoptosis of CD4 and CD8 T cells correlates with lymphocyte activation and with disease progression. J. Immunol. **156:** 3509–.

34. ESTAQUIER, J.T., F. IDZIOREK, F. DE BELS, F. BARRE-SINOUSSI, *et al.* 1994. Programmed cell death and AIDS: significance of T cell apoptosis in pathogenic and nonpathogenic primate lentiviral infections. Proc. Natl. Acad. Sci. USA **91:** 9431.

35. BISHOP, S.A., T.J. GRUFFYDD-JONES, D.A. HARBOUR & C.R. STOKES. 1993. Programmed cell death as a mechanism of cell death in PBMC from cats infected with feline immunodeficiency virus (FIV). Clin. Exp. Immunol. **93:** 65.

36. HOLZNAGEL, E., R. HOFMANN-LEHMANN, C.M. LEUTENEGGER, K. ALLENSPACH, S. HUETTNER, *et al.* 1998. The role of in vitro-induced lymphocyte apoptosis in feline immunodeficiency virus infection: correlation with different markers of disease progression. J. Virol. **72:** 9025–9033.

37. GOUGEON, M.-L., H. LECOEUR, F. BOUDET, E. LEDRU, S. MARZABAL, S. BOULLIER, R. ROUE, S. NAGATA & J. HEENEY. 1997. Lack of chronic immune activation in HIV-infected chimpanzees correlates with the resistance of T cells to Fas/Apo-1 (CD95)-induced apoptosis and preservation of a Th1 phenotype. J. Immunol. **158:** 2964–2976.

38. AKBAR, A.N. & M. SALMON. 1997. Cellular environments and apoptosis: tissue microenvironments control activated T-cell death. Immunol. Today **18:** 72–76.

39. PIGNATA, C., M. FIORE, S. DE FILIPPO, M. CAVALCANTI, L. GAETANIELLO & I. SCOTESE. 1998. Apoptosis as a mechanism of peripheral blood mononuclear cell death after measles and varicella-zoster virus infections in children. Ped. Res. **43:** 77–83.

40. BOFILL, M., W. GOMBERT, N.J. BORTHWICK, A.N. AKBAR, J.E. MCLAUGHLIN, *et al.* 1995. Presence of $CD3^+CD8^+$Bcl-2low lymphocytes undergoing apoptosis and activated macrophages in lymph nodes of HIV-1^+ patients. Am. J. Pathol. **146:** 1542.

41. GIORGI, J.V., Z. LIU, L.E. HULTIN, W.G. CUMBERLAND, K. HENNESSEY & R. DETELS. 1993. Elevated level of $CD38^+CD8^+$ cells in HIV infection add to the prognostic value of low $CD4^+T$ cell level: results of 6 years follow-up. J. Acquired Immune Defic. Syndr. **6:** 904–.
42. BLANCHARD, A., L. MONTAGNIER & M.-L. GOUGEON. 1997. Influence of microbial infections on the progression of HIV disease. Trends Microbiol. **5:** 326–331.
43. MICHEL, P., A. TOURE-BALDE, C. ROUSSILHON, G. ARIBOT, J.-L. SARTHOU & M.-L. GOUGEON. 2000. Reduced immune activation and T cell apoptosis in HIV-2 compared to HIV-1 infected West African patients. Correlation of T cell apoptosis with seric β2 microglobulin and with disease evolution. J. Infect. Dis. **181:** 64.
44. BOUDET, F., H. LECOEUR & M.-L. GOUGEON. 1996. Apoptosis associated with ex vivo down-regulation of bcl-2 and up-regulation of Fas in potential cytotoxic $CD8^+$ T lymphocytes during HIV infection. J. Immunol **156:** 2282.
45. KATSIKIS, P.D., E.S. WUNDERLICH, C.A. SMITH, L.A. HERZENBERG & L.A. HERZENBERG. 1995. Fas antigen stimulation induces marked apoptosis of T lymphocytes in human immunodeficiency virus-infected individuals. J. Exp. Med. **181:** 2029–2036.
46. DEBATIN, K.-M., A. FAHRIG-FAISSNER, S. ENENKEL-STOODT, W. KREUZ, A. BENNER & P.H. KRAMMER. 1994. High expression of APO-1 (CD95) on T lymphocytes from HIV-infected children. Blood **83:** 3101–3103.
47. GOUGEON, M.-L., H. LECOEUR & Y. SASAKI. 1999. The CD95 system in HIV infection. Impact of HAART. Immunol. Lett. **66:** 97–103.
48. MEDRANO, F.J., M. LEAL, D. ARIENTI, C. REY, A. ZAGLIANI, Y. TORRES, A. SANCHEZ-QUIJANO, E. LISSEN & M. CLERICI. 1998. TNFβ and soluble APO-1/Fas independently predict progression to AIDS in HIV-seropositive patients. AIDS Res. Hum. Retroviruses **14:** 835–843.
49. BÄUMLER, C.B., T. BÖHLER, I. HERR, A. BENNER, P.H. KRAMMER & K.-M. DEBATIN. 1996. Activation of the CD95 (APO-1/Fas) system in T cells from human immunodeficiency virus type-1-infected children. Blood **88:** 1741.
50. DOCKRELL, D.H., A.D. BADLEY, J.S. VILLACIAN, C.J. HEPPELMANN, A. ALGECIRAS, S. ZIESMER, H. YAGITA, D.H. LYNCH, P.C. ROCHE, P.J. LEIBSON & C.V. PAYA. 1998. The expression of Fas ligand by macrophages and its upregulation by HIV infection. J. Clin. Invest. **101:** 2394–2405.
51. MITRA, D., M. STEINER, D.H. LYNCH, L. STAIANO-COICO & J. LAURENCE. 1996. HIV-1 upregulates Fas ligand expression in $CD4^+$ T cells in vitro and in vivo: association with Fas-mediated apoptosis and modulation by aurintricarboxylic acid. Immunology **87:** 581–585.
52. BADLEY, A.D., J.A. MCELHINNY, P.J. LEIBSON, D.H. LYNCH, M.R. ALDERSON & C.V. PAYA. 1996. Upregulation of Fas ligand expression by human immunodeficiency virus in human macrophages mediates apoptosis of uninfected T lymphocytes. J. Virol. **70:** 199–.
53. WESTENDORP, M.O., R. FRANK, C. OSCHSENBAUER, K. STRICKER, J. DHEIN, H. WALCZAK, K.-M. DEBATIN & P.H. KRAMMER. 1995. Sensitization of T cells to CD95-mediated apoptosis by HIV-1 Tat and gp120. Nature **375:** 497–500.
54. PIAZZA, C., M.S. MONTANI, S. GILARDINI, S. MORETTI, E. CUNDARI & E. PICCOLELLA. 1997. $CD4^+$ T cells kill $CD8^+$ T cells via Fas/Fas ligand-mediated apoptosis. J. Immunol. **158:** 1503.
55. KATSIKIS, P.D., M.E. GARCIA-OJEDA, J.F. TORRES-ROCA, I.M. TIJOE, C.A. SMITH, L.A. HERZENBERG & L.A. HERZENBERG. 1997. Interleukin-1-beta converting enzyme-like protease involvement in Fas-induced and activation-induced peripheral blood T cell apoptosis in HIV infection-TNF-related apoptosis-inducing ligand can mediate activation-induced T cell death in HIV infection. J. Exp. Med. **186:** 1365–1372.
56. HERBEIN, G., U. MAHLKNECHT, F. BATLIWALLA, *et al.* 1998. Apoptosis of $CD8^+$ T cells is mediated by macrophages through interaction of HIV gp120 with chemokine receptor CXCR4. Nature **395:** 189–194.
57. BERNDT, C., B. MOPPS, S. ANGERMULLER, P. GIERSCHIK & P.H. KRAMMER. 1998. CXCR4 and CD4 mediate a rapid CD95-independent cell death in CD4 T cells. Proc. Natl. Acad. Sci. USA **95:** 12556–12561.

58. HESSELGESSER, J., D. TAUB, P. BASKAR, M. GREENBERG, J. HOXIE, D.L. KOLSON & R. HORUK. 1998. Neuronal apoptosis induced by HIV-1 gp120 and the chemokine SDF-1a is mediated by the chemokine receptor CXCR4. Curr. Biol. **8:** 595–598.

59. MOSMANN, T.R. & S. SAD. 1996. The expanding universe of T-cell subsets: Th1, Th2 and more. Immunol. Today **17:** 138.

60. CLERICI, M. & G.M.SHEARER. 1994. The Th1/Th2 hypothesis of HIV infection: new insights. Immunol. Today, **15:** 575–580.

61. LECOEUR, H., E. LEDRU & M.-L. GOUGEON. 1998. A cytofluorometric method for the simultaneous detection of both intracellular and surface antigens on apoptotic peripheral lymphocytes. J. Immunol. Methods **217:** 11.

62. LEDRU, E., H. LECOEUR, S. GARCIA, T. DEBORD & M.-L. GOUGEON. 1998. Differential susceptibility to activation-induced apoptosis among peripheral Th1 subsets: correlation with Bcl-2 expression and consequences for AIDS pathogenesis. J. Immunol. **160:** 3194–3206.

63. AKBAR, A.N., N.J. BORTHWICK, R.G. WICKREMASINGHE, P. PANAYOITIDIS, D. PILLING, M. BOFILL, S. KRAJEWSKI, J.C. REED & M. SALMON. 1996. Interleukin-2 receptor common gamma-chain signaling cytokines regulate activated T cell apoptosis in response to growth factor withdrawal: selective induction of anti-apoptotic (bcl-2, bcl-xL) but not pro-apoptotic (bax, bcl-xS) gene expression. Eur. J. Immunol. **26:** 294.

64. ADACHI, Y., N. OYAIZU, S. THAN, T.W. MCCLOSKEY & S. PAHWA. 1996. IL-2 rescues in vitro lymphocyte apoptosis in patients with HIV infection: correlation with its ability to block culture-induced down-modulation of Bcl-2. J. Immunol. **157:** 4184.

65. NAORA, H. & M.-L. GOUGEON. 1999. Interleukin-15 is a potent survival factor in the prevention od spontaneous but not CD95-induced apoptosis in CD4 and CD8 T lymphocytes of HIV-infected individuals. Correlation with its ability to increase Bcl-2 expression. Cell Death Diff. **6:** 1002–1011.

66. KOVACS, J.A. *et al.* 1995. Increases in CD4 T lymphocytes with intermittent courses of Interleukin-2 in patients with human immunodeficiency virus infection. A preliminary study. N. Engl. J. Med. **332:** 567–575.

67. GOUGEON, M.L., C. ROUZIOUX, I. LIBERMAN, M. BURGARD, J.-P. VIARD, K. BOUCHENAFA, C. CAPITANT, J.F. DELFRAISSY, THE ANRS 048 GROUP & Y. LEVY. 2000. IL-2 therapy combined with dual nucleoside analogs could lead to long term reduction in the pool of HIV-1 infected cells and preservation of HIV specific immune responses. Submitted for publication.

68. KANAI, T., E.K. THOMAS, Y. YASUTOMI & N.L. LETVIN. 1996. IL-15 stimulates the expansion of AIDS virus-specific CTL. J. Immunol. **157:** 3681–3687.

69. NAORA, H. & M.-L. GOUGEON. 1999. Enhanced survival and potent expansion of the natural killer cell population of HIV-infected individuals by exogenous IL-15. Immunol. Lett. **68:** 359–367.

70. NAORA, H. & M.-L. GOUGEON. 1999. Activation, survival and apoptosis of CD45R0+ and CD45R0– T cells of HIV-infected individuals: effects of IL-15 and comparison with IL-2. Immunology **97:** 181–187.

71. CHEHIMI, J., J.D. MARSHALL, O. SALVUCCI, I. FRANK, S. CHEHIMI, S. KAWECKI, D. BACHELLER, S. RIFAT & S. CHOUAIB. 1997. IL-15 enhances functions during HIV infection. J Immunol. **158:** 5978–5987.

72. GIRI, J.G., M. ADHIEH, J. EISENMANN, K. SHANEBECK, K. GRABSTEIN, S. KUMAKI, A. NAMEN, L.S. PARK, D. COSMAN & D. ANDERSON. 1994. Utilization of the β and γ chains of the IL-2 receptor by the novel cytokine IL-15. EMBO J **13:** 2822–2830.

73. CLERICI, M., A. SARIN, J.A. BERZOFSKY, A.L. LANDAY, H.A. KESSLER, F. HASHEMI, C.W. HENDRIX, S.P. BLATT, J. RUSNAK, M.J. DOLAN, R.L. COFFMAN, P.A. HENKART & G.M. SHEARER. 1996. Antigen-stimulated apoptotic T-cell death in HIV infection is selective for CD4+ T cells, modulated by cytokines and effected by lymphotoxin AIDS **10:** 603–611.

74. ESTAQUIER, J., T. IDZIOREK, E. ZOU, D. EMILIE, C.-M. FARBER, J.-M. BOUREZ & J.C. AMEISEN. 1995. T helper type 1/T helper type 2 cytokines and T cell death: preventive effect of interleukin 12 on activation-induced and CD95 (Fas/APO-1)-mediated apoptosis of CD4+ T cells from human immunodeficiency virus-infected persons. J. Exp. Med. **182:** 1759–1767.
75. BULFONE-PAUS, S., D. UNGUREANU, T. POHL, G. LINDNER, R. PAUS, R. RUCKERT, H. KRAUSE & U. KUNZENDORF. 1997. Interleukin-15 protects from lethal apoptosis in vivo. Nature Med. **3:** 1124.
76. LEVY, Y. *et al.* 1999. Comparison of subcutaneous and intravenous interleukin-2 in asymptomatic HIV-1 infection: a randomised controlled trial. ANRS 048 study group. Lancet **353:** 1923–1929.
77. KELLEHER, A.D. *et al.* 1998. Effects of IL-2 therapy in asymptomatic HIV-infected individuals on proliferative responses to mitogens, recall antigens and HIV-related antigens. Clin. Exp. Immunol. **113:** 85–91.
78. PAKKER *et al.* 1998. Biphasic kinetics of peripheral blood T cells after triple combination therapy in HIV-1 infection: a composite redistribution and proliferation. Nature Med. **4:** 208–214.
79. SLOAND, E.M., P.N. KUMAR, S. KIM, A. CHAUDHURI, F.F. WEICHOLD & N.S. YOUNG. 1999. HIV-1 protease inhibitor modulates activation of peripheral blood CD4+ T cells and decreases their susceptibility to apoptosis in vitro and in vivo. Blood **94:** 1021.
80. ANDRE, P., M. GROETTRUP, P. KLENERMAN, *et al.* 1998. An inhibitor of HIV-1 protease modulates proteasome activity, antigen presentation and T cell responses. Proc. Natl. Acad. Sci. USA **95:** 13120.
81. CARR, A., K. SAMARAS, S. BURTON, *et al.* 1998. A syndrome of peripheral lipodystrophy, hyperlipidaemia and insulin resistance in patients receiving HIV protease inhibitors. AIDS **12:** F51–58.
82. FEINGOLD, K.R., M. SOUED, M.K. SERIO, A.H. MOSER, W. FIERS, C.A. DINARELLO & C. GRUNFELD. 1989. Multiple cytokines stimulate hepatic lipid synthesis in vivo. Endocrinology **125:** 267–274.
83. LEDRU, E., N. CHRISTEFF, O. PATEY, J.C. MELCHIOR & M.L. GOUGEON. 2000. Alteration of TNFα^+ T cell homeostasis following potent antiretroviral therapy. Contribution to the development of HIV-associated lipodystrophy syndrome. Blood **95:** 3191.

Quantitation of CD95 and CD95L mRNA Expression in Chronic and Acute HIV-1 Infection by Competitive RT-PCR

MARCELLO PINTI,[a] MILENA NASI,[a] LAURA MORETTI,[a] CRISTINA MUSSINI,[b] DANIELA PETRUSCA,[c] ROBERTO ESPOSITO,[b] AND ANDREA COSSARIZZA[a,d]

[a]*Department of Biomedical Sciences and* [b]*Infectious Diseases Clinics, University of Modena and Reggio Emilia, Modena 41100, Italy*

[c]*Center of Immunology, Bucharest, Romania*

ABSTRACT: Human immunodeficiency virus-type 1 (HIV-1) infection is characterized by increased immune cell apoptosis. Apoptosis can be triggered by signals that arise from within the cell, or by signals that are elicited by binding of extracellular "death ligands" to their "death receptors," most of which belong to the tumor necrosis factor (TNF)-receptor family, such as CD95 (Fas/Apo-1). In immune cells the oligomerization of CD95, induced by its ligand CD95L, and the recruitment of different intracytoplasmic molecules that in turn activate FLICE/caspase 8 are crucial. To study the role of CD95/CD95L interactions during HIV-1 infection, we developed an original method based upon quantitative-competitive (QC) RT-PCR that allowed us to quantify the amounts of mRNA coding for the total (tCD95) and membrane (mCD95) forms of CD95. We first studied the expression of different forms of CD95 mRNA in a classical model of chronic HIV infection using two infected cell lines of different origin—lymphocytic (ACH-2) or monocytic (U1). We have shown that infected cells of monocytic origin preferentially produce the "protective" (soluble) form of CD95, and no detectable CD95L mRNA, while lymphoid cells produce more mRNA for the membrane form of CD95 (which triggers apoptosis) along with low but detectable amounts of CD95L mRNA. One can hypothesize that a complex balance exists between pro-apoptotic events, perhaps triggered by the host to limit viral production, and anti-apoptotic events likely triggered by the virus to increase its production and survival. In cells of monocytic origin, which act as a reservoir for the virus, the anti-apoptotic molecules are favored; in cells of lymphocytic origin, molecules with an apoptotic meaning are prevalent.

KEYWORDS: HN-1; Apoptosis; CD95; QC-RT-PCR

[d]Address for correspondence: Andrea Cossarizza, M.D., Ph.D., Chair of Immunology, Department of Biomedical Sciences, University of Modena and Reggio Emilia School of Medicine, Via Campi 287, 41100 Modena, Italy. Voice: +39 059 2055.415; fax: +39 059 2055.426.
cossariz@unimo.it

INTRODUCTION

The infection with the human immunodeficiency virus-type 1 (HIV-1) is characterized by an increased tendency to undergo apoptosis that affects immune cells. Apoptosis can be triggered by signals that arise from within the doomed cell, or by signals that are elicited by binding of extracellular "death ligands" to their "death receptors," most of which belong to the tumor necrosis factor (TNF)-receptor family (TNF-RF), such as CD95 (Fas/Apo-1).[1] These type of molecules have an intracellular domain, the "death domain" (DD) that is capable of triggering a cascade of cysteine proteases (caspases), whose activity is essential for almost all pathways to apoptosis. Distinct members of the caspase family are involved in both the initiation and execution phases of cell death, with the initiator caspases being recruited to multicomponent signaling complexes. In immune cells a crucial role is played by the oligomerization of CD95, induced by its ligand CD95L, and the recruitment of different intracytoplasmic molecules that in turn activate FLICE/caspase 8.

After activation, immune cells can express either CD95 or other death receptors such as DR4, DR5, DR6 and Apo-3/DR3, which bind TRAIL (TNF-related-apoptosis-inducing-ligand)/Apo-2L, and Apo-3L, respectively; however, these cells have developed different strategies to protect themselves from apoptosis. In some cases, they can produce membrane receptors that can bind death ligands but do not induce apoptosis as they lack the DD. In other cases, they can produce soluble molecules that provoke a pre-receptor blockage of the death ligand.

In the case of CD95, besides the full-length mRNA that codes for the membrane-bound form of the protein, a variety of cells are able to produce different mRNA variants that derive from alternative splicing of the primary transcript. The variant called FasTMDel, which represents the large majority of soluble variants, produces a protein that can modulate apoptosis by interacting with CD95L. However, the mechanisms that regulate the production of mRNA for the membrane form of CD95 or of different mRNA variants are not known. To get more insight into the role of CD95/CD95L interactions during HIV-1 infection, we have developed an original method based upon quantitative-competitive (QC) RT-PCR that allowed us to quantify the amounts of mRNA coding for the total (tCD95) and membrane (mCD95) forms of CD95. This method has been used to quantify the amount of these forms of CD95 mRNA contained in basal conditions (i.e., without any stimulus) in two HIV-infected cell lines, ACH-2 (derived from A301) and U1 (derived from U937).

MATERIALS AND METHODS

For performing QC RT-PCR, scalar amounts of competitor RNA are added in several PCR tubes containing constant amounts of "wild type" RNA;[2] the first step consists in the preparation of RNA competitors. For this, we followed an approach similar to that described for quantifying CD95L mRNA.[3] RNA was extracted from U937 cells, and used as target for amplification by RT-PCR, using primers called 255D (5′-GGAATTCGTCTGTTGCTAGATTATCGTCCA-3′) and 400R (5′-GCTCTAGATTGACTGTGCAGTCCCTAGCTT-3′), or primers 521D (5′-GGAATTCATTGTGTGATGAAGGACATGGCT-3′) and 710R (5′-GCTCTAGACAAAGCCACCCCAAG-

TTAG-3′). The first pair of primers gives a PCR product of 210 bp, spanning exons 2 and 3 of CD95 mRNA, which are present both in total and membrane forms of CD95. The second pair of primers gives a PCR product of 200 bp spanning exons 4 and 6; exon 6 codes for transmembrane domain of the protein and is present only in mCD95 mRNA. An internal deletion of 20 bp in the PCR product of mCD95 was then carried out with a PCR strategy, using, together with 710R, the primer d57R (5′-GCTCTAGACAAAGCCACCCCAAGTTAGCTTGGTGTTGCT-GGTGAGTT-3′), which anneals inside the fragment and bears a tail that is identical to primer 710R. At the end of this reaction, a 20 bp shorter PCR product was obtained; this product can be amplified at the same rate and with the same primers of the wild type product, but can be easily distinguished as a separate, faster band on agarose gel. The same strategy was used to obtain an internal deletion for the total CD95 form; in this case primer 340R (5′-GCTCTAGATTGACTGTGCAGTCCCTAGCT-TCATGATGCAGGCCTTCAAAGTT-3′) was used to obtain a 40 bp internal deletion. The deleted fragments were cloned "tail to tail" in pGEM 3Z vector and *in vitro* transcribed using T7 external promoter; RNA was purified from DNA used as template by a digestion with 40U of RNase-free DNase, quantified, aliquoted and stored at –80°C.

Identical amounts of total RNA were extracted from samples, and scalar amounts of competitor RNA were added to several tubes containing the mix for reverse transcription, the mix consisting of MMLV RT buffer (50 mM TrisCl pH 8.3,75 mM KCl, 3 mM $MgCl_2$, 10 mM DTT), dNTPs 400 μM, reverse primer (710R for mCD95, 400R for tCD95) in a final volume of 20 μL. RNA was denatured by heating at 70°C for 5 min and immediately cooled on ice. 40 U of MMLV RT and 40 U of Rnasin, diluted in 5 μL of RT mix were added to denatured RNA. Reverse transcription was carried out at 42°C for 45 min. MMLV RT was denatured by heating at 95°C for 5 min. Twenty-five μL of Mg^{++} free-PCR mix (10 mM TrisCl pH 9, 50 mM KCl, 0.1% Triton X-100, primers 710R/521D or 400R/255D 200 μM, *Taq* polymerase 1 U) was added to the RT mix, and PCR was carried out for 50 cycles. Each cycle consists of a denaturation step (94°C, 30 sec), an annealing step (52°C, 30 sec), and an extension step (72°C, 30 sec). The first cycle was preceded by a denaturation step of 5 min at 94°C, and the last one was followed by an extension step of 7 min at 72°C. Ten μL of PCR reaction was loaded on a 3% agarose gel, run for 50 min at 100V stained with EtBr 0.5 mg/ml and acquired on computer with BioRad GelDoc 2000 densitometer. Densitometric analysis was performed using Multianalyst 1.0.1 software. The relative light intensity of competitor band was corrected with a 1.11 (mCD95) or 1.25 (tCD95) factor, to compensate the minor EtBr incorporation. The number of copies of mRNAs for tCD95 or mCD95 were then normalized to 1 μg of RNA. The number of mRNA copies for the soluble forms of CD95 (sCD95) was determined subtracting from the number of mRNA copies for tCD95 the value of mCD95 copies.

RESULTS AND DISCUSSION

We first studied the expression of different forms of CD95 mRNA in a classical model of chronic HIV infection using two infected cell lines of different origin, i.e., lymphocytic (ACH-2) or monocytic (U1), kindly provided by Drs. G. Poli and P.

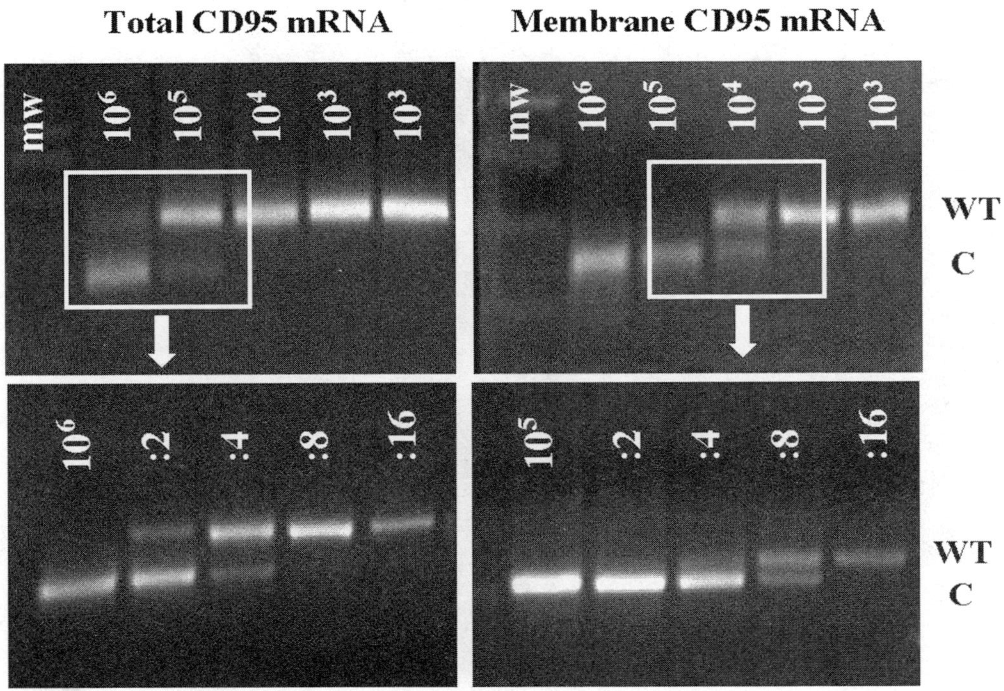

FIGURE 1. Typical example of quantitative–competitive RT PCR for the total form of CD95 (**left panels**) and for the membrane form of CD95 (**right panels**). A first PCR has to be performed to find the range in which the wildtype (WT) molecule is present (i.e., within the *white boxes*), using concentrations of competitor (C) that increase 10-fold from one sample to the other. Then, a second PCR is performed with concentrations of C that decrease by a factor of 2 (**lower panels**) to precisely quantify the number of WT mRNA molecules. Note the different area of competition for total and membrane forms of CD95 mRNA.

TABLE 1. Expression of CD95 mRNA in unstimulated, chronically HIV-infected cell lines

Cell line	tCD95	mCD95	sCD95	Ratio m/s
A301	630	403	232	1.7
ACH-2	100	97	3	32.3
U937	54	47	7	6.7
U1	17	6	11	0.5

NOTE: Data are expressed as number of copies × 10^{-5}/μg RNA.

Biswas (DIBIT-San Raffaele, Milan, Italy). FIGURE 1 shows a typical example of quantitative-competitive RT-PCR. We have previously demonstrated, at the protein level, that A301 cell line and its HIV-1 infected clone ACH-2 expressed the same amount of plasma membrane CD95.[3] Here we show that they contain substantially different ratios between mCD95 and sCD95 mRNA forms (TABLE 1); indeed, the infected cell line of lymphocytic origin has a relatively higher amount of mRNA for the membrane form of CD95. We showed that the infected clone U1 had a consistent decrease in CD95 expression in comparison to parental U937 cells.[3] Here we report that such decrease is mirrored by a decrease in membrane CD95 mRNA. Consistent with this observation is the fact that agonist anti-CD95 mAbs are able to trigger apoptosis and provoke mitochondrial functional alterations in U937 but not in U1 cells (manuscript in preparation). As far as the expression of mRNA for CD95L is concerned, compared to A301, ACH-2 expressed about 1/400 (in basal condition) or 1/10 (after a 4 hour stimulation with TPA and ionomycin) the amount of CD95L mRNA. The CD95L mRNA was undetectable (i.e., less than 2,000 copies RNA per million cells) in U937 and U1 cell lines.

Ex vivo studies showed that most lymphocytes from patients with primary HIV infection were CD95$^+$; at the single cell level, however, the level of protein was not increased. Anti-CD95 mAbs or binding to CD95L$^+$ cells caused rapid apoptosis, but CD95L mRNA was not expressed, and poorly inducible.[4] Studies are still in progress on the expression of the different forms of CD95 mRNA, but preliminary data in lymphocytes and in cell lines indicate that the ratio between total and membrane form of CD95 mRNA does not change after acute infection with HIV-1.

We have shown that infected cells of monocytic origin preferentially produce the "protective" (soluble) form of CD95 and no detectable CD95L mRNA, while lymphoid cells produce more mRNA for the membrane form of CD95 (which triggers apoptosis) along with reduced but easily detectable amounts of CD95L mRNA. It is possible to hypothesize that a complex balance exists between pro-apoptotic events, likely triggered by the host to limit viral production, and anti-apoptotic events likely triggered by the virus to increase its production and survival. In cells of monocytic origin, which act as reservoir for the virus, the anti-apoptotic molecules are favored; in cells of lymphocytic origin, molecules with an apoptotic meaning are prevalent.

REFERENCES

1. NAGATA, S. & P. GOLSTEIN. 1995. Fas death factor. Science **267:** 1449–1456.

2. DIVIACCO, S., P. NORIO, L. ZENTILIN, *et al.* 1992. A novel procedure for quantitative polymerase chain reaction by coamplification of competitive templates. Gene **122:** 313–320.
3. PINTI, M., J. PEDRAZZI, F. BENATTI, *et al.* 1999. Differential down-regulation of CD95 or CD95L in chronically HIV-infected cells of monocytic or lymphocytic: cellular studies and molecular analysis by quantitative, competitive RT-PCR. FEBS Lett. **458:** 209–214.
4. COSSARIZZA, A., G. STENT, C. MUSSINI, *et al.* 2000. Deregulation of the CD95/CD95L system in lymphocytes from patients with primary, acute HIV infection. AIDS **14:** 345–355.

The Complexity of TNF-Related Apoptosis-Inducing Ligand

KARON ABE,[a] ALEXEI KURAKIN,[b] MEHRDAD MOHSENI-MAYBODI,[a] BRIAN KAY,[c] AND ROYA KHOSRAVI-FAR[a,d]

[a]*Department of Pathology, Beth Israel Deaconess Medical Center and Harvard Medical School, Boston, Massachusetts 02215, USA*

[b]*Buck Center For Research in Aging, 8001 Redwood Boulevard, Novato, California 94945, USA*

[c]*Department of Pharmacology, University of Wisconsin-Madison, 1300 University Avenue, Madison, Wisconsin 53706-1532, USA*

ABSTRACT: One of the major goals of researchers in the field of apoptosis is to understand the molecular mechanisms of the various components of the apoptotic pathways, with the hope to identify targets for novel cancer therapies. The discovery of a TNF-related, apoptosis-inducing ligand, TRAIL, that kills transformed cells with great specificity *in vitro*, has provided the hope that TRAIL may be used to induce cell death in tumor cells without affecting normal tissues. However, TRAIL signaling is very complex and a clear understanding of its function is necessary before it can be used in cancer therapy. Complexity of TRAIL-induced signaling is apparent from its ubiquitous expression, its ability to interact with five receptors, and its tumor-selective induction of apoptosis. The signaling events that mediate the tumor selectivity of TRAIL-induced apoptosis and the biological functions of each of the TRAIL receptors are not well characterized. This review will focus on the complexity of TRAIL and the role of c-FLIP in mediating TRAIL function.

KEYWORDS: Cancer therapy; TRAIL; TNF superfamily

INTRODUCTION

Apoptosis, or programmed cell death, is a process fundamental to the development of multicellular organisms and plays a critical role in the immune response.[1–5] The specific activation of apoptosis in tumor cells has been a long sought, but elusive, approach to tumor destruction.[6–8] TNF-Related Apoptosis Inducing Ligand (TRAIL)/

[d]Address for correspondence: Dr. Roya Khosravi-Far, Department of Pathology, Beth Israel Deaconess Medical Center and Harvard Medical School, 99 Brookline Ave., RN 270F, Boston, MA 02215, USA. Voice: 617-667-8526; fax: 617-667-3591.
rkhosrav@caregroup.harvard.edu

ABBREVIATIONS: c, cellular; CrmA, cytokine response modifier A; DD, death domain; DED, death effector domain; DR, death receptor; DcR, decoy receptor; DISC, death inducing signaling complex; EST, expressed sequence tag; FLICE, Fas-associated death domain-like IL-1β-converting enzyme; FLIP, FLICE-inhibitory protein; JNK, Jun amino terminal kinase; NF-κB, nuclear factor kappa B; OPG, osteoprogerin; TNF, tumor necrosis factor; TRAIL, TNF-related apoptosis inducing ligand.

Apo-2 is a member of the TNF superfamily (TNF, Fas, CD40, etc.).[5,9–11] TRAIL has been shown to selectively induce cell death in tumor and transformed— but not in normal—cell lines, suggesting it as a valuable target for cancer therapy.[9,10,12] There are at least five receptors that can be engaged by TRAIL.[9,11,13,14] The existence of multiple receptors for TRAIL suggests that TRAIL-induced signals are complex, and that multiple signaling pathways may mediate its function. Some of these signal transduction pathways may be unique to each of the receptors and some may constitute cell specific/tissue specific/tumor specific pathways. However, the molecular mechanism by which TRAIL and its receptors mediate their function and how the tumor selectivity of TRAIL-induced apoptosis is achieved is not well characterized. TRAIL receptors possess no intrinsic signaling capacity (e.g., kinase activity), suggesting that signal transduction is likely to be mediated by molecules associated with their death domain (protein/protein interaction domain).[9,11,13] In this review, we will summarize the current and often controversial knowledge on the adaptor molecules and signaling pathways that are activated by TRAIL and its receptors. We will also summarize the results of our phage-displayed library screening. Using this approach, we have identified the FLICE Inhibitory Protein, c-FLIP, as a Death Receptor 5-(DR5) binding protein. Finally, we will summarize the current understanding on the role of c-FLIP in regulation of apoptosis.

TNF-RELATED APOPTOSIS INDUCING LIGAND AND ITS RECEPTORS

TRAIL is an apoptotic ligand and a member of the TNF superfamily. Similar to many family members, TRAIL is a type II membrane, having its N-terminus in the cytoplasm and its C-terminus in the extracellular region.[18,19] Both human and murine cDNAs encoding TRAIL were cloned using an expressed sequence tag (EST), originally identified from a region conserved in several TNF ligand family members.[18,19] Although the N-terminal region of TRAIL does not share sequence similarity with other TNF family members, the C-terminal portion of TRAIL is highly homologous to other TNF family members, sharing sequence identity with Fas ligand (28%), TNF-α (23%), lymphotoxin-α (23%) and lymphotoxin-β (22%).[19] Furthermore, crystal structure analysis of TNF-α [22] as well as functional studies utilizing soluble TRAIL[18,19] showed that these ligands form multimeric structures (homotrimers) that are necessary for effective receptor activation, signal transduction, and apoptosis.

Unlike Fas or TNF, which have one or two receptors, five different receptors have been identified as TRAIL receptors (FIG. 1). The receptors that bind to TRAIL include death receptor (DR) 4/TRAIL-R1,[23] DR5/KILLER/TRAIL-R2/TRICK2,[23–27] decoy receptor (DcR) 1/TRAIL-R3/TRID (TRAIL without an extracellular domain)/LIT (Lymphocyte Inhibitor of TRAIL),[23–26,28] DcR2/TRAIL-R4/TRUDD (TRAIL receptor with a truncated death domain)[25,28–30] and a soluble receptor called osteoprotegerin (OPG).[14] DR4, DR5, DcR1, and DcR2 have been shown to bind TRAIL with similar affinities.[13] However, only two of the TRAIL receptors, DR4 and DR5, contain functional death domains and are capable of inducing apoptosis.[24,25,31]

Despite having multiple receptors, TRAIL can selectively target tumor cells to undergo apoptosis while leaving normal cells unaffected.[9,12,15,16] The selectivity of

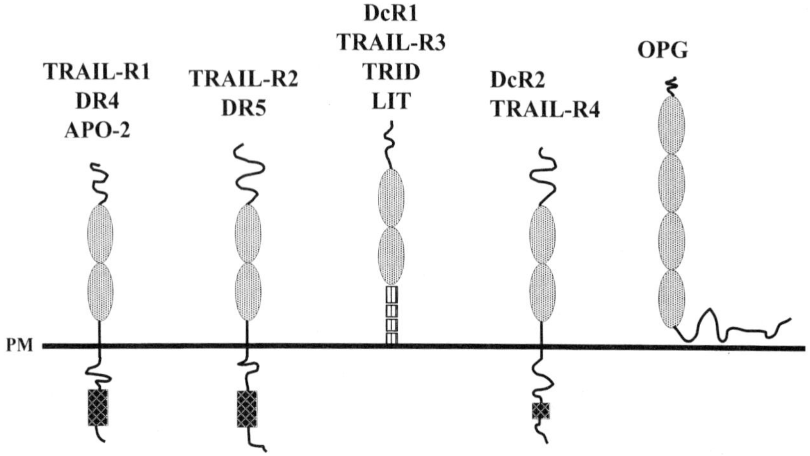

FIGURE 1. TRAIL Receptors. Extracellular cysteine rich domains characterize TRAIL receptors. The apoptotic TRAIL receptors DR4 and DR5 contain intracellular DD, the DcR2 contains a truncated non-functional DD, and the DcR1 and OPG lack the death domain.

TRAIL-induced apoptosis has been demonstrated by systemic administration of TRAIL, which reduces tumor growth *in vivo,* without the severe toxic effects often seen in TNF or Fas treated mice.[15,20] However, the recent observation that human primary hepatocytes are sensitive to TRAIL-induced apoptosis[21] suggests that TRAIL may cause hepatotoxicity. A better understanding of the TRAIL-mediated apoptotic signaling pathways will provide valuble information on the molecular mechanism by which tumor selectivity of TRAIL is mediated.

APOPTOTIC SIGNALING BY TRAIL RECEPTORS

Although TRAIL shares many similarities with the other TNF-superfamily members, the molecular components and signaling pathways that mediate TRAIL-induced apoptosis are more complex. The ability of TNF family of receptors to mediate apoptosis is attributed to a conserved region of 80 amino acids within the cytoplasmic region of the receptors termed the death domain (DD).[32,33] The role of DD in TNFR1-, 2-, and FasR-induced apoptosis has been well characterized.[9,33] For these death receptors, ligand binding causes receptor oligomerization and the formation of DISC (Death Inducing Signaling Complex).[34] DISC formation involves the recruitment of caspase-8 via an adaptor protein (FADD, TRADD, and RIP) to the DD of the activated receptor (FIG. 2). Caspase-8 initiates the activation of a cascade of caspases causing the proteolytic cleavage of a number of substrates, leading to the subsequent cell death.[2,9,35]

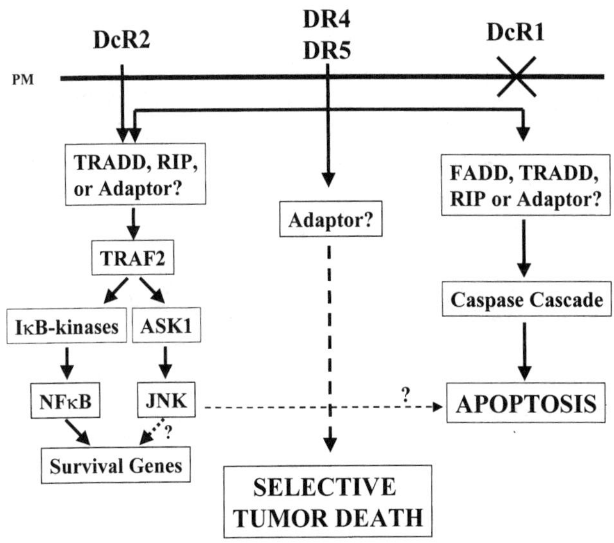

FIGURE 2. Apoptotic signaling by TRAIL receptors. The recruitment of FADD, TRADD and RIP to DR4 and/or DR5 results in the initiation of the caspase machinery as well as the NF-κB and JNK pathways. The adaptors involved in and the mechanism by which tumor selectivity of TRAIL is achieved are not yet characterized.

Although initial studies utilizing caspase inhibitors (cytokine response modifier A (CrmA) or carbobenzyloxy (z) VAD) showed that TRAIL-induced apoptosis was mediated via the activation of caspases,[23,25,31] the upstream effectors involved in the activation of caspases are now being delineated. Studies using overexpression approaches showed that TRAIL receptors DR4 and DR5 did not bind known adaptor proteins FADD, TRADD, or RIP.[23,25,31] In agreement with this, the dominant negative mutant of FADD efficiently blocked TNF-α and Fas-induced apoptosis, but failed to do so in the case of TRAIL receptor signaling.[23] Moreover, experiments with cells from FADD knockout mice indicated that even though the FADD adaptor was indispensable in signaling death from the TNF and Fas receptors, the absence of FADD did not impair the TRAIL death pathway.[41] These studies suggest differences in signaling and/or adaptor redundancy in the TRAIL system.[41]

In contrast, several other groups have shown an association between DR4 and DR5 with FADD, TRAF2, RIP and TRADD as well as inhibition of TRAIL-induced apoptosis upon over-expression of the dominant-negative forms of these adaptor proteins.[24,42–44] Furthermore, recent studies, utilizing mutant Jurkat cells that are defective in FADD or caspase-8, or analysis of the DISC components in a variety of cells, have demonstrated that both FADD and caspase-8 are required for mediating TRAIL-induced apoptosis.[45–47]

Finally, in certain cell types, TRAIL activates nuclear factor kappa B (NF-κB) and the Jun N-terminal kinase (JNK) pathways (FIG. 2).[37,41,44] Activation of these pathways in TRAIL-mediated apoptosis is believed to be through the interaction of its receptors with TRADD and/or RIP (FIG. 2).[41] NF-κB is a transcription factor that can prevent cells from undergoingTNF-induced apoptosis by transcriptional regulation of anti-apoptotic and/or survival genes.[36] On the other hand, the involvement of JNK in mediating apoptosis is more controversial. In some cell systems, JNK has been shown to mediate apoptotic signals from Fas and/or TNF, whereas in other cell systems it is anti-apoptotic or has no effect.[37–40] Although both NF-κB and JNK are activated by TRAIL, the exact functional role of these proteins in mediating TRAIL function remains to be characterized.

TUMOR SELECTIVITY OF TRAIL-INDUCED APOPTOSIS

The presence of multiple TRAIL receptors as well as the tissue distribution of TRAIL expression suggests a higher degree of complexity and regulation in TRAIL-mediated apoptosis and signal transduction. TRAIL mRNA transcripts have been detected in the adult spleen, thymus, lung, prostate, ovary, small intestine, peripheral blood lymphocytes, colon, heart, placenta, skeletal muscle, and kidney, but not in the adult liver, brain, and testis.[18,19] The broad range of TRAIL expression suggests that these cells may have developed mechanisms to regulate cellular activation and apoptosis induced by TRAIL. These cellular mechanisms may thus be deregulated in tumor cells to allow selective tumor cell death by TRAIL. One such mechanism of regulation may be at the receptor level, where the amount of receptor or the type of receptor expressed on the cell surface is tightly controlled. Another mechanism may be the differential expression of inhibitory intracellular proteins that either block an apoptotic signaling pathway or activate a proliferative signaling pathway.

Expression of a tissue specific receptor may account for the ability of some cells to resist an apoptotic signal by TRAIL. Studies on the tissue distributions of DR4, DR5, and TRAIL mRNA demonstrate similar patterns of expression. Therefore, TRAIL resistance is not likely to be regulated by restricted expression of these receptors.[28,29] However, sensitivity to TRAIL-induced apoptosis by tumor cells may be attributed in part to the levels of expression of the remaining receptors, all of which have been shown to possess the ability to inhibit TRAIL-induced apoptosis. OPG is a secreted TNF family protein that can bind and inhibit TRAIL-induced apoptosis.[14] DcR1 and DcR2 bind TRAIL with comparable affinities as DR4 and DR5. However, DcR1 and DcR2 are incapable of transducing an apoptotic signal. DcR1 lacks transmembrane and cytoplasmic residues and is glycosylphosphatidylinositol (GPI)–linked to the cell membrane. DcR2 contains only a partial DD.[28–30] Thus, it has been proposed that these receptors act as decoy receptors on normal cells, competing with DR4 and DR5 for binding to TRAIL and providing a mechanism of resistance from TRAIL-mediated apoptosis.[28–30] In support of this idea, several studies have demonstrated that over-expression of DcR1 or DcR2 in TRAIL-sensitive tumor cells converts them to TRAIL-resistant cells.[28–30,48] In addition, there are observations that demonstrate lower mRNA transcripts for DcR1 in some transformed cell lines in comparison to normal human tissue.[31] Thus, expression level of these decoy receptors might protect TRAIL sensitive cells from undergoing apoptosis.

In contrast, other recent studies have suggested that decoy receptors may not be involved in selectivity of tumor death by TRAIL in all tumors. Several groups have screened panels of human tumors without finding a correlation between DcR1/DcR2 mRNA expression and TRAIL resistance.[49,50] In another study, similar levels of receptor surface expression between DcR1, DcR2, and OPG in normal and transformed keratinocytes was observed.[51] Finally, experiments using melanoma cells showed that the levels of various TRAIL receptors did not correlate with their sensitivity or resistance toward TRAIL-induced apoptosis.[49]

Several groups have demonstrated that resistance to TRAIL-mediated apoptosis in some cases correlates better with expression of intracellular inhibitors. Studies on intracellular inhibitors of apoptosis were prompted by experiments showing an increase in sensitivity to TRAIL-mediated apoptosis upon treatment of resistant melanoma cells with protein synthesis inhibitor cycloheximide or actinomycin D.[49] These studies indicated the presence of a protein that could regulate the effects of TRAIL-induced apoptosis. Subsequently, after screening known inhibitors of apoptosis, it was shown that the TRAIL-resistant melanoma cells possessed high levels of the FADD-like IL-1β-converting enzyme (FLICE)-like inhibitory protein (c-FLIP).[49] Furthermore, studies in both melanomas and keratinocytes have demonstrated a reduction in c-FLIP expression upon actinomycin D or cycloheximide treatment.[49,51] However, other studies did not find any correlation between TRAIL-induced apoptosis and c-FLIP expression in melanoma cells.[52] Therefore, the role of c-FLIP expression in TRAIL-mediated apoptosis is not well established.

Finally, other intracellular anti-apoptotic pathways may be involved in determining the tumor selectivity of TRAIL-induced apoptosis. For example, TRAIL, DR4, DR5 and DcR2 activate NF-κB. The activation of NF-κB by DR4 and DR5 indicated a possible dual function for DR4 and DR5 as apoptotic as well as anti-apoptotic activators depending on extra- and intracellular signals[44,53] and may contribute to the ability of normal cells to resist TRAIL-induced apoptosis.[36,54]

Taken together, these studies indicate that TRAIL activation resulting in apoptosis or resistance to apoptosis may be cell-type specific requiring: (1) the up-regulation or down-regulation of one of the TRAIL receptors; (2) the assembly of different receptors; (3) involvement of decoy receptors; and (4) the function of intracellular inhibitors or intracellular anti-apoptotic/survival pathways working in concert. Furthermore, the ubiquitous expression of TRAIL suggests that it may have functions other than apoptosis in some cell types. As for Fas and TNF receptors, stepwise identification of the components of additional signaling pathways that are initiated by one or more of the TRAIL receptors could provide valuable information on the molecular mechanism of TRAIL-induced apoptosis. These studies could allow a better understanding of selectivity/specificity of TRAIL signals in normal versus tumor cells and the possible specific functions of each of the TRAIL receptors. Hence, we have utilized the phage-displayed library screening approach to identify DR4- and/or DR5-interacting proteins.

IDENTIFICATION OF NOVEL PROTEIN-PROTEIN INTERACTIONS BY PHAGE-DISPLAYED LIBRARY SCREENING

Combinatorial peptide libraries displayed on bacteriophage M13 have been a rich source of ligands for various proteins.[55,56] In a typical screening experiment, the target protein is immobilized on a microtiter plate or sepharose beads and an aliquot of the phage library is allowed to interact with the target. An aliquot of the library typically contains 10^{12} phages displaying on their surface about 10^{9} different peptides genetically fused to one of the phage coat proteins. Phage that displays a peptide which binds to the target with affinity high enough to resist the following washing step are retained in the complex with the target. After the elution step, retained phages are amplified by infecting bacteria and applied for a second round of selection on the same target. Thus, by repeating this selection-amplification cycle several times, binder phage is selected at the expense of non-binding phage. The identity of the selected peptides is then deduced by sequencing the phage DNA.

The ability of peptides presented on the phage to home in on "functional" surfaces of a protein target such as protein-protein interaction interfaces allows mimicking protein-protein interactions in a biochemically characterized system *in vitro*. In this situation, the peptide ligand plays the role of a "surrogate" of the target's natural binding partner. Results of successful screens indicate that a surprising number of peptides isolated from random libraries are highly similar in their primary structure to stretches of amino acid sequences present in natural binding partners of the proteins used as targets.[57] Thus, phage display can be used as a predictive tool to identify novel protein-protein interactions. In other words, the primary sequence of peptides isolated from combinatorial libraries can be successfully used in homology searches of protein databases to identify candidate interacting partners of the target protein.

We used the death domain of the DR5 receptor as a target to screen a 11-mer combinatorial peptide library displayed at the N-terminus of the pIII coat protein of M13 phage.[55,58] The amino acid sequence of the selected peptide resembled a short stretch of amino acids present in the sequence of c-FLIP, an enzymatically inactive homologue of caspase-8. Moreover, the corresponding peptide derived from c-FLIP bound specifically to the DR5 death domain *in vitro*. These facts prompted us to test the interaction between DR5 and c-FLIP in cultured cells. GST-pull down and co-immunoprecipitation experiments confirmed the ligand-dependent association of the DR5 death domain and c-FLIP (to be presented elsewhere).

c-FLIP AND REGULATION OF APOPTOSIS

The access to ever-growing databases of genomic sequences and to software that allows the fast analysis of nucleotide and protein sequences through web browsers resulted in the virtually simultaneous identification of c-FLIP as an enzymatically inactive caspase homologue in eight different laboratories. It has been named, accordingly, Casper,[59] c-FLIP,[60] CASH,[61] CLARP,[62] MRIT,[63] Usurpin,[64] FLAME,[65] and I-FLICE.[66]

This parallel discovery created a unique situation, an experiment in itself, where a novel protein was characterized simultaneously in different systems and by different approaches, in a presumably unbiased fashion. As a result, four groups[59,61–63] reported it as being proapoptotic protein and five[60,61,64–66] as a negative regulator of programmed cell death. The disagreement was also in respect to whether c-FLIP, being a close homologue of caspase-8, interacts with FADD or not. All reports agreed on its ability to bind caspase-8. These contradictory results either put into question the reliability of some methods routinely used in the field, or indicate the unexpected complexity of c-FLIP function, or both.

At present there are several lines of evidence indicating that c-FLIP is an important negative regulator of the TNF family of cytokine receptors signaling. First, it is highly expressed in transformed cells, T lymphocytes and healthy, but not in injured, myocytes and is downregulated following lymphocyte activation.[60,64,67,68] Second, analysis of EBV-positive Burkitt's lymphoma cell lines resistant to Fas-mediated apoptosis revealed the ratio of caspase 8/c-FLIP mRNA levels as a main correlate of susceptibility to Fas killing.[69] Third, c-FLIP introduced into B lymphoma cells acted as a tumor progression factor by suppressing death receptor signaling.[70] In another report,[71] tumor cells with higher expression levels of c-FLIP demonstrated a selective advantage in escape from T-cell dependent killing *in vivo*. Furthermore, B cell antigen receptor (BCR) cross-linking led to elevated levels of c-FLIP in primary and transformed B cells. In addition, BCR cross-linking or c-FLIP over-expression protected B cells from Fas and TRAIL-induced apoptosis indicating a possible involvement of c-FLIP in the selection of antigen-specific B cells.[72] It was also suggested that pathological destruction of endothelial tissue by oxidized lipids might involve sensitization of endothelial cells to Fas-mediated suicide through down-regulation of c-FLIP levels.[73] Finally, several reports indicated that the resistance of primary and transformed cells to TRAIL-mediated killing correlated with expression levels of c-FLIP rather than with the ratio of TRAIL receptors to their decoys.[12,51,74]

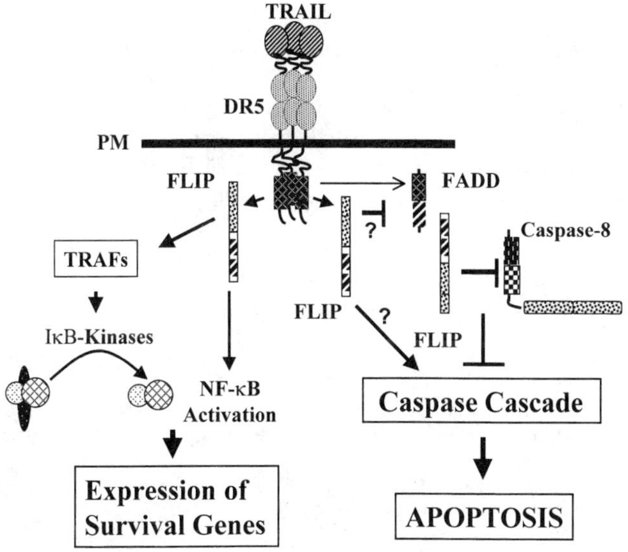

FIGURE 3. The regulation of TRAIL-induced apoptosis by FLIP. FLIP can inhibit the apoptotic machinery by either preventing the formation of a functional DISC, or activating the antiapoptotic pathways such as the NF-κB pathway.

The mechanism by which c-FLIP protects from death receptor-mediated apoptosis still remains a subject of controversy. Two splice isoforms of c-FLIP detected on the protein level,[60,64] c-FLIP(L) and c-FLIP(S), both contain death effector domain (DED) repeats within their N termini. The long (L) isoform includes an additional C-terminal sequence resembling the p20 and p10 subunits of caspases, and is most closely related to caspase-8 in its primary sequence. C-FLIP(L) represents an enzymatically inactive caspase homologue carrying mutations that affect amino acids critical for proteolytic activity.[61,64–66] c-FLIP has been proposed to function in a fashion similar to the previously described viral FLIPs, which are essentially a tandem of DED domains, a structure resembling the N-terminal part of caspase-8 and c-FLIP(S).[60,75] They have been postulated to prevent the formation of a competent death receptor signaling complex (DISC) by binding to the FADD adaptor protein through DED-DED association and competing off caspase-8.[75] Another possible mechanism for this c-FLIP-induced protection is the ability of c-FLIP to activate the NF-κB and ERK pathways, both of which mediate the transcription activation of survival and proliferative genes.[36,76,77]

In case of c-FLIP at least three groups[59,62,64] failed to demonstrate the interaction of FADD with c-FLIP, while others[60,61,63,65,66] confirmed the interaction. In some reports,[64,72] c-FLIP was shown to prevent the recruitment of caspase-8 to the DISC. While Scaffidi *et al.*[78] showed that even though the overexpression of c-FLIP(L) prevented caspase-8 recruitment, it did not prevent its initial processing. Furthermore, several groups could not detect any c-FLIP/FADD or c-FLIP/caspase-8 complexes in the cytoplasm.[60,61,63,65,66] The reasons for this controversy are not clear at present and ask for more work.

We find that c-FLIP(L) may directly associate with the death domain of DR5 through its p10 subunit suggesting that, in fact, c-FLIP may function as an adaptor protein and recruit caspase-8 to the ligated death receptor, thus promoting the transduction of a death signal in the absence of FADD. In this scenario, c-FLIP can function as either promoter or inhibitor of apoptosis depending on the outcome of the complex interplay between FADD, caspase-8 and c-FLIP(L) (FIG. 3). The availability, local concentrations of FADD, caspase-8 and c-FLIP(L) as well as their affinities for each other would be critical parameters defining the outcome in any particular situation.

REFERENCES

1. SCHULZE-OSTHOFF, K., D. FERRARI, *et al.* 1998. Apoptosis signaling by death receptors. Eur. J. Biochem. **254**: 439–459.
2. NAGATA, S. 1999. Fas ligand-induced apoptosis. Annu. Rev. Genet. **33:** 29–55.
3. KRAMMER, P.H. 1999. CD95(APO-1/Fas)-mediated apoptosis: live and let die. Adv. Immunol. **71:** 163–210.
4. WALLACH, D., E.E. VARFOLOMEEV, *et al.* 1999. Tumor necrosis factor receptor and Fas signaling mechanisms. Annu. Rev. Immunol. **17:** 331–367.
5. WALCZAK, H. & P.H. KRAMMER. 2000. The CD95 (APO-1/Fas) and the TRAIL (APO-2L) apoptosis systems. Exp. Cell Res. **256:** 58–66.
6. THOMPSON, C.B. 1995. Apoptosis in the pathogenesis and treatment of disease. Science **267:** 1456–1462.
7. MCGILL, G. 1997. Apoptosis in tumorigenesis and cancer therapy. Front. Biosci. **2:** d353–379.

8. OLIFF, A, & S.H. FRIEND. 1997. Molecular targets for drug development. *In* Cancer: Principles and Practice of Oncology. V.T. DeVita, S. Hellman, S.A. Rosenberg, Eds.: 3115–3125. Lippincott-Raven. Philadelphia, PA.
9. ASHKENAZI, A. & V.M. DIXIT. 1998. Death receptors: signaling and modulation. Science **281:** 1305–1308.
10. FRENCH, L.E. & J. TSCHOPP. 1999. The TRAIL to selective tumor death. Nat. Med. **5:** 146–147.
11. ASHKENAZI, A. & V.M. DIXIT. 1999. Apoptosis control by death and decoy receptors. Curr. Opin. Cell Biol. **11:** 255–260.
12. GRIFFITH, T.S. & D.H. LYNCH. 1998. TRAIL: a molecule with multiple receptors and control mechanisms. Curr. Opin. Immunol. **10:** 559–563.
13. DEGLI-ESPOSTI, M. 1999. To die or not to die—the quest of the TRAIL receptors. J. Leuk. Biol. **65:** 535–542.
14. EMERY, J.G., P. MCDONNELL, *et al.* 1998. Osteoprotegerin is a receptor for the cytotoxic ligand TRAIL. J. Biol. Chem. **273:** 14363–14367.
15. ASHKENAZI, A., R.C. PAI, *et al.* 1999. Safety and antitumor activity of recombinant soluble Apo2 ligand. J. Clin. Invest. **104:** 155–162.
16. MARSTERS, S.A., R.A. PITTI, *et al.* 1999. Control of apoptosis signaling by Apo2 ligand. Recent Prog. Horm. Res. **54:** 225–234.
17. MARSTERS, S.A., R.M. PITTI, *et al.* 1996. Activation of apoptosis by Apo-2 ligand is independent of FADD but blocked by CrmA. Curr. Biol. **6:** 750–752.
18. PITTI, R.M., S.A. MARSTERS, *et al.* 1996. Induction of apoptosis by Apo-2 ligand, a new member of the tumor necrosis factor cytokine family. J. Biol. Chem. **271:** 12687–12690.
19. WILEY, S.R., K. SCHOOLEY, *et al.* 1995. Identification and characterization of a new member of the TNF family that induces apoptosis. Immunity **3:** 673–682.
20. WALCZAK, H., R.E. MILLER, *et al.* 1999. Tumoricidal activity of tumor necrosis factor-related apoptosis-inducing ligand in vivo. Nat. Med. **5:** 157–163.
21. JO, M., T.H. KIM, *et al.* 2000. Apoptosis induced in normal human hepatocytes by tumor necrosis factor-related apoptosis-inducing ligand. Nat. Med. **6:** 564–567.
22. BANNER, D.W., A. D'ARCY, *et al.* 1993. Crystal structure of the soluble human 55 kd TNF receptor-human TNF beta complex: implications for TNF receptor activation. Cell **73:** 431–445.
23. PAN, G., K. O'ROURKE, *et al.* 1997. The receptor for the cytotoxic ligand TRAIL. Science **276:** 111–113.
24. SCHNEIDER, P., J.L. BODMER, *et al.* 1997. Characterization of two receptors for TRAIL. FEBS Lett. **416:** 329–334.
25. SHERIDAN, J.P., S.A. MARSTERS, *et al.* 1997. Control of TRAIL-induced apoptosis by a family of signaling and decoy receptors. Science **277:** 818–821.
26. MACFARLANE, M., M. AHMAD, *et al.* 1997. Identification and molecular cloning of two novel receptors for the cytotoxic ligand TRAIL. J. Biol. Chem. **272:** 25417–25420.
27. SCREATON, G.R., J. MONGKOLSAPAYA, *et al.* 1997. TRICK2, a new alternatively spliced receptor that transduces the cytotoxic signal from TRAIL. Curr. Biol. **7:** 693–696.
28. DEGLI-ESPOSTI, M.A., P.J. SMOLAK, *et al.* 1997. Cloning and characterization of TRAIL-R3, a novel member of the emerging TRAIL receptor family. J. Exp. Med. **186:** 1165–1170.
29. MARSTERS, S.A., J.P. SHERIDAN, *et al.* 1997. A novel receptor for Apo2L/TRAIL contains a truncated death domain. Curr. Biol. **7:** 1003–1006.
30. PAN, G., J. NI, *et al.* 1998. TRUNDD, a new member of the TRAIL receptor family that antagonizes TRAIL signalling. FEBS Lett. **424:** 41–45.
31. PAN, G., J. NI, *et al.* 1997. An antagonist decoy receptor and a death domain-containing receptor for TRAIL. Science **277:** 815–818.
32. TARTAGLIA, L.A., T.M. AYRES, *et al.* 1993. A novel domain within the 55 kD TNF receptor signals cell death. Cell **74:** 845–853.
33. NAGATA, S. 1997. Apoptosis by death factor. Cell **88:** 355–365.

34. Kischkel, F.C., S. Hellbardt, *et al.* 1995. Cytotoxicity-dependent APO-1 (Fas/CD95)-associated proteins form a death-inducing signaling complex (DISC) with the receptor. EMBO J. **14:** 5579–5588.
35. Thornberry, N.A. & Y. Lazebnik. 1998. Caspases: enemies within. Science **281:** 1312–1316.
36. Beg, A. & D. Baltimore. 1996. An essential role for NF-κB in preventing TNF-α-induced cell death. Science **274:** 782–784.
37. Liu, Z.G., H. Hsu, *et al.* 1996. Dissection of TNF receptor 1 effector functions: JNK activation is not linked to apoptosis while NF-kappaB activation prevents cell death. Cell **87:** 565–576.
38. Xia, Z., M. Dickens, *et al.* 1995. Opposing effects of ERK and JNK-p38 MAP kinases on apoptosis. Science **270:** 1326–1331.
39. Verheij, M., R. Bose, *et al.* 1996. Requirement for ceramide-initiated SAPK/JNK signalling in stress-induced apoptosis. Nature **380:** 75–79.
40. Yang, X., F.R. Khosravi, *et al.* 1997. Daxx, a novel Fas-binding protein that activates JNK and apoptosis. Cell **89:** 1067–1076.
41. Yeh, W.C., J.L. Pompa, *et al.* 1998. FADD: essential for embryo development and signaling from some, but not all, inducers of apoptosis. Science **279:** 1954–1958.
42. Walczak, H., E.M. Degli, *et al.* 1997. TRAIL-R2: a novel apoptosis-mediating receptor for TRAIL. EMBO J. **16:** 5386–5397.
43. Wu, G.S., T.F. Burns, *et al.* 1997. KILLER/DR5 is a DNA damage-inducible p53-regulated death receptor gene. Nat. Genet. **17:** 141–143.
44. Chaudhary, P.M., M. Eby, *et al.* 1997. Death receptor 5, a new member of the TNFR family, and DR4 induce FADD-dependent apoptosis and activate the NF-kappaB pathway. Immunity **7:** 821–830.
45. Bodmer, J.L., N. Holler, *et al.* 2000. TRAIL receptor-2 signals apoptosis through FADD and caspase-8. Nat. Cell Biol. **2:** 241–243.
46. Kischkel, F.C., D.A. Lawrence, *et al.* 2000. Apo2L/TRAIL-dependent recruitment of endogenous FADD and capase-8 to death receptors 4 and 5. Immunity **12:** 611–620.
47. Sprick, M.R., M.A. Weigand, *et al.* 2000. FADD/MORT1 and capase-8 are recruited to TRAIL receptors 1 and 2 and are essential for apoptosis mediated by TRAIL receptor 2. Immunity **12:** 599–609.
48. Mongkolsapaya, J., A.E. Cowper, *et al.* 1998. Lymphocyte inhibitor of TRAIL (TNF-related apoptosis-inducing ligand): a new receptor protecting lymphocytes from the death ligand TRAIL. J. Immunol. **160:** 3–6.
49. Griffith, T.S., W.A. Chin, *et al.* 1998. Intracellular regulation of TRAIL-induced apoptosis in human melanoma cells. J. Immunol. **161:** 2833–2840.
50. Degli-Esposti, M.A., W.C. Dougall, *et al.* 1997. The novel receptor TRAIL-R4 induces NF-kappaB and protects against TRAIL-mediated apoptosis, yet retains an incomplete death domain. Immunity **7:** 813–820.
51. Leverkus, M., M. Neumann, *et al.* 2000. Regulation of tumor necrosis factor-related apoptosis-inducing ligand sensitivity in primary and transformed human keratinocytes. Cancer Res. **60:** 553–559.
52. Zhang, X.D., A. Franco, *et al.* 1999. Relation of TNF-related apoptosis-inducing ligand (TRAIL) receptor and FLICE-inhibitory protein expression to TRAIL-induced apoptosis of melanoma. Cancer Res. **59:** 2747–2753.
53. Schneider, P., M. Thome, *et al.* 1997. TRAIL receptors 1 (DR4) and 2 (DR5) signal FADD-dependent apoptosis and activate NF-kappaB. Immunity **7:** 831–836.
54. Van, A.D., S.J. Martin, *et al.* 1996. Suppression of TNF-alpha-induced apoptosis by NF-kappaB. Science **274:** 787–789.
55. Kay, B.K., A.V. Kurakin, *et al.* 1998. From peptides to drugs via phage display. Drug Discovery Today **3:** 370–378.
56. Smith, G.P. & V.A. Petrenko. 1997. Phage display. Chem. Rev. **97:** 391–410.
57. Kay, B.K., J. Kasanov, *et al.* 2000. Convergent evolution with combinatorial peptides. FEBS. Submitted.

58. Kurakin, A., N.G. Hoffman, *et al.* 1998. Molecular recognition properties of the C-terminal Sh3 domain of the Cbl associated protein, Cap. J. Pept. Res. **52:** 331–337.
59. Shu, H.B., D.R. Halpin, *et al.* 1997. Casper is a FADD- and caspase-related inducer of apoptosis. Immunity **6:** 751–763.
60. Irmler, M., M. Thome, *et al.* 1997. Inhibition of death receptor signals by cellular FLIP. Nature **388:** 190–195.
61. Goltsev, Y.V., A.V. Kovalenko, *et al.* 1997. CASH, a novel caspase homologue with death effector domains. J. Biol. Chem. **272:** 19641–19644.
62. Inohara, N., T. Koseki, *et al.* 1997. CLARP, a death effector domain-containing protein interacts with caspase-8 and regulates apoptosis. Proc. Natl. Acad. Sci. USA **94:** 10717–10722.
63. Han, D.K., P.M. Chaudhary, *et al.* 1997. MRIT, a novel death-effector domain-containing protein, interacts with caspases and BclXL and initiates cell death. Proc. Natl. Acad. Sci. USA **94:** 11333–11338.
64. Rasper, D.M., J.P. Vaillancourt, *et al.* 1998. Cell death attenuation by 'Usurpin', a mammalian DED-caspase homologue that precludes caspase-8 recruitment and activation by the CD-95 (Fas, APO-1) receptor complex. Cell Death Differ. **5:** 271–288.
65. Srinivasula, S.M., M. Ahmad, *et al.* 1997. FLAME-1, a novel FADD-like anti-apoptotic molecule that regulates Fas/TNFR1-induced apoptosis. J. Biol. Chem. **272:** 18542–18545.
66. Hu, S., C. Vincenz, *et al.* 1997. I-FLICE, a novel inhibitor of tumor necrosis factor receptor-1- and CD- 95-induced apoptosis. J. Biol. Chem. **272:** 17255–17257.
67. Tschopp, J., M. Irmler, *et al.* 1998. Inhibition of fas death signals by FLIPs. Curr. Opin Immunol. **10:** 552–558.
68. Van Parijs, L., Y. Refaeli, *et al.* 1999. Autoimmunity as a consequence of retrovirus-mediated expression of C-FLIP in lymphocytes. Immunity **11:** 763–770.
69. Tepper, C.G. & M.F. Seldin. 1999. Modulation of caspase-8 and FLICE-inhibitory protein expression as a potential mechanism of Epstein-Barr virus tumorigenesis in Burkitt's lymphoma. Blood **94:** 1727–1737.
70. Djerbi, M., V. Screpanti, *et al.* 1999. The inhibitor of death receptor signaling, FLICE-inhibitory protein defines a new class of tumor progression factors. J. Exp. Med. **190:** 1025–1032.
71. Medema, J.P., J. de Jong, *et al.* 1999. Immune escape of tumors in vivo by expression of cellular FLICE- inhibitory protein. J. Exp. Med. **190:** 1033–1038.
72. Wang, J., A.A. Lobito, *et al.* 2000. Inhibition of Fas-mediated apoptosis by the B cell antigen receptor through c-FLIP. Eur. J. Immunol. **30:** 155–163.
73. Sata, M. & K. Walsh. 1998. Endothelial cell apoptosis induced by oxidized LDL is associated with the down-regulation of the cellular caspase inhibitor FLIP. J. Biol. Chem. **273:** 33103–33106.
74. Kim, K., M.J. Fisher, *et al.* 2000. Molecular determinants of response to TRAIL in killing of normal and cancer cells. Clin. Cancer Res. **6:** 335–346.
75. Thome, M., P. Schneider, *et al.* 1997. Viral FLICE-inhibitory proteins (FLIPs) prevent apoptosis induced by death receptors. Nature **386:** 517–521.
76. Hu, W.H., H. Johnson, *et al.* 2000. Activation of NF-kappaB by FADD, Casper, and caspase-8. J. Biol. Chem. **275:** 10838–10844.
77. Kataoka, T., R.C. Budd, *et al.* 2000. The caspase-8 inhibitor FLIP promotes activation of NF-kappaB and erk signaling pathways. Curr. Biol. **10:** 640–648.
78. Scaffidi, C., I. Schmitz, *et al.* 1999. The role of c-FLIP in modulation of CD95-induced apoptosis. J. Biol. Chem. **274:** 1541–1548.

Tissue Biology of Apoptosis

Ref-1 and Cell Differentiation in the Developing Retina

LUCIANA B. CHIARINI AND RAFAEL LINDEN[a]

Instituto de Biofísica da UFRJ, Rio de Janeiro, Brazil

ABSTRACT: Programmed cell death by apoptosis plays a major role in neurogenesis. The sensitivity to apoptosis in developing nervous tissue is strongly dependent on cell interactions taking place within a highly structured environment, composed of various cell types at distinct stages of differentiation. In this article, we review evidence gathered both *in vivo* and in a histotypical retinal explant preparation *in vitro* that the bifunctional AP endonuclease/redox factor Ref-1 (HAP1, APE, APEX) may be an anti-apoptotic protein associated with cell differentiation in the developing retina.

KEYWORDS: Programmed cell death; Apoptosis; Retina; Ref-1; TNF receptors; NGF family

ASSOCIATIVE CONTROL OF APOPTOSIS

Mechanisms of apoptosis have usually been studied in either cell lines or primary cultures of isolated cells. These preparations allow for both biochemical measurements and genetic modification of a homogeneous cell population in a controlled environment. Together with cell-free systems, homogeneous cell cultures provided a wealth of information regarding the molecular mechanisms of programmed cell death.[1–5]

Briefly, the execution of the apoptotic cell death program has been ascribed to proteolysis driven by the cysteine-aspartate proteases of the caspase family, which appear to explain every major morphological trait of apoptosis.[6] The execution program may be activated either through the engagement of membrane receptors such as CD95/Fas and tumor necrosis factor (TNF) receptors, or by mitochondrial release of cytochrome *c*, and these two major pathways may interact.[1,2] Intracellular modulators of the apoptotic pathway include several members of the Bcl-2 protein family,[5] as well as transcription factors such as p53 and c-Myc, which are associated with apoptosis in various conditions, including cell cycle checkpoints.[7–10]

Beside the direct initiation of apoptosis induced by the ligands of the CD95 or TNF membrane receptors, other extracellular messengers modulate apoptosis. In the nervous system, the neurotrophins of the nerve growth factor (NGF) family, as well as other growth factors, control neuronal survival.[11–14] Depending on circumstances, which include both the concentration of the neuroactive substance and the target cell type, neurotransmitters such as dopamine and glutamate may either trigger or

[a]Address for correspondence: Dr. Rafael Linden, Instituto de Biofísica da UFRJ, CCS, bloco G, Cidade Universitária, 21949-900, Rio de Janeiro, Brazil. Fax: 55-21-2808193.
rlinden@biof.ufrj.br

block apoptosis.[15–17] In addition, the extracellular matrix decisively affects the control of apoptosis,[18] with important consequences both for the metastatic spread of cancer, and for the regulation of programmed cell death in developing tissues and organs.

Extracellular modulation of cell death implies that the sensitivity to apoptosis of any cell is strongly dependent on interactions with their neighbors within living tissues. Therefore, histotypical preparations are likely to represent more closely than dissociated cell cultures the behavior of normal cells vis à vis apoptotic stimuli.

Mechanisms of apoptosis also vary according to the stage of cell differentiation.[19,20] During embryogenesis of the nervous system, there coexist proliferating neuroblasts and glioblasts, immature neurons and glial cells, and differentiated cells of both types, either intermixed or geometrically arranged in ways that are characteristic of each area. Prior to the establishment of mature synaptic connections, released neuroactive substances, such as neurotransmitters, may influence developing neurons.[21,22] In addition, intimate contact between glia and neurons is a major characteristic of the embryonic nervous tissue. This complex and dynamic environment imposes additional constraints upon the control of apoptosis.

RETINAL EXPLANTS AS A MODEL OF NERVOUS TISSUE FROM THE CNS

The retina of newborn rats has a thickness of 150–200 μm and is composed of two major cellular strata separated by an emerging layer of processes (FIG. 1). The innermost cellular stratum is the ganglion cell layer (GCL), and immediately after birth (at postnatal day 0–P0) this layer contains only ganglion cells, the axons of which form the optic nerve. Displaced amacrine (axonless) cells, which comprise roughly one-half of all neurons in the GCL of adult rodents, migrate postnatally into the GCL.[23] The outer cellular stratum of the neonatal rat retina contains elongated

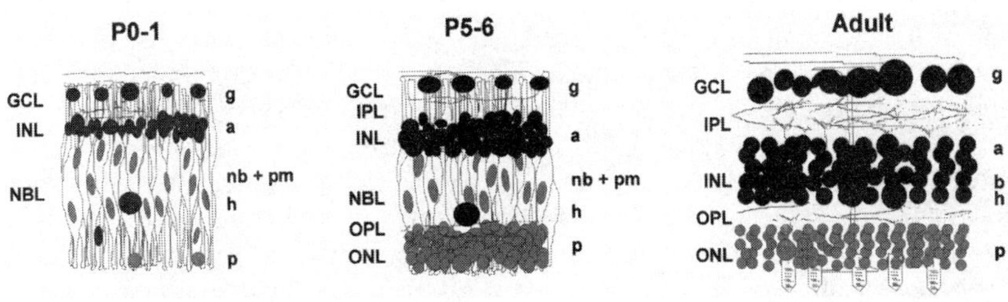

FIGURE 1. Organization of the retina in the rat, and distribution of Ref-1 at various ages. Major retinal layers are indicated to the left, and cell types are indicated to the right of each diagram, depicting newborn (postnatal days 0–1, P0–1), P5–6 and adult rats. *Shading* represents the relative intensity of immunoreactivity with an antibody to Ref-1.[36] ABBREVIATIONS: GCL = ganglion cell layer; INL = inner nuclear layer; NBL = neuroblastic layer; IPL = inner plexiform layer; OPL = outer plexiform layer; ONL = outer nuclear layer; g = ganglion cells; a = amacrine cells; nb = neuroblasts; pm = undifferentiated postmitotic cells; h = horizontal cells; p = photoreceptors; b = bipolar cells.

neuroblasts delimiting a strip of retinal tissue in which extensive cell division occurs postnatally. This corresponds to the ventricular zone[24] and is called the neuroblastic layer (NBL). The NBL in the newborn rat is sandwiched between a few rows of early differentiating amacrine cells, which will form the inner part of the inner nuclear layer (INL) adjacent to the inner plexiform layer (IPL), and a small number of early developing photoreceptors already found close to the outer margin of the neuroretina adjoining the pigment epithelium.

Progressive postnatal development of the rodent retina occurs through the addition of postmitotic cells generated in the NBL. The nuclei of the proliferating neuroblasts undergo interkinetic migration[25] back and forth through the full thickness of this layer, during each cell cycle. Following the last round of cell division, newly formed neurons migrate through the NBL towards their definitive positions, where they finally acquire mature phenotypic markers and connect with their partners through the elaboration of dendrites and, in some cell types, axons. Thus, at any time during development, the NBL contains not only proliferating cells, but also a number of early postmitotic neurons *en route* to their definitive positions. In addition, the early developing horizontal cells can be identified at about the middle of the NBL at birth.

At 4–5 days after birth (P4–P5), the outer plexiform layer (OPL) appears[26] and separates an outer nuclear layer (ONL) containing increasing numbers of differentiating photoreceptor cell somas, from the NBL, which is now internally bound by several rows of amacrine cells (FIG. 1). The photoreceptors will eventually comprise the largest population of neurons in the retina. Cell divisions continue within the NBL at least until the end of the first postnatal week,[27] while the INL and ONL further mature.

During at least the first few days of postnatal development in the rat, waves of naturally occurring cell death[23,28] overlap the progressive events of development that occur within the complex and highly structured environment of the retina.

To preserve cell interactions of the developing nervous tissue we devised a preparation of free-floating explants taken from the retina of neonatal rats, which has been used for studies both of the production of intra-retinal neurotrophic factors[29–31] and of the mechanisms of programmed cell death within the retinal tissue.[32] Fragments of the neural retina excluding the pigment epithelium of approximately 1 mm^2 are cut and placed in culture at 70–90 rpm in an orbital shaker.[29] The multi-layered arrangement, components of the extracellular matrix, and presumably normal cell interactions are essentially maintained in the unattached explants.[32] After various treatments, the explants are fixed with aldehyde mixtures and transverse sections are cut through the retinal layers in a cryostat. The sections are then subject to either histo- or immunocytochemical procedures. Alternatively, the tissue is harvested for various biochemical techniques. The maintenance of histotypical structure *in vitro* allows comparison with the normal course of retinal development *in situ*, as well as controlled experiments upon the microenvironment of retinal cells.

The explants maintain the histotypical organization of the retinal layers for several days in culture. An outer plexiform layer appears at about 4 days after plating explants of newborn rat retinae, with a time course similar to that observed *in vivo*. By the end of the second week in culture the explants tend to form rosettes and undergo massive degeneration. However, during the first week in culture the small amount of cell death within the explant tissue is comparable to the retina *in situ*.[29] The only exception is caused by the unavoidable lesion of optic axons at enucleation, and therefore the axotomized ganglion cells die within 2–3 days in explants prepared from newborn rats.[33] This time course is, nevertheless, identical to that observed *in situ* after damage to the optic nerve.[28,34] Cell proliferation and interkinetic nuclear migration are also preserved in the explants at 24 hours *in vitro*.[32,35] The migration of displaced amacrine cells through the IPL into the ganglion cell layer also continues in the explants.[33]

Thus, both progressive events such as cell and nuclear migration and the emergence of plexiform layers through neuritogenesis, and regressive events such as the death of axotomized ganglion cells are reproduced within the explants. All stages of neural development, from proliferation to relatively advanced stages of cell differentiation, are represented in single cross sections of the explants, similar to the situation *in situ*. The cells within the explant are surrounded by normal neighbors and their neurites relate to normal partners. Each cell is subject to physiological concentrations of extracellular messengers released by either connecting or neighboring cells, and lives and dies within a neighborhood very similar to its natural microenvironment *in situ*. Most of these features are lost when the tissue is dissociated, which is likely to severely affect the metabolic pathways directly related to the control of apoptosis.

Retinal cells at distinct stages of development were shown to be differentially sensitive to various inducers of apoptosis. Thus, both ganglion cells and photoreceptors are more sensitive to apoptosis induced by either the Ca^{2+}-ATPase inhibitor thapsigargin or the phosphatase inhibitor okadaic acid than are proliferating neuroblasts.[36,37] Proliferating cells are, in turn, particularly sensitive to the topoisomerase II inhibitor etoposide, heat stress and gamma irradiation, which nonetheless may also kill recent postmitotic cells.[35,37] In contrast, excitotoxic concentrations of glutamate kill only relatively differentiated retinal neurons, particularly amacrine cells in neonatal rat retinae.[38] Whereas the selective effects of excitatory aminoacids may be a trivial consequence of the progressive expression of specific membrane receptors, it is unclear why drugs that have widespread effects upon cell metabolism, such as thapsigargin and okadaic acid, preferentially kill photoreceptors.

We showed that cells at various stages of differentiation harbor distinct mechanisms of apoptosis regarding the requirement of protein synthesis.[39,40] Thus, apoptosis of axotomized ganglion cells requires newly translated proteins, whereas inhibition of protein synthesis induced apoptosis within the NBL.[40] Further experiments showed that the cells sensitive to apoptosis induced by inhibition of protein synthesis are mainly, if not exclusively, undifferentiated postmitotic cells within 3–4 days after withdrawal from the cell cycle.[41] These cells are likely *en route* to their final destinations among the multi-layered structure of the mature retina.[42]

Apoptosis induced by translation blockade in the neuroblastic layer was prevented by an increase in the intracellular concentration of cAMP.[40,43,44] Apoptosis of photoreceptors induced by either thapsigargin or okadaic acid was also prevented by an increase in cAMP.[37] In contrast, the degeneration of proliferating cells following treatment with either etoposide or heat stress was not sensitive to the intracellular concentration of cAMP, similar to ganglion cells, which also did not resist axotomy-induced apoptosis upon activation of adenylyl-cyclase in retinal explants. It should be noted, however, that in dissociated retinal cell cultures, ganglion cells often respond with increased survival to the activation of adenylyl-cyclase,[45,46] which further emphasizes the distinct behavior of neurons in either histotypical or dissociated cell cultures.

Retinal cell differentiation is consistently associated with increased resistance of retinal neurons to apoptosis induced by various means, such as gamma irradiation, heat stress, exposure to etoposide, and inhibition of protein synthesis. The data reviewed below indicate that the bifunctional redox factor/AP endonuclease Ref-1/HAP1 is an anti-apoptotic protein associated with retinal cell differentiation, which may explain the progressive resistance of post-mitotic retinal cells to programmed cell death.

REF-1: AN ANTI-APOPTOTIC PROTEIN ASSOCIATED WITH RETINAL CELL DIFFERENTIATION

Ref-1 (also known as HAP1, APE or APEX) is a bifunctional enzyme, in which the carboxy terminal domain has an apurinic/apyrimidinic endonuclease activity, while the amino terminal domain has redox activity.[47–50] As an endonuclease, Ref-1 is involved in mechanisms of DNA base excision repair.[51] As a redox factor, Ref-1 modulates the activity of various transcription factors, such as c-Jun, c-Fos, NF-κB, CREB, Myb and p53.[47,52,53]

Ref-1 was distinctly described in the nucleus, the cytoplasm, or both in various cell types.[54] The extra-nuclear location of Ref-1 in certain cells may be associated with a mitochondrial function, since these organelles possess AP endonucleases, and removal of AP sites occur in damaged mitochondrial DNA.[55,56] A possible role of Ref-1 in ribosome function has also been suggested, based on the spatial coincidence of Ref-1 immunoreactivity and staining of neuronal Nissl substance with cresyl-violet.[57,58]

The activities of Ref-1 can be modulated both by phosphorylation and by redox reactions. Phosphorylation *in vitro* by both casein kinases I and II and PKC have been described.[59,60] The endonuclease activity was reportedly blocked *in vitro* by phosphorylation mediated by casein kinase II,[59] although this finding has been disputed.[60] On the other hand, the AP-1 regulatory activity of Ref-1 was activated by casein kinase II.[60] In addition the direct interaction of Ref-1 and c-Jun *in vitro* depends on the reduction of Cys 65,[52,61] which is likely to affect its role as a redox factor.

In the retina *in situ*, the expression of Ref-1 is associated with cell differentiation.[36] In newborn rats, differentiated retinal cells, such as ganglion cells, early-developing amacrines and horizontal cells showed high amounts of Ref-1 in the nucleus, while the neuroblastic layer contained little immunoreactivity. The total amount of Ref-1 protein increases postnatally (FIG. 1 and ref. 36). Ref-1 mRNA also increases in the testis after withdrawal from the cell cycle.[62] In contrast, relatively high levels of Ref-1 are expressed in the proliferative zone of various regious of the brain, followed by a steady decrease associated with development.[36,63]

Control of Ref-1 expression is poorly understood. However, increased intracellular levels of cAMP increase the amount of Ref-1 mRNA,[64] and it was shown that

CREB is necessary for the transcription of *ref-1*.[65] Since increased cAMP leads to retinal cell differentiation,[66] a link between cAMP, differentiation and expression of Ref-1 may be warranted.

Immunoreactivity for Ref-1 was found associated with retinal cell survival. Apoptotic cells lose imunoreactivity for Ref-1 (FIG. 2A–C). This is particularly striking in ganglion cells, which show heavy nuclear immunoreactivity for Ref-1 either before or immediately after axotomy. The loss of Ref-1 immunoreactivity in the nucleus of axotomized ganglion cells precedes the condensation of chromatin. In addition, preservation of Ref-1 in the nuclei of ganglion cells is associated with blockade of apoptosis by an inhibitor of protein synthesis.[36]

Photoreceptors are considered the most vulnerable cells of the retina.[67] These cells have the least content of Ref-1 among differentiated retinal neurons, and this low content was associated with a high sensitivity to apoptosis induced by either thapsigargin or okadaic acid.[36,37] Apoptosis of photoreceptors was also associated with loss of Ref-1, and blockade of apoptosis was accompanied by preservation of Ref-1 in their nuclei.[36]

We showed that anisomycin induces apoptosis of cells that have recently left the cell cycle.[41] Undifferentiated cells have a low content of Ref-1 when compared with amacrine and ganglion cells, which are resistant to inhibition of protein synthesis (FIG. 2D–F). Even the differentiated photoreceptors have a slightly higher immunoreactivity for Ref-1 than proliferating cells. Since our data indicate that Ref-1 content is inversely correlated with sensitivity to apoptosis, inhibition of Ref-1 synthesis upon treatment with anisomycin may be a major determinant of the death of the recent postmitotic cells, as is shown later in FIGURE 4. Proliferating cells are relatively resistant to inhibition of protein synthesis,[41] but also contain little Ref-1. Thus, the dependence on protein synthesis in the retina correlates mainly with a stage of dynamic increase in Ref-1 content, associated with extensive changes in the pattern of gene expression typical of cell differentiation. The repair endonuclease APE/Ref-1 may be particularly important for differentiated nerve cells, because AP sites are the most common form of DNA damage under physiological conditions.[68]

Other reports are also consistent with the hypothesis that Ref-1 is a cell-protective protein. Loss of Ref-1 was associated with neuronal death,[36,69–73] while increased levels of Ref-1 were found in surviving neurons[36,70,71,74] in various scenarios. Finally, down-regulation of Ref-1 by transfection of antisense cDNA was followed by increased sensitivity to genotoxic insult.[75,76]

Both increased nuclear levels of Ref-1 and increased numbers of labeled cells were reported in uterine cervical pre-malignant and malignant tissue when compared with normal tissue.[77] Increased levels of Ref-1 also correlated with decreased radiosensitivity in human cervical carcinomas.[78] Thus, increased Ref-1 expression seems to be associated with enhanced survival and radioresistance of cancer cells *in situ*. However, Ref-1 was not detected in some tumors,[54,77] and the levels of endogenous AP endonuclease activities in various cell lines were not limiting for efficient DNA repair following exposure to either ionizing radiation or methylating agents.[79]

We have previously shown that cells within the neuroblastic layer degenerate by apoptosis following either gamma-irradiation or treatment with 2-aminopurine.[35,43] In both cases, dying cells lose their content of Ref-1, while neighboring cells spared

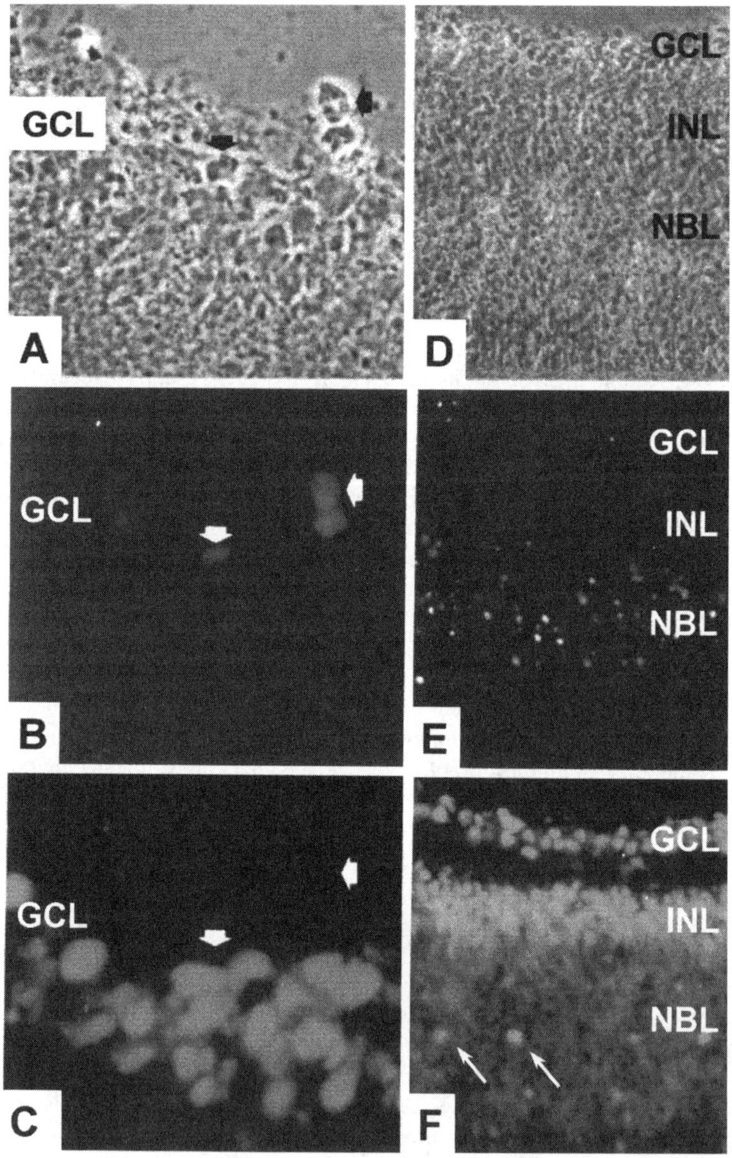
GCL
A
GCL
INL
NBL
D
GCL
B
GCL
INL
NBL
E
GCL
C
GCL
INL
NBL
F

from degeneration present an apparently increased content of Ref-1 (FIG. 3). The increased expression of Ref-1 is likely to be associated with cell resistance to insult in the retina, consistent with the data in tumor cells *in situ*.[77,78]

Various reports showed that stresses such as hypoxia, genotoxic agents, nontoxic levels of reactive oxygen species (ROS), and heat stress induce an increase in the expression of Ref-1.[74,80–82] Ref-1 has been previously associated with hypoxia-induced apoptosis,[83] but in this case the increased level of Ref-1 detected by Western blot may reflect increased levels in surviving cells, as is observed both at the periphery of cold injury-induced brain lesions[71] and following certain insults to the retinal tissue (FIG. 3).

Ref-1 has been shown to activate the tumor supressor p53,[53] which can induce either cell cycle arrest or apoptosis. In apparent contradiction to the proposed protective role, Ref-1 increased the ability of p53 to induce apoptosis in a cotransfection assay, while downregulation of Ref-1 reduced this ability.[84] Nonetheless, the evidence from that study does not demonstrate clearly whether Ref-1 is relevant for p53-mediated apoptosis with normal levels of expression of the tumor supressor gene. Activation of native p53 by Ref-1 may alternatively result in cell cycle arrest associated with DNA repair. The roles of the interaction of Ref-1 and p53 upon distinct responses to DNA damage warrant further examination. Nonetheless, Ref-1 was shown to selectively potentiate the lethal effects of certain chemotherapeutic drugs that are activated by reduction.[85]

The functions of Ref-1 both as an AP endonuclease and as a redox factor may be related to adaptive response to oxidative stress. As an AP endonuclease, Ref-1 is directly involved in the repair of oxidative DNA damage. As a redox factor Ref-1 may lead to AP1- or NF-κB-mediated activation of genes associated with detoxification of ROS (reviewed in refs. 82,86,87). It has also been shown that Ref-1 is directly involved in calcium-induced gene repression.[88] We found that inhibition of protein synthesis blocks photoreceptor cell death induced by thapsigargin.[37] These data are consistent with the hypothesis that the low Ref-1 content in photoreceptors renders this protein unable to counteract the calcium-induced activation of pro-apoptotic genes.

FIGURE 2. Association of Ref-1 with apoptosis and differentiation in developing retinal tissue. Transverse sections from retinal explants are shown in phase contrast (**A**, **D**), and were double-labeled with the TUNEL method developed with fluorescein (**B**, **E**) and immunohistochemistry for Ref-1 using antibody sc-334 from Santa Cruz Biotechnology and secondaries with rhodamin (**C**, **F**). **A–C:** The ganglion cell layer (GCL) of an explant from the retina of a P5 rat, maintained in control culture medium for one day. *Arrowheads* show apoptotic cells. Notice that the TUNEL-labeled apoptotic ganglion cells (**B**) are not labeled for Ref-1 (**C**), while all the remaining, healthy cell nuclei are strongly labeled for the enzyme. **D–F:** Explant from the retina of a P5 rat, incubated for 1 day with 1 μg/ml anisomycin, which induces apoptosis of recent postmitotic cells within the neuroblastic layer.[41] TUNEL-labeled cells (**E**) are restricted to the undifferentiated neuroblastic layer, while Ref-1 staining (**F**) is the most intense within the more differentiated ganglion cell and inner nuclear layers (GCL, INL, respectively). The pattern of Ref-1 labeling in the explant is similar to the staining at P5 in the corresponding layers of the retina *in situ*. *Arrows* in **F** point to early differentiating horizontal cells, located at this age in the outer edge of the neuroblastic layer.

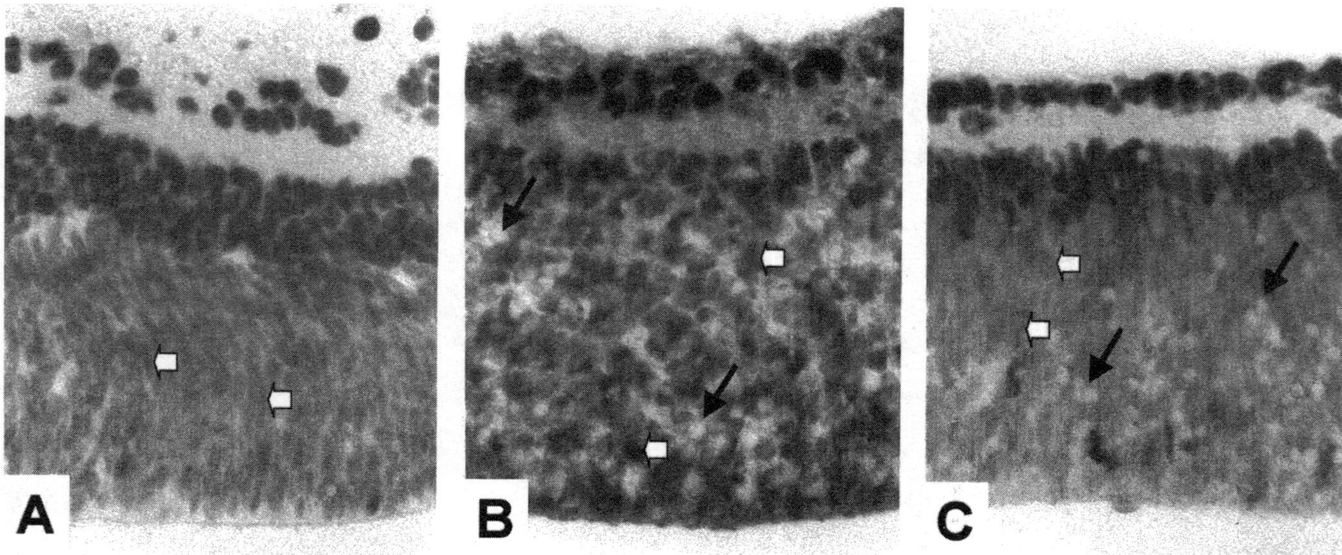

FIGURE 3. Upregulation of Ref-1 in surviving cells following induction of apoptosis by 2-aminopurine. Transverse sections from retinal explants maintained for 24 hours in control medium (**A**), or incubated either with the kinase inhibitor 2-aminopurine at 10 mM (**B**), or 1 μg/ml anisomycin, an inhibitor of protein synthesis (**C**). Notice that the round apoptotic bodies (*arrows*) are devoid of Ref-1 staining, while live cell nuclei (*arrowheads*) treated with 2-aminopurine (**B**) are more intensely stained than either in the control explant or following incubation with anisomycin.

GENERAL CONCLUSIONS AND FUTURE DIRECTIONS

Our studies in the developing retina have uncovered a striking association of Ref-1 with both cell differentiation and resistance to apoptosis. The use of the retinal explant preparations *in vitro* allowed a correlation of the stage of development with responses to apoptotic insults to be studied within the context of the organized neural tissue. Although direct evidence for the proposed antiapoptotic role of Ref-1 is lacking owing to the early lethal effect of *ref-1* gene deletion,[89] the overall data suggest a strong link of Ref-1 with neuroprotective mechanisms, in keeping with current data obtained in various cell types, including tumors *in situ*. The level of Ref-1 may be a major determinant of sensitivity to apoptosis at various stages of retinal development (FIG. 4).

Increased intracellular cAMP blocks apoptosis induced either by anisomycin in undiffferentiated postmitotic retinal cells,[40,43] or by thapsigargin in developing photoreceptors.[37] The report that the expression of Ref-1 is enhanced following an increase of intracellular cAMP[64] raises the hypothesis that photoreceptor protection by increased intracellular cAMP may be due at least in part to induced synthesis of Ref-1. It remains to be tested whether Ref-1 in retinal cells also responds to increased concentrations of cAMP.

On the other hand, altered expression of *ref-1* cannot explain the protective effects of cAMP upon apoptosis of undifferentiated retinal cells induced by the inhibitor of protein synthesis. It has been reported that Ref-1 is not a target of phosphorylation by protein kinase A,[60] therefore further work should be directed at testing effects of cAMP upon the stability of Ref-1 protein.

The relative timing of Ref-1 down-regulation and caspase activation has yet to be determined for retinal apoptosis. These data are needed to understand the exact relationship between the activity of Ref-1 and the control of apoptosis execution pathways.

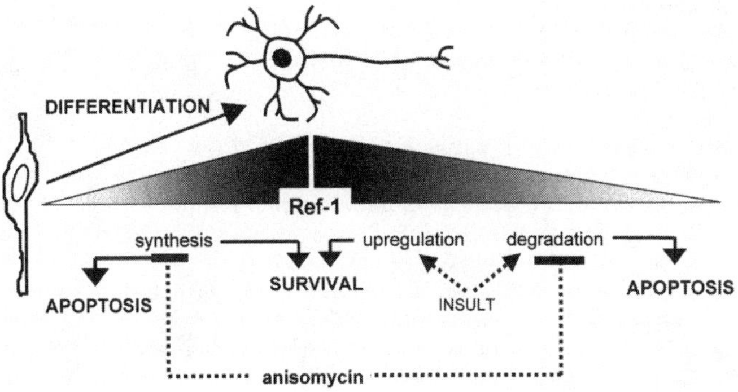

FIGURE 4. A general hypothesis for the association of Ref-1 with both cell differentiation and control of apoptosis in developing retina. Expression of Ref-1 increases with differentiation following withdrawal from the cell cycle, and imparts a progressively lower sensitivity to apoptosis in retinal cells. Lack of Ref-1 synthesis leads to apoptosis in recent postmitotic cells treated with anisomycin. Turnover of Ref-1 helps the execution of apoptosis, while anisomycin prevents cell death partly due to blockade of Ref-1 degradation. Upregulation of Ref-1 leads to resistance of cells subject to certain insults.

REFERENCES

1. Ashkenazi, A. & V.M. Dixit. 1998. Death receptors: signalling and modulation. Science **281:** 1305–1308.
2. Green, D.R. & J.C. Reed. 1998. Mitochondria and apoptosis. Science **281:** 1309–1312.
3. Thornberry, N.A. & Y.A. Lazebnik. 1998. Caspases: enemies within. Science **281:** 1312–1317.
4. Evan, G. & T. Littlewood. 1998. A matter of life and cell death. Science **281:** 1317–1322.
5. Adams, J.M. & S. Cory. 1998. The Bcl-2 protein family: arbiters of cell survival. Science **281:** 1322–1326.
6. Stroh, C. & K. Schulze-Osthoff. 1998. Death by a thousand cuts: an ever increasing list of caspase substrates. Cell Death Diff. **5:** 997–1000.
7. Lane, D.P., X. Lu, T. Hupp & P.A. Hall. 1994. The role of the p53 protein in the apoptotic response. Philos. Trans. R. Soc. Lond. [Biol]. **345:** 277–280.
8. Chiarugi, V. & M. Ruggiero. 1996. Role of three cancer "master genes" p53, bcl2 and c-myc on the apoptotic process. Tumori **82:** 205–209.
9. Liebermann, D.A., B. Hoffman & R.A. Steinman. 1995. Molecular controls of growth arrest and apoptosis: p53-dependent and independent pathways. Oncogene **11:** 199–210.
10. Strasser, A., D.C. Huang & D.L. Vaux. 1997. The role of the bcl-2/ced-9 gene family in cancer and general implications of defects in cell death control for tumourigenesis and resistance to chemotherapy. Biochim. Biophys. Acta **1333:** F151–78.
11. Kaplan, D.R. & R.M. Stephens. 1994. Neurotrophin signal transduction by the Trk receptor. J. Neurobiol. **25:** 1404–1417.
12. Segal, R.A. & M.E. Greenberg. 1996. Intracellular signaling pathways activated by neurotrophic factors. Annu. Rev. Neurosci. **19:** 463–489.
13. Henderson, C.E., H.S. Phillips, R.A. Pollock, A.M. Davies, C. Lemeulle, M. Armanini, L.C. Simpson, B. Moffet, R.A. Vandlen, V.E. Koliatsos & A. Rosenthal. 1994. GDNF: a potent survival factor for motoneurons present in peripheral nerve and muscle. Science **266:** 1062–1064.
14. Ohgoh, M., M. Kimura, H. Ogura, K. Katayama & Y. Nishizawa. 1998. Apoptotic cell death of cultured cerebral cortical neurons induced by withdrawal of astroglial trophic support. Exp. Neurol. **149:** 51–63.
15. Luo, Y., H. Umegaki, X. Wang, R. Abe & G.S. Roth. 1998. Dopamine induces apoptosis through an oxidation-involved SAPK/JNK activation pathway. J. Biol. Chem. **273:** 3756–3764.
16. Ankarcrona, M., J.M. Dypbukt, E. Bonfoco, B. Zhivotovsky, S. Orrenius, S.A. Lipton & P. Nicotera. 1995. Glutamate-induced neuronal death: a succession of necrosis or apoptosis depending on mitochondrial function. Neuron **15:** 961–973.
17. Linden, R. 2000. The anti-death league: associative control of apoptosis in developing retinal tissue. Brain Res. Rev. **32:** 146–158.
18. Meredith, J.E., B. Fazeli & M.A. Schwartz. 1993. The extracellular matrix as a cell survival factor. Mol. Biol. Cell **4:** 953–961.
19. Stolen, C.M., M.W. Jackson & A.E. Griep. 1997. Overexpression of FGF-2 modulates fiber cell differentiation and survival in the mouse lens. Development **124:** 4009–4017.
20. Szondy, Z. 1997. Methylprednisolone and 2-chloroadenosine induce DNA fragmentation at different stages of human T-lymphocyte development. Immunol. Lett. **58:** 59–65.
21. Young, S.H. & M.M. Poo. 1983. Spontaneous release of transmitter from growth cones of embryonic neurones. Nature **305:** 634–637.
22. Xie, Z.P. & M.M. Poo. 1986. Initial events in the formation of neuromuscular synapse: rapid induction of acetylcholine release from embryonic neuron. Proc. Natl. Acad. Sci. USA **83:** 7069–7073.

23. PERRY, V.H., Z. HENDERSON & R. LINDEN. 1983. Postnatal changes in retinal ganglion cell and optic axon populations in the pigmented rat. J. Comp. Neurol. **219:** 356–368.
24. ALEXIADES, M.R. & C.L. CEPKO. 1997. Subsets of retinal progenitors display temporally regulated and distinct biases in the fates of their progeny. Development **124:** 1119–1131.
25. SIDMAN, R.L. 1961. Histogenesis of mouse retina studied with thymidine-H. *In* The Structure of the Eye. G.K. Smelser, Ed.: 487–506. Academic Press. New York.
26. CRAFT, J.L., A.B. FULTON, J. SILVER & D.M. ALBERT. 1982. Development of the outer plexiform layer in albino rats. Curr. Eye Res. **2:** 295–299.
27. ALEXIADES, M.R. & C.L. CEPKO. 1996. Quantitative analysis of proliferation and cell cycle length during development of the rat retina. Dev, Dyn. **205:** 293–307.
28. BEAZLEY, L.D., V.H. PERRY, B. BAKER & J.E. DARBY. 1987. An investigation into the role of ganglion cells in the regulation of division and death of other retinal cells. Dev. Brain Res. **33:** 169–184.
29. ARAUJO, E.G. & R. LINDEN. 1993. Trophic factors produced by retinal cells increase the survival of retinal ganglion cells in vitro. Eur. J. Neurosci. **5:** 1181–1188.
30. ARY PIRES, R., M. NAKATANI, S.K. REHEN & R. LINDEN. 1997. Developmentally regulated release of intraretinal neurotrophic factors in vitro. Int. J. Dev. Neurosci. **15:** 239–255.
31. ARY-PIRES, R., S.K. REHEN & R. LINDEN. 1992. Evidence for a developmentally regulated neurotrophic activity from chick retina affecting the survival of retinal ganglion cells from newborn rats. Braz. J. Med. Biol. Res. **25:** 845–848.
32. LINDEN, R., S.K. REHEN & L.B. CHIARINI. 1999. Apoptosis in developing retinal tissue. Progr. Retinal Eye Res. **18:** 133–165.
33. REHEN, S.K., R. ARY-PIRES & R. LINDEN. 1993. Intraretinal neurotrophic activity prevents the degeneration of ganglion cells in retinal explants. Braz. J. Med. Biol. Res. **26:** 955–959.
34. MILLER, N.M. & M. OBERDORFER. 1981. Neuronal and neuroglial responses following retinal lesions in the neonatal rats. J. Comp. Neurol. **202:** 493–504.
35. BORGES, H.L. & R. LINDEN. 1999. Gamma irradiation leads to two waves of apoptosis in distinct cell populations of the retina of newborn rats. J. Cell Sci. **112:** 4315–4324.
36. CHIARINI, L.B., F.G. FREITAS & R. LINDEN. 2000. Evidence that the bifunctional redox factor/AP endonuclease Ref-1 is an anti-apoptotic protein associated with differentiation in the developing retina. Cell Death Diff. **7:** 272–281.
37. CHIARINI, L.B., F.G. FREITAS, M.L. LEAL-FERREIRA & R. LINDEN. 2000. Major changes in the sensitivity to cells death during the transition from the cell cycle to the differentiated state in retinal photoreceptors. Invest. Ophthalmol. Vis. Sci. Submitted.
38. ROCHA, M., R.A.P. MARTINS & R. LINDEN. 1999. Activation of NMDA receptors protects against glutamate neurotoxicity in the retina: evidence for the involvement of neurotrophins. Brain Res. **827:** 79–92.
39. REHEN, S.K. & R. LINDEN. 1994. Apoptosis in the developing retina: paradoxical effects of protein synthesis inhibition. Braz. J. Med. Biol. Res. **27:** 1647–1651.
40. REHEN, S.K., M.H. VARELLA, F.G. FREITAS, M.O. MORAES & R. LINDEN. 1996. Contrasting effects of protein synthesis inhibition and of cyclic AMP on apoptosis in the developing retina. Development **122:** 1439–1448.
41. REHEN, S.K., D.M. DINIZ, L.F. MADEIRA, L.R.G. BRITO & R. LINDEN. 1999. Selective sensitivity of early postmitotic cells to apoptosis induced by inhibition of protein synthesis. Eur. J. Neurosci. **11:** 349–356.
42. GALLI-RESTA, L. & M. ENSINI. 1996. An intrinsic time limit between the genesis and death of individual neurons in the developing retinal ganglion cell layer. J. Neurosci. **16:** 2318–2324.
43. VARELLA, M.H., D.F. CORREA, C.B.L. CAMPOS, L.B. CHIARINI & R. LINDEN. 1997. Protein kinases selectively modulate apoptosis in the developing retina in vitro. Neurochem. Int. **31:** 217–227.

44. VARELLA, M.H., F.G. DE MELLO & R. LINDEN. 1999. Evidence for an anti-apoptotic role of dopamine in developing retinal tissue. J. Neurochem. **73:** 485–492.
45. KAISER, P.K. & S.A. LIPTON. 1990. VIP-mediated increase in cAMP prevents tetrodotoxin-induced retinal ganglion cell death in vitro. Neuron **5:** 373–381.
46. MEYERFRANKE, A., M.R. KAPLAN, F.W. PFRIEGER & B.A. BARRES. 1995. Characterization of the signaling interactions that promote the survival and growth of developing retinal ganglion cells in culture. Neuron **15:** 805–819.
47. XANTHOUDAKIS, S. & T. CURRAN. 1992. Identification and characterization of Ref-1, a nuclear protein that facilitates AP-1 DNA-binding activity. EMBO J. **11:** 653–665.
48. XANTHOUDAKIS, S., G. MIAO & T. CURRAN. 1994. The redox and DNA-repair activities of Ref-1 are encoded by nonoverlapping domains. Proc. Natl. Acad. Sci. USA **91:** 23–27.
49. ROBSON, C.N. & I.D. HICKSON. 1991. Isolation of cDNA clones encoding a human apurinic/apyrimidinic endonuclease that corrects DNA repair and mutagenesis defects in *E. coli* xth (endonuclease III) mutants. Nucleic Acids Res. **19:** 5519–5523.
50. DEMPLE, B., T. HERMAN & D.S. CHEN. 1991. Cloning and expression of APE, the cDNA encoding the major human apurinic endonuclease: definition of a family of DNA repair enzymes. Proc. Natl. Acad. Sci. USA **88:** 11450–11454.
51. ROTHWELL, D.G., G. BARZILAY, M. GORMAN, S. MORERA, P. FREEMONT & I.D. HICKSON. 1997. The structure and function of the HAP1/Ref-1 protein. Oncol. Res. **9:** 275–280.
52. XANTHOUDAKIS, S., G. MIAO, F. WANG, Y.C.E. PAN & T. CURRAN. 1992. Redox activation of Fos-Jun DNA binding activity is mediated by a DNA repair enzyme. EMBO J. **11:** 3323–3335.
53. JAYARAMAN, L., K.G.K. MURTHY, C. ZHU, T. CURRAN, S. XANTHOUDAKIS & C. PRIVES. 1997. Identification of redox/repair protein Ref-1 as a potent activator of p53. Genes Dev. **11:** 558–570.
54. KAKOLYRIS, S., L. KAKLAMANIS, A. GIATROMANOLAKI, M.I. KOUKOURAKIS, I.D. HICKSON, G. BARZILAY, H. TURLEY, R.D. LEEK, P. KANAVAROS, V. GEORGOULIAS, K.C. GATTER & A.L. HARRIS. 1998. Expression and subcellular localization of human AP endonuclease 1 (HAP1/Ref-1) protein: a basis for its role in human disease. Histopathology **33:** 561–569.
55. DRIGGERS, W.J., V.I. GRISHKO, S.P. LEDOUX & G.L. WILSON. 1996. Defective repair of oxidative damage in the mitochondrial DNA of a xeroderma pigmentosum group A cell line. Cancer Res. **56:** 1262–1266.
56. TOMKINSON, A.E., R.T. BONK & S. LINN. 1988. Mitochondrial endonuclease activities specific for apurinic/apyrimidinic sites in DNA from mouse cells. J. Biol. Chem. **263:** 12532–12537.
57. DUGUID, J.R., J.N. EBLE, T.M. WILSON & M.R. KELLEY. 1995. Differential cellular and subcellular expression of the human multifunctional apurinic/apyrimidinic endonuclease (APE/ref-1) DNA repair enzyme. Cancer Res. **55:** 6097–6102.
58. WILSON, D.M., W.A. DEUTSCH & M.R. KELLEY. 1994. Drosophila ribosomal protein S3 contains an activity that cleaves DNA at apurinic/apyrimidinic sites. J. Biol. Chem. **269:** 25359–25366.
59. YACOUB, A., M.R. KELLEY & W.A. DEUTSCH. 1997. The DNA repair activity of human redox/repair protein APE/Ref-1 is inactivated by phosphorylation. Cancer Res. **57:** 5457–5459.
60. FRITZ, G. & B. KAINA. 1999. Phosphorylation of the DNA repair protein APE/REF-1 by CKII affects redox regulation of AP-1. Oncogene **18:** 1033–1040.
61. WALKER, L.J., C.N. ROBSON, E. BLACK, D. GILLESPIE & I.D. HICKSON. 1993. Identification of residues in the human DNA repair enzyme HAP1 (Ref-1) that are essential for redox regulation of Jun DNA binding. Mol. Cell Biol. **13:** 5370–5376.
62. WILSON, T.M., S.A. RIVKEES, W.A. DEUTSCH & M.R. KELLEY. 1996. Differential expression of the apurinic/apyrimidinic endonuclease (APE/ref-1) multifunctional DNA base excision repair gene during fetal development and in the adult rat brain and testis. Mutation Res. **362:** 237–248.

63. Ono, Y., M. Watanabe, Y. Inoue, T. Ohmoto, K. Akiyama, K. Tsutsu & S. Seki. 1995. Developmental expression of APEX nuclease, a multifunctional DNA repair enzyme, in mouse brains. Dev. Brain Res. **86:** 1–5.
64. Asai, T., F. Kambe, T. Kikumori & H. Seo. 1997. Increase in Ref-1 mRNA and protein by thyrotropin in rat thyroid FRTL-5 cells. Biochem. Biophys. Res. Commun. **236:** 71–74.
65. Grosch, S. & B. Kaina. 1999. Transcriptional activation of apurinic/apyrimidinic endonuclease (Ape, Ref-1) by oxidative stress requires CREB. Biochem. Biophys. Res. Commun. **261:** 859–863.
66. Taylor, M. & T.A. Reh. 1990. Induction of differentiation of rat retinal, germinal, neuroepithelial cells by dbcAMP. J. Neurobiol. **21:** 470–481.
67. Stone, J., J. Maslim, K. Valter-Kocsi, K. Mervin, F. Bowers, Y. Chu, N. Barnett, J. Provis, G. Lewis, S.K. Fisher, S. Bisti, C. Gargini, L. Cervetto, S. Merin & J. Peler. 1999. Mechanisms of photoreceptor death and survival in mammalian retina. Prog. Retinal Eye Res. **18:** 689–735.
68. Lindhal, T. & B. Nyberg. 1972. Rate of depurination of native deoxyribonucleic acid. Biochemistry **11:** 3810–3818.
69. Robertson, K.A., D.P. Hill, Y. Xu, L. Liu, S. Van Epps, D.M. Hockenbery, J.R. Park, T.M. Wilson & M.R. Kelley. 1997. Down-regulation of apurinic/apyrimidinic endonuclease expression is associated with the induction of apoptosis in differentiating myeloid leukemia cells. Cell Growth Diff. **8:** 443–449.
70. Gillardon, F., B. Bottiger & K.A. Hossmann. 1997. Expression of nuclear redox factor ref-1 in the rat hippocampus following global ischemia induced by cardiac arrest. Mol. Brain Res. **52:** 194–200.
71. Morita-Fujimura, Y., M. Fujimura, M. Kawase & P.H. Chan. 1999. Early decrease in apurinic/apyrimidinic endonuclease is followed by DNA fragmentation after cold injury-induced brain trauma in mice. Neuroscience **93:** 1465–1473.
72. Walton, M., P. Lawlor, E. Sirimanne, C. Williams, P. Gluckman, M. Dragunow & R. Takahashi. 1997. Loss of Ref-1 protein expression precedes DNA fragmentation in apoptotic neurons. Brain Res. Mol. Brain Res. **24:** 1381–1385.
73. Kawase, M., M. Fujimura, Y. Morita-Fujimura & P.H. Chan. 1999. Reduction of apurinic/apyrimidinic endonuclease expression after transient global cerebral ischemia in rats. Stroke **30:** 441–449.
74. Edwards, M., D.K. Rassin, T. Izumi, S. Mitra & J.R. Perez-Polo. 1998. APE/Ref-1 responses to oxidative stress in aged rats. J. Neurosci. Res. **54:** 635–638.
75. Walker, L.J., R.B. Craig, A.L. Harris & I.D. Hickson. 1994. A role for the human DNA repair enzyme HAP1 in cellular protection against DNA damaging agents and hypoxic stress. Nucleic Acids Res. **22:** 4884–4889.
76. Ono, Y., T. Furuta, T. Ohmoto, K. Akiyama & S. Seki. 1994. Stable expression in rat glioma cells of sense and antisense nucleic acids to a human multifunctional DNA repair enzyme, APEX nuclease. Mutation Res. **315:** 55–63.
77. Xu, Y., D.H. Moore, J. Broshearws, L. Liu, T.M. Wilson & M.R. Kelley. 1997. The apurinic/apyrimidinic endonuclease (APE/ref-1) DNA repair enzyme is elevated in premalignant and malignant cervical cancer. Anticancer Res. **17:** 3713–3719.
78. Herring, C.J., C.M. West, D.P. Wilks, S.E. Davidson, R.D. Hunter, P. Berry, G. Forster, J. MacKinnon, J.A. Rafferty, R.H. Elder, J.H. Hendry & G.P. Margison. 1998. Levels of the DNA repair enzyme human apurinic/apyrimidinic endonuclease (APE1, APEX, Ref-1) are associated with the intrinsic radiosensitivity of cervical cancers. Br. J. Cancer **78:** 1128–1133.
79. Herring, C.J., B. Deans, D.J. Elder, J.A. Rafferty, J. MacKinnon, G. Barzilay, I.D. Hickson, J.H. Hendry & G.P. Margison. 1999. Expression levels of the DNA repair enzyme HAP1 do not correlate with the radiosensitivities of human or HAP1-transfected rat cell lines. Br. J. Cancer **80:** 940–945.
80. Grosch, S., G. Fritz & B. Kaina. 1998. Apurinic endonuclease (Ref-1) is induced in mammalian cells by oxidative stress and involved in clastogenic adaptation. Cancer Res. **58:** 4410–4416.

81. RAMANA, C.V., I. BOLDOGH, T. IZUMI & S. MITRA. 1998. Activation of apurinic/apyrimidinic endonuclease in human cells by reactive oxygen species and its correlation with their adaptive response to genotoxicity of free radicals. Proc. Natl. Acad. Sci. USA **95:** 5061–5066.
82. DIAMOND, D.A., A. PARSIAN, C.R. HUNT, S. LOFGREN, D.R. SPITZ, P.C. GOSWAMI & D. GIUS. 1999. Redox factor-1 (Ref-1) mediates the activation of AP-1 in HeLa and NIH 3T3 cells in response to heat shock. J. Biol. Chem. **274:** 16959–16964.
83. YAO, K.S., M. CLAYTON & P.J. O'DWYER. 1995. Apoptosis in human adenocarcinoma HT29 cels induced by exposure to hypoxia. J. Natl. Cancer Inst. **87:** 117–122.
84. GAIDDON, C., N.C. MOORTHY & C. PRIVES. 1999. Ref-1 regulates the transactivation and pro-apoptotic functions of p53 in vivo. EMBO J. **18:** 5609–5621.
85. PRIETO-ALAMO, M.J. & F. LAVAL. 1999. Overexpression of the human HAP1 protein sensitizes cells to the lethal effect of bioreductive drugs. Carcinogenesis **20:** 415–419.
86. CASTAGNE, V., M. GAUTSCHI, K. LEFEVRE, A. POSADA & P.G.H. CLARKE. 1999. Relationships between neuronal death and the cellular redox status: focus on the developing nervous system. Progr. Neurobiol. **59:** 397–423.
87. YAO, K.S., S. XANTHOUDAKIS, T. CURRAN & P.J. O'DWYER. 1994. Activation of AP-1 and of a nuclear redox factor, Ref-1, in the response of HT29 colon cancer cells to hypoxia. Mol. Cell. Biol. **14:** 5997–6003.
88. OKAZAKI, T., U.I. CHUNG, T. NISHISHITA, S. EBISU, S. USUDA, S. MISHIRO, S. XANTHOUDAKIS, T. IGARASHI & E. OGATA, 1994. A redox factor protein, Ref-1 is involved in negative gene regulation by extracellular calcium. J. Biol. Chem. **269:** 27855–27862.
89. XANTHOUDAKIS, S., R.J. SMEYNE, J.D. WALLACE & T. CURRAN. 1996. The redox/DNA repair protein Ref-1 is esential for early embryonic development in mice. Proc. Natl. Acad. Sci. USA **93:** 8919–8923.

Modulation of Cell Death in the Rat Thymus

Light and Electron Microscopic Investigations

DANIELA QUAGLINO,[a] MIRIAM CAPRI, AND IVONNE PASQUALI RONCHETTI

Department of Biomedical Sciences, Division of General Pathology, University of Modena and Reggio Emilia, 41100 Modena, Italy

ABSTRACT: In mammals, the thymus is the primary central organ of the lymphoid system; after birth, it progressively diminishes in size, undergoing gradual atrophy. Physiological maturation and/or involution of the thymus may be accelerated by endogenous or exogenous factors. Exposure to extremely low frequency EMF seems to interfere with thymic cell death. Data suggest that, in the rat model, a prolonged exposure to 50 Hz electric and magnetic fields, independently from field strength, seems to affect thymic cell death and possibly thymic physiology, since alterations in the balance of cell death and other parameters such as mitoses might interfere with the positive and negative selection of thymocytes and with the immunosurveillance properties of the thymus.

KEYWORDS: EMF; Thymus; Necrosis; Apoptosis; Clustered cell death

In mammals, the thymus is the primary central organ of the lymphoid system, where lymphocytes, produced by the bone marrow, enter and differentiate into antigen-recognizing cells.[1,2] After birth, the thymus progressively diminishes in size, undergoing gradual atrophy and replacement by fat.[3] Since apoptosis is responsible for negative selection of thymocytes, this parameter as well as other types of cell death have been investigated in the thymus of male Sprague Dawley rats from birth up to the one year of age, as this period is crucial for developmental, maturational and perhaps early involutive processes of the thymus and of the whole immune system.[4] Moreover, since the physiological maturation and/or involution of the thymus may be accelerated by endogenous or exogenous factors[5] and the thymus may be regarded as a possible target of 50 Hz electric and magnetic fields,[6] male Sprague Dawley rats at birth or two months-old have been exposed, for periods up to eight months, to 50 Hz-sinusoidal electric and magnetic fields at two different intensities (1kV/m-5μT and 5kV/m-100μT) in a dark-light cycle or under continuous light. Their thymuses were analyzed by light and electron microscopy.

Data indicate that cell death within the normal rat thymus may be recognized as: (1) apoptosis, characterized by condensed and clumped chromatin (FIG. 1 A), which is maximal at one month from birth (FIG. 2 A); (2) necrosis, comprised of cell debris in the extracellular space as well as within macrophages (FIG. 1 B), which progressively increases with time (FIG. 2 E); and (3) clustered cell death, characterized by

[a]Address for correspondence: Dr. Daniela Quaglino, Department of Biomedical Sciences, Division of General Pathology, University of Modena and Reggio Emilia, Via Campi 287, 41100 Modena, Italy. Voice: 0039-059-205.5442; fax: 0039-059-205.5426.

quaglino.daniela@unimo.it

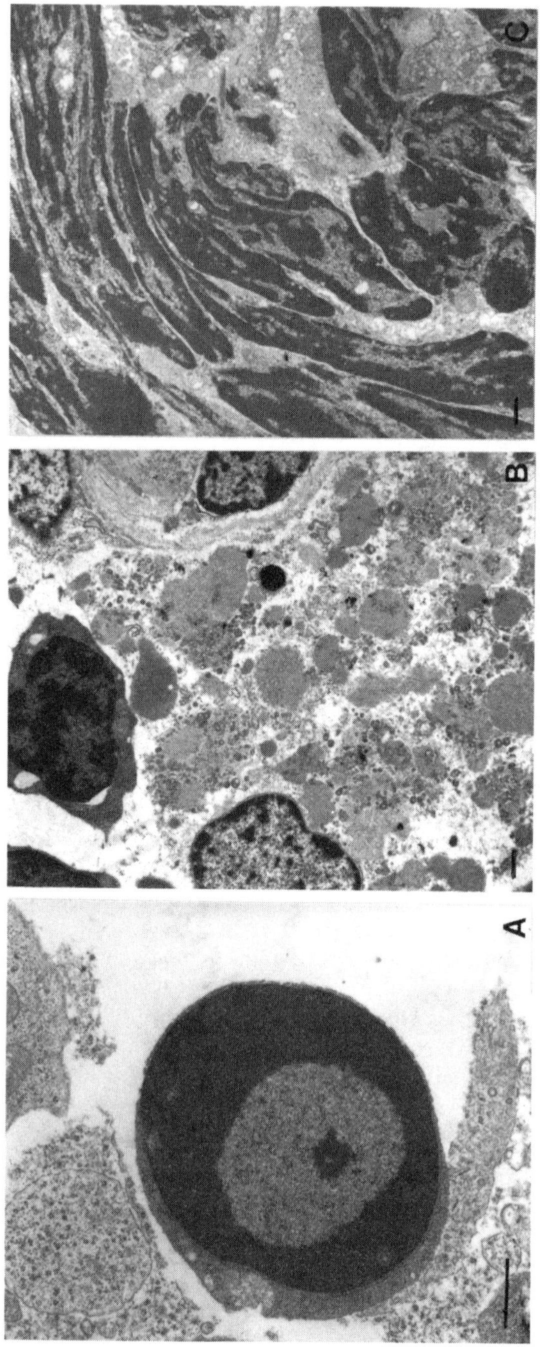

FIGURE 1. Electron microscopy exhibiting the various types of cell death within the rat thymus: apoptosis (**A**), necrosis (**B**) and clustered cell death (**C**). Bar = 1 μm.

groups of lengthened cells with elongated nuclei and dense chromatin (FIG. 1C), which is more represented in old animals (FIG. 2I). In some cases, exposure to extremely low frequency EMF seems to interfere with thymic cell death, depending on field intensity, on the age at which animals start to be exposed, and on the presence of other stress conditions, such as continuous light (FIG. 2).

In conclusion, thymocytes may die in different ways after birth;[4] besides classical apoptosis and necrosis, clustered cell death can be recognized. Further investigation is needed to clarify the nature of this type of cell death, which seems consistent with that described as "thread-like lymphocytes" in lymph nodes.[7] Moreover, data

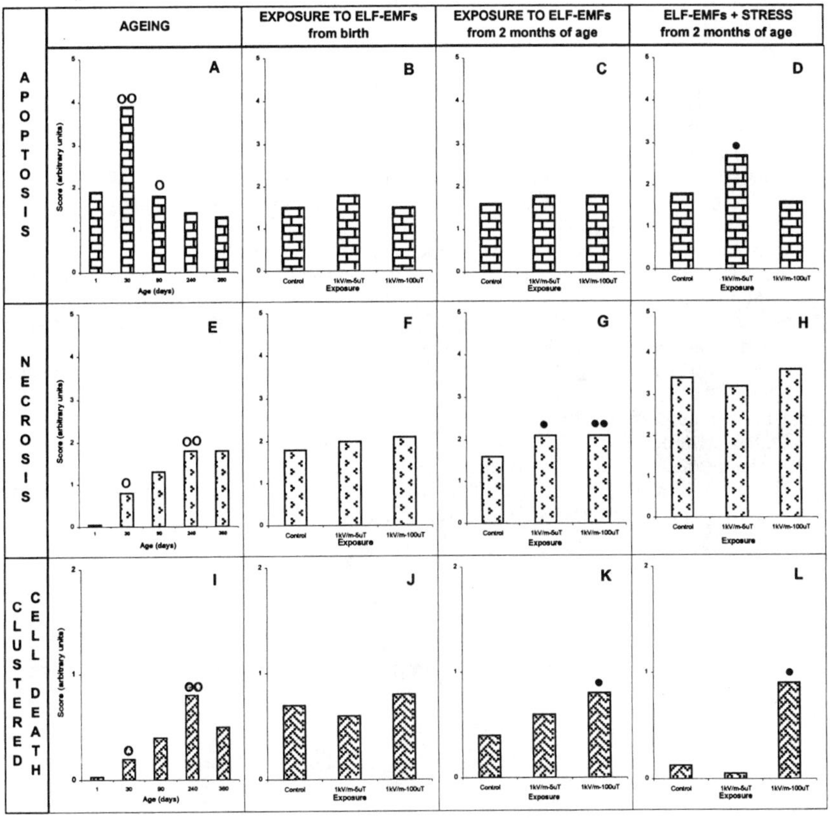

FIGURE 2. Morphometric analysis on semithin sections at 40× magnification. Apoptosis (**A–D**), necrosis (**E–H**) and clustered cell death (**I–L**) were graded semi-quantitatively on a scale from 0 to 5 (0, absent; 1, very scarce but present; 2, scarce; 3, moderately represented; 4, relatively abundant; 5, exuberant). Changes of cell death in aging rats (**A, E, I**) and in rats exposed for 8 months since birth (**B, F, J**) or from 2 months of age in a dark-light cycle (**C, G, K**) or under continuous light (**D, H, L**) are shown. Non-parametric statistical analyses were performed in order to assess the significance of data. ○, $p < 0.05$; ○○, $p < 0.01$ exposed vs.control; ●, $p < 0.05$; ●●, $p < 0.01$ vs. previous age.

suggest that in the rat model a prolonged exposure to 50 Hz electric and magnetic fields, independently of field strength, seems to affect thymic cell death and possibly thymic physiology, since alterations in the balance of cell death and other parameters such as mitoses[6] might interfere with the positive and negative selection of thymocytes and with the immunosurveillance properties of the thymus.[8]

ACKNOWLEDGMENTS

This work was supported by grants from CNR-ENEL and MURST (ex 60%). The authors gratefully acknowledge the contribution of Dr. L. Zecca during ELF-EMF experiments.

REFERENCES

1. KENDALL, M. 1991. Functional anatomy of the thymic microenvironment. J. Anat. **177:** 1–29.
2. SPRENT, J. 1996. The thymus and central tolerance. Horm. Metab. Res. **28:** 294–295.
3. MAJEED, N. *et al.* 1991. Reappraisal of age-related morphological changes in the rat thymus-morphometric study. Func. Develop. Morphol. **1:** 57–60.
4. QUAGLINO, D. JR. *et al.* 1998. Age-dependent remodeling of rat thymus. Morphological and cytofluorimetric analysis from birth up to one year of age. Eur. J. Cell Biol. **76:** 156–166.
5. SCHUURMAN, H.J. *et al.* 1994. Histopathology of the immune system as a tool to assess immunotoxicity. Toxicology. **86:** 187–212.
6. CAPRI, M. *et al.* 1999. Thymus as a possible target of 50Hz electric and magnetic fields. *In* Electricity and Magnetism in Biology and Medicine. F. Bersani, Ed.: 195–198. Plenum. New York.
7. SAINTE MARIE, G. & F.S. PENG. 1990. Atrophy of compartments of rat lymph nodes related to an entry of lethally altered lymphocytes. Cell Tissue Res. **262:** 263–271.
8. URBAN, J.L. & H. SCHREIBER. 1988. Host-tumor interactions in immunosurveillance against cancer. Progr. Exp.Tumor Res. **32:** 17–68.

The Role of the Pro-Apoptotic Bcl-2 Family Member Bim in Physiological Cell Death

PHILIPPE BOUILLET, DAVID C.S. HUANG, LORRAINE A. O'REILLY, HAMSA PUTHALAKATH, LIAM O'CONNOR, SUZANNE CORY, JERRY M. ADAMS, AND ANDREAS STRASSER[a]

The Walter and Eliza Hall Institute, Melbourne, Australia

ABSTRACT: Apoptosis, an evolutionarily conserved process for killing unwanted cells in multicellular organisms, is essential for normal development, tissue homeostasis and as a defense against pathogens. The control of apoptosis is of considerable importance for clinical medicine, as its deregulation can lead to cancer, autoimmunity or degenerative diseases. We have disrupted the Bim gene in the mouse and demonstrated that it plays a major and non-redundant role in embryogenesis, in the control of hematopoietic cell death, and as a barrier against autoimmunity.

KEYWORDS: Bcl-2 family; Bim; Apoptosis

INTRODUCTION

Apoptosis is an evolutionarily conserved process for killing unwanted cells in multicellular organisms. [1] It is essential for normal development, tissue homeostasis and as a defense against pathogens.[2,3] The control of apoptosis is of considerable importance for clinical medicine, as its deregulation can lead to cancer, autoimmunity or degenerative diseases.[4–6]

Our present understanding of the molecular control of apoptosis comes largely from the convergence of genetic studies in the worm *C. elegans* and biochemical and genetic analysis in mammalian cells.[7, 8] Three families of proteins are necessary for the execution of the death program: cysteine proteases termed caspases,[9] their adaptors,[10] and the pro-apoptotic members of the Bcl-2 family.[11] Pro-survival members of the Bcl-2 family inhibit the activation of the process and are required for cell survival. Pro apoptotic Bcl-2 family members can be subdivided into two subgroups. Bax, Bak, Bok/Mtd and Bcl-x_S all possess two or three Bcl-2 homology regions (BH1 to 4). By contrast, Bim/Bod, Bad, Bid, Bik, Blk, Hrk/DP5 and *C. elegans* EGL-1 share homology with each other and with the Bcl-2 family in general only within a short 9–16 amino acid domain referred to as the BH3 region.[11] These proteins are now known as BH3-only members of the Bcl-2 family. The BH3 region

[a]Address for correspondence: Dr. Andreas Strasser, The Walter and Eliza Hall Institute, Post Office, The Royal Melbourne Hospital, VIC 3050, Australia. Voice: +61-3-9345-2624; fax: +61-3-9347-0852.

strasser@wehi.edu.au

within pro-apoptotic Bcl-2 family members is essential for heterodimerization with the pro-survival Bcl-2 relatives and for proapoptotic function.[12]

We have disrupted the Bim gene in the mouse and shown that it is essential for the response to certain apoptotic stimuli, to maintain homeostasis in the hematopoietic compartment and as a barrier against autoimmunity.[13]

THE *C. ELEGANS* BH3-ONLY PROTEIN EGL-1

The BH3-only protein EGL-1 is necessary for all developmentally programmed deaths of somatic cells in *C. elegans.*[14] Loss-of-function (lf) mutations in *egl-1* cause a phenotype that is very similar to lf mutations in the caspase CED-3 or its adaptor CED-4, and gain-of-function (gf) mutations in the Bcl-2 homologue CED-9.[14] Gf mutations in *egl-1* induce the death of a group of neurons that normally survive in hermaphrodites and are required for egg laying (HSN).[15] Cell deaths induced by ectopic expression of EGL-1 or by *egl-1* gf mutations can be suppressed by lf mutations in either *ced-3* or *ced 4* or by gf mutations in *ced-9*, indicating that *egl-1* acts upstream of these genes. EGL-1 protein contains a 9 amino acid BH3 domain, but is otherwise unrelated to the mammalian Bcl-2-like proteins. The BH3 region of EGL-1 is predicted to adopt an amphipathic α-helical structure that binds to a cleft in CED-9 formed by the BH1 and BH2 domains. It has been proposed that the interaction between CED-9 and CED-4 prevents CED-4 from activating CED-3 and thereby promotes cell survival.[16–18] Binding of EGL-1 to CED-9 is thought to release CED-4, thereby allowing CED-4 to activate CED-3 and trigger downstream events required for the killing process.

PRO-APOPTOTIC BCL-2 FAMILY MEMBERS IN MAMMALS

It has been speculated that the Bax subfamily of proapoptotic Bcl-2-related molecules may induce cell death by more than one mechanism.[19] Although some studies have shown that these molecules can function by antagonizing Bcl-2-like pro-survival molecules, other studies have demonstrated ability of these molecules to form channels in mitochondrial membranes,[20] either directly or by interacting with mitochondrial pore forming proteins.[21, 22] It has also been reported that Bax can kill cells by a caspase-independent mechanism.[23] It must however be emphasized that many of these observations stem from experiments in which non-physiologically high levels of Bax were expressed. It therefore remains unclear which mechanisms of cell killing are activated by Bax and its closest relatives under normal conditions. The absence of similarity outside the BH3 domain makes it likely that the BH3-only proteins induce cell killing exclusively by inactivating Bcl-2-like pro-survival molecules.

BIM, A MAMMALIAN BH3-ONLY PROTEIN

Bim was first cloned in our laboratory as a Bcl-2-interacting protein using λ phage expression library screening.[24] Analysis of its sequence showed that it is a BH3-only protein with a transmembrane domain (TM). Three isoforms of Bim, Bim_{EL}, Bim_L and Bim_S, can be generated by alternative splicing. They all contain the BH3 and TM regions, but their N-terminal regions are different. The BH3 and TM domains are both required for the pro-apoptotic function in all Bim isoforms. Interestingly, the Bim isoforms differ substantially in their killing potency, the shorter one (Bim_S) being the most potent and the longer (Bim_L, Bim_{EL}) being less active. This indicated that the extra N-terminal sequences may attenuate the pro-apoptotic activity of Bim.

EXPRESSION OF BIM

The tissue distribution of Bim has been investigated by IP/western blotting, immunohistochemistry and *in situ* hybridization. These studies have demonstrated that Bim is expressed in hematopoietic, epithelial, neuronal and germ cells. Bim_L and Bim_{EL} were co-expressed at similar levels in many cell types, but Bim_S was never detected.[25, 26] This may indicate that Bim_S is not expressed under physiological conditions or that it is expressed only in a limited number of cells that need to be killed rapidly. Microscopic examination revealed a punctate pattern of Bim_L and Bim_{EL} immunostaining, indicating association with cytoplasmic structures.

BIM_L AND BIM_{EL} ARE KEPT INACTIVE BY BINDING TO THE CYTOSKELETON

Screening of a HeLa cell library with the yeast two-hybrid system using Bim_L as bait revealed an interaction between Bim and the LC8 dynein light chain.[27] In yeast and in mammalian cells, LC8 interacted with Bim_{EL} and Bim_L, but not with Bim_S. The sequence required to bind to LC8 was shown to be localized within the N-terminal region that is absent in Bim_S. Single amino acid mutations that disrupt the interaction of Bim_L with LC8 increased its killing potency to that of Bim_S, indicating that the binding to LC8 is a physiological regulator of Bim pro-apoptotic activity. In healthy cells, Bim_{EL} and Bim_L are associated with LC8 and thereby sequestered to the microtubule associated dynein motor complex. Upon an apoptotic stimulus, Bim is released from this complex together with LC8 and can then translocate to and inactivate pro-survival Bcl-2-like molecules.

BIM PLAYS AN IMPORTANT ROLE DURING EMBRYONIC DEVELOPMENT

To investigate the physiological role of Bim, we have inactivated the gene in the mouse by replacing the critical BH3-containing exon with a *neo* cassette.[13] No Bim protein could be seen by IP/Western blotting in cells from homozygous *bim*–/– mice, demonstrating that a null-mutation had been created. Intercrosses of *bim*+/– heterozygotes yielded only one third of the expected homozygous offspring, demonstrating an important role of Bim during embryogenesis. Embryonic lethality occurred prior to E9.5 (for a presently unknown reason) and was observed on a mixed genetic background as well as on inbred C57BL/6 or 129 backgrounds. Interestingly, a significant proportion of *bim*+/– heterozygotes also died during development, indicating that there is a gene dosage effect (see below).

BIM IS REQUIRED FOR LEUKOCYTE HOMEOSTASIS

The *bim*–/– mice that were born looked outwardly normal and were fertile. Although the production of hematopoietic cells was normal, *bim*–/– mice had a significant elevation in blood leukocytes. Immunofluorescent staining and flow cytometric analysis revealed 2- to 4-fold increased numbers of B cells, T cells, granulocytes and macrophages compared to wt littermates. Numbers of erythrocytes and megakaryocytes were normal, but interestingly, platelet numbers were reduced to 50% of the levels observed in wt mice. This may indicate that platelet shedding from megakaryocytes occurs by an apoptosis-like mechanism that requires Bim and perhaps also other Bcl-2 family members. The spleen and lymph nodes of young adult *bim*–/– mice contained 2- to 3- times more B and T lymphocytes than wt littermates, and the lymphadenopathy progressed with age. These results demonstrate that Bim is a critical regulator of homeostasis in several hematopoietic cell types.

BIM IS REQUIRED FOR THE EXECUTION OF CERTAIN APOPTOTIC STIMULI

In tissue culture, *bim*–/– lymphocytes ($CD4^+8^+$ thymocytes, $B220^+sIg$ pre B cells, mature, resting and activated B and T cells) were resistant to certain Bcl-2 inhibitable apoptotic stimuli, such as cytokine withdrawal and treatment with a calcium ionophore.[28] Interestingly, sensitivity of *bim*+/– cells was always intermediate between that of wt and *bim*–/– cells, demonstrating that the concentration of Bim is a central factor in these pathways of apoptosis. Bim-deficient cells were almost normally sensitive to other stimuli that can be inhibited by Bcl-2, including treatment with glucocorticoids, γ-radiation, or PKC activation. Consistent with the observation that Bcl-2 does not block death by receptor-induced apoptosis in lymphocytes,[29] *bim*–/– lymphocytes were normally sensitive to Fas-ligand. These results demonstrate that different death stimuli require different BH3-only proteins for apoptosis induction.

BIM IMPOSES A CRITICAL BARRIER AGAINST AUTOIMMUNE DISEASE

Lymphadenopathy progressed with age and many older *bim–/–* mice accumulated enormous numbers of immunoglobulin (Ig)-secreting cells. IgG antibody–forming cells reached levels 50- to 200-fold above those in wt mice. About 60% of *bim–/–* mice died before 1 year of age, and 85% of those were diagnosed with fatal immune complex glomerulonephritis. High titers of IgG autoantibodies to nuclear antigens and deposits of immune complexes in the kidneys were observed in most of these mice. Interestingly, about 30% of *bim+/–* heterozygotes developed the same symptoms, reinforcing the notion that there is a gene dosage effect for Bim. A very similar systemic lupus erythematosus (SLE)-like disease had been observed in mice expressing a *bcl*-2 transgene in B cells on a (C57BL/6×SJL)F2 genetic background.[5]

CONCLUSIONS

These results demonstrate that Bim plays a major and non-redundant role in embryogenesis, in the control of hematopoietic cell death, and as a barrier against autoimmunity. Several tissues in which Bim is normally expressed appear normal in *bim–/–* mice. It is likely that in these cell types, other BH3-only molecules can compensate for Bim and that they can trigger the same downstream apoptotic events as Bim does.

Unlike EGL-1, which appears to be regulated mainly at the transcriptional level,[14] Bim and also some other mammalian BH3-only proteins (e.g., Bad, Bid) are regulated post-translationally by sequestration to sites where they cannot interact with Bcl-2-like pro-survival molecules.[27,30–32] They also seem to be required differentially for the execution of distinct death stimuli. The systematic disruption of all genes coding for BH3-only proteins and intercrossing of the resulting mice should reveal more fascinating insights into the functions of these essential cell death initiators.

ACKNOWLEDGMENTS

We are grateful to all of our past and present colleagues at the Walter and Eliza Hall Institute, particularly Drs. A. Harris, D. Vaux, D. Metcalf, F. Koentgen, D. Tarlinton, S. Bath, L. Harrison, T. Kay, and G. Hausmann, for help with experiments and insightful discussions. We thank L. Barnett, L. Zhang, S. Novakovic, L. Cullen and J. Beaumont for technical assistance and G. Filby for editorial help. Our laboratories are supported by the NHMRC (Reg. Key 973002), the US NCI (CA 43540 and CA80188), the Anti-Cancer Council of Victoria (Melbourne), The Australian Research Council, the Dr. Josef Steiner Cancer Research Foundation (Bern) the Leukemia and Lymphoma Society of America, and the Cancer Research Institute (New York).

REFERENCES

1. Kerr, J.F.R. *et al.* 1972. Apoptosis: a basic biological phenomenon with wide-ranging implications in tissue kinetics. Br. J. Cancer **26:** 239–257.
2. Vaux, D.L. *et al.* 1994. An evolutionary perspective on apoptosis. Cell **76:** 777–779.
3. Jacobson, M.D. *et al.* 1997. Programmed cell death in animal development. Cell **88:** 347–354.
4. Strasser, A. *et al.* 1990 Novel primitive lymphoid tumors induced in transgenic mice by cooperation between myc and bcl-2. Nature **348:** 331–333.
5. Strasser, A. *et al.* 1991. Enforced BCL-2 expression in B-lymphoid cells prolongs antibody responses and elicits autoimmune disease. Proc. Natl. Acad. Sci. USA **88:** 8661–8665.
6. Thompson, C.B. 1995. Apoptosis in the pathogenesis and treatment of disease. Science **267:** 1456–1462.
7. Hengartner, M.O. *et al.* 1994. Programmed cell death in *Caenorhabditis elegans*. Curr. Opin. Genet. Dev. **4:** 581–586.
8. Vaux, D.L. *et al.* 1996. The molecular biology of apoptosis. Proc. Natl. Acad. Sci. USA **93:** 2239–2244.
9. Thornberry, N.A. *et al.* 1998. Caspases: enemies within. Science **281:** 1312–1316.
10. Vaux, D.L. 1997. CED-4: the third horseman of apoptosis. Cell **90:** 389–390.
11. Adams, J.M. *et al.* 1998. The Bcl-2 family: arbiters of cell survival. Science **281:** 1322–1326.
12. Sattler, M. *et al.* 1997. Structure of Bcl-x_L-Bak peptide complex: recognition between regulators of apoptosis. Science **275:** 983–986.
13. Bouillet, P. *et al.* 1999. Proapoptotic Bcl-2 relative Bim required for certain apoptotis responses, leukocyte homeostasis, and to preclude autoimmunity. Science **286:** 1735–1738.
14. Conradt, B. *et al.* 1998. The *C. elegans* protein EGL-1 is required for programmed cell death and interacts with the Bcl-2-like protein CED-9. Cell **93:** 519–529.
15. Ellis, H.M. *et al.* 1986. Genetic control of programmed cell death in the nematode *C. elegans*. Cell **44:** 817–829.
16. Spector, M.S. *et al.* 1997. Interactions between the *C. elegans* death regulators CED-9 and CED-4. Nature **385:** 653–656.
17. Chinnaiyan, A.M. *et al.* 1997. Interaction of CED-4 with CED-3 and CED-9: a molecular framework for cell death. Science **275:** 1122–1126.
18. Wu, D. *et al.* 1997. Interaction and regulation of subcellular localization of CED-4 and CED-9. Science **275:** 1126–1129.
19. Xiang, J. *et al.* 1996. Bax-induced cell death may not require interleukin 1β-converting enzyme-like proteases. Proc. Natl. Acad. Sci. USA **93:** 14559–14563.
20. Antonsson, B. *et al.* 1997. Inhibition of Bax channel-forming activity by Bcl-2. Science **277:** 370–372.
21. Marzo, I. *et al.* 1998. Bax and adenine nucleotide translocator cooperate in the mitochondrial control of apoptosis. Science **281:** 2027–2031.
22. Shimizu, S. *et al.* 1999. Bcl-2 family of proteins regulate the release of apoptogenic cytochrome c by the mitochondrial channel VDAC. Nature **399:** 483–487.
23. Knudson, C.M. *et al.* 1997. Bcl-2 and Bax function independently to regulate cell death. Science **270:** 96–99.
24. O'Connor, L. *et al.* 1998. Bim, a novel member of the Bcl-2 family that promotes apoptosis. EMBO J. **17:** 384–395.
25. O'Reilly, L.A. *et al.* 1998. Rapid hybridoma screening method for the identification of monoclonal antibodies to low abundance cytoplasmic proteins. BioTechniques **25:** 824–830.
26. O'Reilly, L.A. *et al.* 2000. The pro-apoptotic BH3-only protein Bim is expressed in hematopoietic, epithelial, neuronal and germ cells. Am. J. Pathol. **157:** 449–461.
27. Puthalakath, H. *et al.* 1999. The pro-apoptotic activity of the Bcl-2 family member Bim is regulated by interaction with the dynein motor complex. Mol. Cell. **3:** 287–296.

28. STRASSER, A. *et al.* 1991. Bcl-2 transgene inhibits T cell death and perturbs thymic self-censorship. Cell **67:** 889–899.
29. STRASSER, A. *et al.* 1995. Bcl-2 and Fas/APO-1 regulate distinct pathways to lymphocyte apoptosis. EMBO J. **14:** 6136–6147.
30. ZHA, J. *et al.* 1996. Serine phosphorylation of death agonist Bad in response to survival factor results in binding to 14.3.3 not Bcl-x_L. Cell **87:** 619–628.
31. LI, H. *et al.* 1998. Cleavage of Bid by caspase-8 mediates the mitochondrial damage in the Fas pathway of apoptosis. Cell **94:** 491–501.
32. LUO, X. *et al.* 1998. Bid, a Bcl-2 interacting protein, mediates cytochrome c release from mitochondria in response to activation of cell surface death receptors. Cell **94:** 481–490.

Evolution of Functions within the p53/p63/p73 Family

VINCENZO DE LAURENZI AND GERRY MELINO[a]

IDI-IRCCS Biochemistry Lab., c/o Department of Experimental Medicine and Biochemical Sciences, University Tor Vergata, Rome, Italy

ABSTRACT: Even though the tumor suppressor gene p53 is highly important in human cancer, as indicated by the fact that it is mutated in about 50% of cases, up to a few years ago no similar proteins had been identified. Recently, two p53 homologues have been identified, p73 and p63, with high aminoacid identity suggesting similar functions. Indeed, like p53, p73 as well (i) can bind mdmX, mdm2, p300/CAF and adenovirus E4-orf6 proteins, (ii) can trigger several promoters including p21, bax, mdm2, gadd45, cyclin G, IGFBP3, 14-3-3σ, (iii) is able to trigger cell death, (iv) is involved in the DNA damage response, although through a different pathway. Here we analyze the data present in the literature in search of diverging pathways among the p53, p63, p73 family. Both p63 and p73 present two significant structural peculiarities: the presence of an extended non-conserved C-terminus containing a sterile alpha motive (SAM), typical of developmental proteins, and the presence of number of different splicing isoforms differing in the N-terminus or in the absence of the transactivation domain (ΔN forms), acting as dominant negative. The mouse knockout of p63 and p73, unlike the ones for p53, shows developmental abnormalities; p63 and p73 are rarely mutated in human cancers; both genes are regulated in different differentiation models. This strongly suggests the involvement of p63 and p73 in development. A picture is emerging showing a gradient of function among p53, p73, p63 ranging from tumor suppression to development.

KEYWORDS: p53; p63; p73; c-Abl; Apoptosis; DNA damage; Differentiation; Development

p53 Is the most frequently mutated gene in all forms of human cancers.[1,2] The p53 gene encodes for a transcription factor that is activated in response to genotoxic damage.[3–8] Increased p53 levels lead to cell-cycle arrest or apoptosis preventing replication of damaged cells. Inactivation of p53 allows the replication of mutated cells eventually leading to cancer development.

Recently two homologues of p53, p63 and p73, have been identified (reviewed by Arrowsmith[9] and Levrero[10,11]). The proteins encoded by these two genes share similar, but not identical, structure with p53,[12,13] they also share some of the functions with p53 but seem to have different physiological functions in cells. In this paper we

[a]Address for correspondence: Dr. Gerry Melino, IDI-IRCCS Biochemistry Lab., c/o Department of Experimental Medicine and Biochemical Sciences, Room D26/F153, University Tor Vergata, via Tor Vergata 135, 00133 Rome, Italy. Voice: +39-06-20427299; fax +39-06-20427290.

gerry.melino@uniroma2.it

summarize the current knowledge on the p63 and the p73 genes, with a focus on the differences among the three members in terms of regulation; physical interaction with one another, and with other cellular and viral proteins; and physiological role in development and response to DNA damage.

STRUCTURE OF p73 AND p63 GENES AND THEIR PRODUCTS

The three members of the p53 family share a very high homology and all contain a transactivation domain (TAD),[b] a DNA binding domain (DBD) and an oligomerization domain (OD). The highest level of homology is reached in the DBD (63% aminoacid identity between p53 and p73 and 60% identity between p53 and p63), suggesting that the three genes can transactivate the same target promoters. The high conservation of the OD suggests that these family members can form hetero-oligomers as well as homo-oligomers. However the data published so far summarized in FIGURE 1 show that homo-oligomers are favored.[12,14–16]

In addition to the three domains that they share with p53 both p63 and p73 contain in their long C-terminus a protein-protein interaction domain known as "sterile alpha motive" (SAM).[17] The SAM domain is a globular domain composed of four alpha-helices and a small 3_{10} helix that is found in proteins involved in the regulation of development. Unlike other SAM domains the ones found in p73 and p63 do not form homo-dimers or hetero-dimers with each other,[17] indicating that they may indeed be important for interactions with other proteins.

Different from p53, both the p63 and the p73 genes present a variety of alternative splicing that results in an equal number of different proteins. Most of the splices are found at the 3′ end of both genes in a part of the sequence that is not present in p53, generating proteins with different C-termini. At least six different p73 proteins, named with Greek Letters from α to ζ, are generated by alternative splicing of the 3′ end in normal cells.[12,14,18,19] The β and the δ isoforms are truncated forms (respectively at amino acid 399 and 494 of the full length p73-α. The ζ isoform lacks amino acids 400 to 496 of the α isoform, but contains the remaining carboxi-terminal part. The γ isoform is probably one of the more interesting variants since it contains a long alternative reading frame and therefore has a completely different C-terminus. The ε isoform uses of part of the γ reading frame and then goes back to the α reading frame.

Similarly to p73, p63 gives rise to different splice variants at the 3′ end. However only 3 different variants at the 3′ end of this gene have been identified so far: a full length α form, a truncated β form and a γ form that splices out exons from 11 to 14 and uses an additional exon 15 not found in p73.

The transcription of both p63 (human and mouse) and mouse p73 can start from exon 1 or from an additional exon (exon 3′) located between exons 3 and 4. The use of exon 3′ produces proteins that lack the TA domain (ΔN isoforms) and act as dominant negatives inhibiting the transactivation functions of the p63/p73 isoforms that have a complete N-terminus (TA isoforms) and of p53.[13] At present, N-terminal

[b]ABBREVIATIONS: TAD, transactivation domain; OD, oligomerization domain; DBD, DNA binding domain; TA, transactivating forms; ΔN, non transactivating N-terminal deleted forms; UV, ultraviolet light; IR, ionizing irradiation; MMS, methylmethane sulfonate; CDDP, cisplatin.

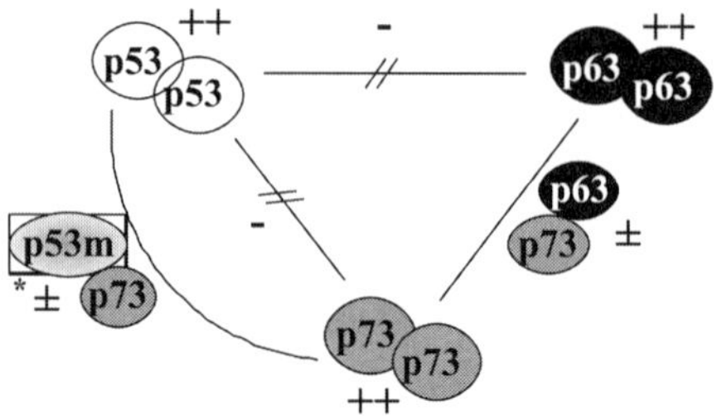

FIGURE 1. Physical interaction between p53, p63 and p73. Symbols: ++ indicates strong interaction; ± indicates weak interaction; – indicates no interaction detected; * = interaction revealed between p53 R175H and p73, and between p53R248W and p73. See text for details.

deletion variants of p73 have been found only in mice where they seem to be the most abundant forms but have not been yet found in humans. Nevertheless their existence in humans cannot be excluded.

Recently the upstream promoter region of the human p73 gene has been cloned and partially characterized[20] (FIG. 2). Unlike the p53 gene promoter it contains a TATA-like box. Preliminary studies show that the region between nucleotides –119 to +19 relative to the start of exon 1 contains the region required for the basal transcription of p73. This region contains putative SP1, AP2 and Egr-1, 2, 3 sites (FIG. 2) and several stretches of CpG dinucleotides. The region located between position –119 and –2714 contain additional regulatory sites: SP1 sites, AP-2, E2F sites, NFAT sites, c-Myb sites, N-Myc sites, and an homeobox.

GENES REGULATED BY p63 AND p73

As expected several genes that are known to respond to p53 can be regulated also by p63 and p73. However the data published so far show important differences

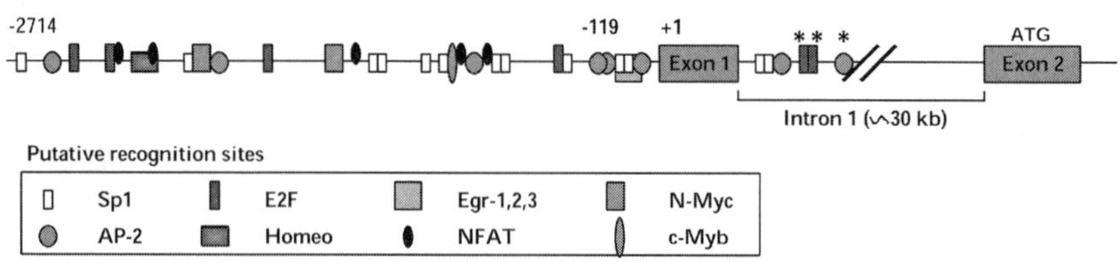

FIGURE 2. Structure of the p73 promoter. Sequence data are from Ding *et al.*, 1999 (nucleotides −1873 to +82), from Genbank AB 031234 (nucleotides −2714 to −1874 and from + 83 to +290). The putative binding sites for transcription factors have been identified using the MATINSPECTOR 2.2 software. All nucleotide positions refer to the start of exon 1. Putative recognition sites are indicated in the legend. See text for explanation.

among the transactivation activities of these three genes. Indeed, although p63 and p73 have been shown to activate the promoters of several p53-responsive genes in cotransfection experiments, including p21, bax, mdm2, cyclin-G, gadd45, IGFBP3 (A) and IGFBP3,[12–14,21–26] many target genes respond differently to the different family members. Zhu *et al.* (1998), using cell lines that inducibly over-produced either p53, p73-α or p73-β, have detected quantitative differences between the effects of p53 and p73 isoforms on the induction of endogenous genes such as p21, mdm2, GADD45 and have identified genes that were selectively induced by p53 but not by p73. Yu *et al.*[27] have made similar observations and, using a Serial Analysis of Gene Expression (SAGE), have found that only 6 out of 36 p53-induced transcripts are significantly induced by p73-α. In addition we have shown that p73 and p63 but not p53 can transactivate promoters of proteins associated with neuronal[28] or keratinocyte[29] differentiation. In conclusion, current data indicate the existence of transcription specificity among the p53-family members.

ROLE OF p73 AND p63 IN CANCER

Since p73 is highly homologous to p53 and since it maps in a region frequently deleted in neuroblastoma[12,30] it was initially suggested that p73 could be a tumor suppressor. However mutation studies performed on large numbers of different human tumors detected only 3 missense mutations. Similarly, in the few mutation studies on p63 only 3 mutations were detected.[23,31] It has also been suggested that p73 may be an imprinted gene expressed by only one allele and therefore the loss of a single copy would be sufficient to lose the function and to predispose to cancer development.[12] Recent studies however seem to show that monoallelic expression of this gene is rare.[32–35] Various tumors express higher levels of p73 when compared to the tissue of origin,[31,33,34,36,38–40] suggesting that an altered expression of p73 rather than a loss of function might be involved in tumorigenesis.

The role of p53 in cancer development is due to its ability to respond to DNA damage.[3,4,41–43] As schematically shown in FIGURE 3, activation of p53 may result in a cell cycle arrest to allow for DNA repair to occur before replication. When cells are irreparably damaged p53 activation leads to their elimination by apoptotic cell death.

Initial observations showing that p73 mRNA levels were not influenced by DNA damage had suggested that p73 did not play a role in the response to genotoxic stress. However. recent studies show that the p73 protein accumulates and is phosphorylated in response to DNA damaging agents in a c-Abl dependent pathway (reviewed by Levrero[10,11] and Shaul[44]). Indeed results from three independent groups show that the nuclear tyrosine kinase c-Abl is activated in response to different DNA damaging agents.[44–47] cAbl can bind and phosphorylate p73. The interaction between p73 and cAbl results in a stabilization of p73 protein and in an increase in its apoptotic potential.

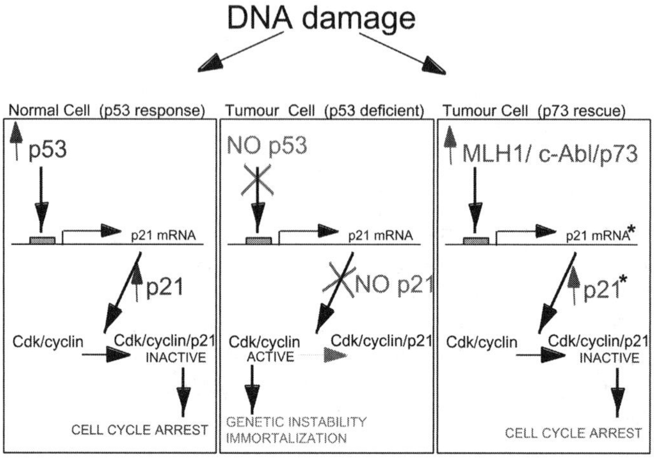

FIGURE 3. Pathways leading to cell death elicited by DNA damage. In normal cells the main "guardian of the genome" gene induced is p53 which by transactivating the cyclin kinase inhibitor p21 inactivates the cyclin kinase/cyclin complex, leading to cell cycle arrest (*left scheme*). For this reason, p53 has been found mutated or inactivated in about 50% of all human cancers. In cancer cells, the absence of, or mutations in, p53 inactivate this pathway, thus allowing an abnormal progression in the cell cycle, which by replicating DNA damage-induced mutations, leads to genetic instability and immortalization (*central panel*). Recently we and others have identified a parallel rescue pathway mediated by p73 (*right panel*). The mismatch repair MLH1 gene induced by DNA damage can physically activate the catalytic enzyme activity of the tyrosine kinase c-Abl which through a physical interaction with p73 transactivates p21, restoring cell cycle arrest. Defined defects in MLH1, c-Abl, p73 block this alternative rescue pathway. Both MLH1 and c-Abl were previously linked to cancer development, but since c-Abl does not interact with p53, the molecular mechanisms were not known. * = p21 might not be the only mechanism involved.

INTERACTIONS OF p73 WITH VIRAL AND CELLULAR PROTEINS

Viruses produce a number of proteins that interact and with p53 and inhibit its function thereby preventing death or growth arrest of infected cells (see for review Ko and Prives[3]; Levine[4]; Choisy-Rossi & Yonish-Rouach[5]; El-Deiry[6]; Prives and Hall[7]; Oren[8]). FIGURE 4A summarizes the interaction between p53 and viral proteins. The adenovirus E1B 55-kD protein and the SV40 T antigen bind to p53 and sequester it into an inactive complex. The E1A and the E1B 55-kD proteins increase the steady state levels of p53 by activating p19Arf which interferes with the Mdm2-mediated degradation of p53.[48,49] The human papillomavirus E6 protein interacts with p53 and promotes its ubiquitin-dependent degradation. It is therefore conceivable to think that these proteins have similar interactions with p73 and p63. The adenovirus E1B 55-kD protein and The SV40 T-Antigen however do not interact with p73.[22,50–52] E1B 55-kD proteins increase the steady state levels p73,[52] but this is probably not mediated by p19Arf since Mdm2 does not affect the stability of p73. E6 has no effect on the stability of p73 and cannot block growth arrest and apoptosis

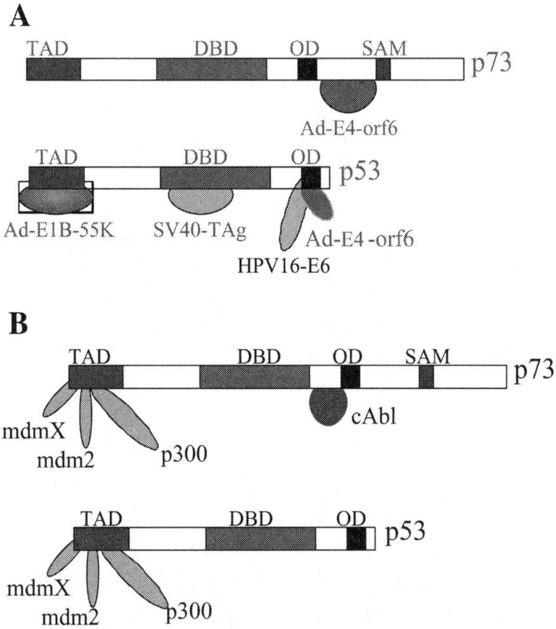

FIGURE 4. Interactions of p53 and p73 with other proteins. **A:** Interactions of p53 and p73 with viral oncoproteins. See text for details. **B:** Physical interaction of p53 and p73 with cellular proteins. See text for details.

induced by p73.[51,53] On the contrary the adenovirus E4orf6 binds to a C-terminal region common to p73 α and β and inhibits the transactivation and the apoptotic functions of p73.[50,52] It seems therefore that viral proteins can clearly distinguish between the different p53 family members and use different systems to block their action.

Similar to what observed for viral proteins, the interactions between p73 and cellular proteins that interact and modulate p53 have shown both similarities and differences, suggesting once again that only certain functions are conserved among the different family members. FIGURE 4B summarizes the known interactions between p73 and cellular proteins. The mouse double minute 2 gene product (Mdm2) and its human homologue (Hdm2)[8,54] sequester and promote the degradation of p53. Since this gene is upregulated by p53 this constitutes an important feedback loop for the control of p53 levels. p73 N-terminal region binds to Mdm2 and MdmX.[25,55,56] The interaction with Mdm2 or MdmX leads to the inactivation of transcription and apoptosis functions of p73-α and p73-β[25] but does not result in the rapid degradation of the p73 protein by the proteasome.[24,25,55,56] The interaction of p53 with the transcription adapters/co-activators p300 and CBP regulates the ability of p53 to bind specific DNA sequences and to activate transcription. p73 binds the CH3 region of p300 through its N-terminal region[24,57] (amino acids 1 to 54). The interaction with p300/CBP is required for p73 to activate transcription and Mdm2 inhibits the transactivating function of p73 by the exclusion of p300/CBP.[25,57]

ROLE OF p63 AND p73 IN DIFFERENTIATION

Studies on the phenotype of knockout mice for the different p53 family members have shown that they have different physiological functions.

p53-Defective mice develop normally but show an increased number of spontaneous and induced tumors.[58,59]

The p63-deficient mice have defects of the apical ectodermal ridge essential to limb development and the mice have truncated limbs. In addition, p63–/– mice have no hair follicles, no teeth, and no mammary, lachrymal or salivary glands.[60,61] Similar defects are found in children affected by ectrodactyly, ectodermal dysplasia and facial clefts (EEC syndrome). Recently mutations in p63 gene have been shown to be responsible for this syndrome.[62] We have recently observed that the over-expression of p63 in normal human epidermal keratinocytes upregulates markers specific for keratinocyte differentiation[29] (i.e., loricrin, involucrin and transglutaminase 1).

p73-Deficient mice show developmental defects that include: hippocampal dysgenesis, hydrocephalus, and abnormalities in pheromone receptors that lead to loss of interest in animals of the opposite sex. The p73–/– mice also showed a number of chronic infections, the origin of which is still unclear since the mice had normal lymphoid and granulocyte populations.[63] Unlike p53-deficient mice, p73–/– animals did not have an increased number of spontaneous tumors. Their susceptibility to cancer-inducing agents however remains to be tested. In line with these results we have found that induction of differentiation in neuroblastoma cells is accompanied by upregulation of p73 and that ectopic expression of this gene is capable of triggering neuronal differentiation in these cells. Indeed, transfection of these cells with p73 results in neurite outgrowth and increased expression of neuronal differentiation markers such as neurofilaments and neuronal cell adhesion molecule (NCAM).[28] Moreover transfection of a dominant negative form of p73 can block some of the differentiation features induced in neuroblastoma cells by retinoic acid.

Altogether the data available so far suggest that the different members of the p53 family, despite their structural similarity, share fewer functional activities than expected. Indeed, while p53 clearly has a main role in DNA damage response and p63 seems to be essential for the development of certain tissues but seems not associated with tumor suppression, p73 might have both functions, being involved in differentiation and participating in DNA damage response in a pathway parallel to the one that involves p53. In addition the ability of the different members of the family to interact with each other suggests that a complicated pattern of crosstalk carefully regulates the activity of these transcription factors.

ACKNOWLEDGMENTS

The work was in part supported by grants: EU grant QLG1-1999-00739, Telethon-E872, AIRC, MURST-Cofin, Min Sanitá, Associazione Neuroblastoma to G.M.

REFERENCES

1. HOLLSTEIN, M., D. SIDRANSKY & C.C. VOGELSTEIN. 1991. p53 mutations in human cancers. Science **253:** 49–53.
2. HOLLSTEIN, M., B. SHOMER, M. GREENBLATT, T. SOUSSI, E. HOVIG, R. MONTESANO & C.C. HARRIS. 1996. Somatic point mutations in the p53 gene of human tumors and cell lines: updated compilation. Nucleic Acids Res. **24:** 141–146.
3. KO, L.J. & C. PRIVES. 1996. p53: puzzle and paradigm. Genes Devel. **10:** 1054–1072.
4. Levine, A.J. 1997. p53, the cellular gatekeeper for growth and division. Cell **88:** 323–331.
5. CHOISY-ROSSI, C. & E. YONISH-ROUACH. 1998. Apoptosis and the cell cycle: the p53 connection. Cell Death Differ. **5:** 129–131.
6. EL-DEIRY, W.S. 1998. Regulation of p53 downstream genes. Semin Cancer Biol **8:** 345–357.
7. PRIVES, C. & P.A. HALL. 1999. The p53 pathway. J. Pathol. **187:** 112–126.
8. OREN, M. 1999. Regulation of the p53 tumor suppressor protein. J. Biol. Chem. **274:** 36031–36034.
9. ARROWSMITH, C.H. 1999. Structure and function in the p53 family. Cell Death Differ. **6:** 1169–1173.
10. LEVRERO, M., V. DE LAURENZI, A. COSTANZO, J. GONG, G. MELINO & J.Y.J. WANG. 1999. Structure, function and regulation of p63 and p73. Cell Death Differ. **6:** 1146–1153.
11. LEVRERO, M., V. DE LAURENZI, A. COSTANZO, J. GONG, J.Y.J. WANG & G. MELINO. 2000. The p53/p63/p73 family of transcription factors: overlapping and distinct functions. J. Cell Sci. **113:** 1661–1670
12. KAGHAD, M., H. BONNET, A. YANG, L. CREACIER, J.C. BISCAN, A. VALENT, A. MINTY, P. CHALON, J.M. LELLAS, X. DUMONT, P. FERRARA, F. MCKEON & D. CAPUT. 1997. Monoallelically expressed gene related to p53 at 1p36, a region frequently deleted in neuroblastoma and other human cancers. Cell **90:** 809–819.
13. YANG, A., M. KAGHAD, Y. WANG, E. GILLETT, M.D. FLEMING, V. DOTSCH, N.C. ANDREWS, D. CAPUT & F. MCKEON. 1998. p63, a p53 homolog at 3q27-29, encodes multiple products with transactivating, death-inducing, and dominant-negative activities. Mol. Cell. **2:** 305–316.
14. DE LAURENZI, V., A. COSTANZO, D. BARCAROLI, A. TERRINONI, M. FALCO, M. ANNICCHIARICO-PETRUZZELLI, M. LEVRERO & G. MELINO. 1998. Two new p73 splice variants, gamma and delta, with different transcriptional activity. J. Exp. Med. **188:** 1763–1768.
15. DAVISON, T.S., C. VAGNER, M. KAGHAD, A. AYED, D. CAPUT & C.H. ARROWSMITH. 1999. p73 and p63 are homotetramers capable of weak heterotypic interactions with each other but not with p53. J. Biol. Chem. **274:** 18709–18714.
16. DI COMO, C.J., C. GAIDDON & C. PRIVES. 1999. p73 function is inhibited by tumor-derived p53 mutants in mammalian cells. Mol. Cell. Biol. **19:** 1438–1449.
17. CHI, S-W., A. AYEDA & C.H. ARROWSMITH. 1999. Solution structure of a conserved C-terminal domain of p73 with structural homology to the SAM domain. EMBO J. **18:** 4438–4445.
18. DE LAURENZI, V., M.V. CATANI, A. COSTANZO, A. TERRINONI, M. CORAZZARI, M. LEVRERO, R.A. KNIGHT & G. MELINO. 1999. Additional complexity in p73: induction by mitogens in lymphoid cells and identification of two new splicing variants epsilon and zeta. Cell Death Differ. **6:** 389–390.
19. UEDA, Y., M. HIJIKATA, S. TAKAGI, T. CHIBA & K. SHIMOTOHNO. 1999. New p73 variants with altered C-terminal structures have varied transcriptional activities. Oncogene **18:** 4993–4998.
20. DING, Y., T. INOUE, J. KAMIYAMA, Y. TAMURA, N. OHTANI-FUJITA, E. IGATA & T. SAKAI. 1999. Molecular cloning and functional characterization of the upstream promoter region of the human p73 gene. DNA Res. **6:** 347–351.
21. JOST, C.A., M.C. MARIN & W.G. KAELIN, JR. 1997. p73 is a human p53 related protein that can induce apoptosis. Nature **389:** 191–194.

22. DOBBELSTEIN, M. & J. ROTH. 1998. The large T antigen of simian virus 40 binds and inactivates p53 but not p73. J. Gen. Virol. **79:** 3079–3083.
23. OSADA, M., M. OHBA, C. KAWAHARA, C. ISHIOKA, R. KANAMARU, I. KATOH, Y. IKAWA. Y. NIMURA, A. NAKAGAWARA, M. OBINATA & S. IKAWA. 1998. Cloning and functional analysis of human p51, which structurally and functionally resembles p53. Nat Med. **4:** 839–843.
24. LEE, C.W. & N.B. LA THANGUE. 1999. Promoter specificity and stability control of the p53-related protein p73. Oncogene **18:** 44171–44181.
25. ZENG, X., L. CHEN, C.A. JOST, R. MAYA, D. KELLER, X. WANG, W.G. KAELIN, JR., M. OREN, J. CHEN & H. LU. 1999. MDM2 suppresses p73 function without promoting p73 degradation. Mol. Cell. Biol. **5:** 3257–3266.
26. ZHU, J., J. JIANG, W. ZHOU & X. CHEN. 1988. The potential tumor suppressor p73 differentially regulates cellular p53 target genes. Cancer Res. **58:** 5061–5065.
27. YU, J., L. ZHANG. P.M. HWANG, C. RAGO, K.W. KINLZER & B. VOGELSTEIN. 1999. Identification and classification of p53-regulated genes. Proc. Natl. Acad. Sci. USA **96:** 14517–14522.
28. DE LAURENZI, V., G. RASCHELLÁ, D. BARCAROLI, M. ANNICCHIARICO-PETRUZZELLI, M. RANALLI, M.V. CATANI, B. TANNO, A. COSTANZO, M. LEVRERO & G. MELINO. 2000. Induction of neuronal differentiation by p73, in a neuroblastoma cell line. J. Biol. Chem. **275:** 15226–15231.
29. DE LAURENZI, V., A. ROSSI, A. TERRINONI, D. BARCAROLI, M. LEVRERO, A. COSTANZO, R.A. KNIGHT, P. GUERRIERI & G. MELINO. 2000. p63 and p73 transactivate differentiation gene promoters in human keratinocytes. Biochem. Biophys. Res. Commun. In press
30. IKAWA, S., A. NAKAGAWARA & Y. IKAWA. 1999. p53 family genes: structural comparison, expression and mutation. Cell Death Differ. **6:** 1154–1161.
31. SUNAHARA, M., S. ICHIMIYA, Y. NIMURA, N. TAKADA, S. SAKIYAMA, Y. SATO, S. TODO, W. ADACHI, J. AMANO & A. NAKAGAWARA. 1998. Mutational analysis of the p73 gene localized at chromosome 1p36.3 in colorectal carcinomas. Int. J. Oncol. **13:** 319–323.
32. NOMOTO, S., N. HARUKI, M. KONDO, H. KONISHI, T. TAKAHASHI, T. TAKAHASHI & T. TAKAHASHI. 1998. Search for mutations and examination of allelic expression imbalance of the p73 gene at 1p36.33 in human lung cancers. Cancer Res. **58:** 1380–1383.
33. YOKOMIZO, A., M. MAI, D.J. TINDALL, L. CHENG, D.G. BOSTWICK, S. NAITO, D.I. SMITH, & W. LIU. 1999. Overexpression of the wild type p73 gene in human bladder cancer. Oncogene **18:** 1629–1633.
34. ZAIKA, A.I., S. KOVALEV, N.D. MARCHENKO & U.M. MOLL. 1999. Overexpression of the wild type p73 gene in breast cancer tissues and cell lines. Cancer Res. **59:** 3257–3263.
35. YOKOZAKI, H., Y. SHITARA, J. FUJIMOTO, T. HIYAMA, W. YASUI & E. TAHARA. 1999. Alterations of p73 preferentially occur in gastric adenocarcinomas with foveolar epithelial phenotype. Int. J. Cancer **83:** 192–196.
36. KOVALEV, S., N. MARCHENKO, S. SWENDEMAN, M. LAQUAGLIA & U.M. MOLL. 1998. Expression level, allelic origin, and mutation analysis of the p73 gene in neuroblastoma tumors and cell lines. Cell Growth Differ **9:** 897–903.
37. MAI, M., C. QIAN, A. YOKOMIZO, D.J. TINDALL, D. BOSTWICK, C. POLYCHRONAKOS, D.I. SMITH & W. LIU. 1998. Loss of imprinting and allele switching of p73 in renal cell carcinoma. Oncogene **17:** 1739–1741.
38. TANNAPFEL, A., M. WASNER, K. KRAUSE, F. GEISSLER, A. KATALINIC, G. HAUSS, J. MOSSNER, K. ENGELAND & C. WITTEKIND. 1999. Expression of p73 and its relation to histopathology and prognosis in hepatocellular carcinoma. J. Natl. Cancer Inst. **91:** 1154–1158.
39. TANNAPFEL, A., K. ENGELAND, L. WEINANS, A. KATALINIC, J. HAUSS, J. MOSSNER & C. WITTEKIND. 1999. Expression of p73, a novel protein related to the p53 tumour suppressor p53 and apoptosis in cholangiocellular carcinoma of the liver. Br. J. Cancer **80:** 1069–1074.

40. CASCIANO, I., M. PONZONI, C. CUNSOLO, G. TONINI & M. ROMANI. 1999. Different p73 splicing variants are expressed in distinct tumour areas of a multifocal neuroblastoma. Cell Death Differ. **6:** 391–393.
41. GOTTLIEB, T.M. & M. OREN. 1996. p53 and apoptosis. Semin. Cancer Biol. **8:** 359–368.
42. AMUNDSON, S.A., T.G. MYERS & A.J. FORNACE, JR. 1998. Roles for p53 in growth arrest and apoptosis: putting on the brakes after genotoxic stress. Oncogene **17:** 3287–3299.
43. GIACCIA, A.J. & M.B. KASTAN. 1998. The complexity of p53 modulation: emerging patterns from divergent signals. Genes Devel. **12:** 2973–2983.
44. SHAUL, Y. 2000. c-Abl: activation and nuclear targets. Cell Death Differ. **7:** 10–16
45. GONG, J.G.,A. COSTANZO, H.Q. YANG, G. MELINO, W.G. KAELIN, M. LEVRERO & J.Y.J. WANG. 1999. The tyrosine kinase c-Abl regulates p73 in apoptotic response to cisplatin-induced DNA damage. Nature **39:** 806–809.
46. AGAMI, R., G. BLANDINO, M. OREN & Y. SHAUL. 1999. Interaction of c-Abl and p73 and their collaboration to induce apoptosis. Nature **399:** 809–813.
47. YUAN, Z.M., H. SHIOYA, T. ISHIKO, X. SUN, J. GU, Y.Y. HUANG, H. LU, S. KHARBANDA, R. WEICHSELBAUM & D. KUFE. 1999. p73 is regulated by tyrosine kinase c-Abl in the apoptotic response to DNA damage. Nature **399:** 814–817.
48. TAO, W. & A.J. LEVINE. 1999. P19(ARF) stabilizes p53 by blocking nucleo-cytoplasmic shuttling of Mdm2. Proc. Natl. Acad. Sci. USA **96:** 6937–6941.
49. ZHANG, Y. & Y. XIONG. 1999. Mutations in human ARF exon 2 disrupt its nucleolar localization and impair its ability to block nuclear export of MDM2 and p53. Mol. Cell. **5:** 579–591.
50. HIGASHINO, F., J.M. PIPAS & T. SHENK. 1998 Adenovirus E4orf6 oncoprotein modulates the function of the p53-related protein, p73. Proc. Natl. Acad. Sci. USA **95:** 15683–15687.
51. MARIN, M.C., C.A. JOST, M.S. IRWIN, J.A. DECAPRIO, D. CAPUT & W.G. KAELIN, JR. 1998. Viral oncoproteins discriminate between p53 and the p53 homolog p73. Mol. Cell. Biol. **18:** 6316–6324.
52. STEEGENGA, W.T., A. SHVARTS, N. RITECO, J.L. BOS & A.G. JOCHEMSEN. 1999. Distinct regulation of p53 and p73 activity by adenovirus E1A, E1B, and E4orf6 proteins. Mol. Cell. Biol. **9:** 3885–3894.
53. PRABHU, N.S., K. SOMASUNDARAM, K. SATYAMOORTHY, M. HERLYN & W.S. EL-DEIRY. 1998. p73beta, unlike p53, suppresses growth and induces apoptosis of human papillomavirus E6-expressing cancer cells. Int. J. Oncol. **13:** 5–9.
54. LOHRUM, M.A.E. & K.H. VOUSDEN. 1999. Regulation and activation of p53 and its family members. Cell Death Differ. **6:** 1162–1168.
55. BALINT, E., S. BATES. & K.H. VOUSDEN. 1999. Mdm2 binds p73 alpha without targeting degradation. Oncogene **18:** 3923–3929.
56. ONGKEKO, W.M., X.Q. WANG, W.Y. SIU, A.W. LAU, K. YAMASHITA, A.L. HARRIS, L.S. COX & R.Y. POON. 1999. MDM2 and MDMX bind and stabilize the p53-related protein p73. Curr. Biol. **15:** 829–833.
57. ZENG, X., X. LI, A. MILLER, Z. YUAN, W. YUAN, R.P. KWOK, R. GOODMAN & H. LU. 2000 The N-terminal domain of p73 interacts with the CH1 domain of p300/CREB binding protein and mediates transcriptional activation and apoptosis. Mol. Cell. Biol. **20:** 1299–1310.
58. DONEHOWER, L.A., M. HARVEY, B.L. SLAGLE, M.J. MCARTHUR, C.A. MONTGOMERY, JR. J.S. BUTEL & A. BRADLEY. 1992. Mice deficient for p53 are developmentally normal but susceptible to spontaneous tumours. Nature **356:** 215–221.
59. DONEHOWER, L.A., L.A. GODLEY, C.M. ALDAZ, R. PYLE, Y.P. SHI, D. PINKEL, J. GRAY, A. BRADLEY, D. MEDINA & H.E. VARMUS. 1995 Deficiency of p53 accelerates mammary tumorigenesis in Wnt-1 transgenic mice and promotes chromosomal instability. Genes Devel. **9:** 882–895.
60. MILLS, A.A., B. ZHENG, X.J. WANG, H. VOGEL, D.R. ROOP & A. BRADLEY. 1999. p63 is a p53 homologue required for limb and epidermal morphogenesis. Nature **398:** 708–713.

61. YANG, A., M. KAGHAD, Y. WANG, E. GILLETT, M.D. FLEMING, V. DOETSCH, N.C. ANDREWS, D. CAPUT & F. MCKEON. 1999. p63 is essential for regenerative proliferation in limb, craniofacial and epithelial development. Nature **398:** 714–718.
62. CELLI, J., P. DUIJF, B.C.J. HAMEL, M. BAMSHAD, B. KRAMER, A.P.T. SMITS, R. NEWBURY-ECOB, R.C.M. HENNEKAM, G. VAN BUGGENHOUT, A. VAN HAERINGEN, C.G. WOODS, A.J. VAN ESSEN, R. DE WAAL, G. VRIEND, D.A. HABER, A. YANG, F. MCKEON, H.G. BRUNNER & V. H. BOKHOVEN. 1999. Heterozygous germline mutations in the p53 homolog p63 are the cause of EEC syndrome. Cell **99:** 143–153.
63. YANG, A., N. WALKER, R. BRONSON, M., KAGHAD, M. OOSTERWEGEL, J. BONNIN, C. VAGNER, H. BONNET, P. DIKKES, A. SHARPE, F. MCKEON & D. CAPUT. 2000. p73-deficient mice have neurological, pherormonal and inflammatory defects but lack spontaneous tumours. Nature **404:** 99–103.

First and Second Messenger Role of Calcium

Survival versus Apoptosis in Serum-Free Cultured Granulosa Explants

SYLVIE MUSSCHE,[a,b] LUC LEYBAERT,[c] AND KATHARINA D'HERDE[a]

[a]*Department of Anatomy, Embryology, Histology and Medical Physics Faculty of Medicine and Health Sciences, Ghent University, B-9000 Ghent, Belgium*

[c] *Department of Physiology and Pathophysiology, Faculty of Medicine and Health Sciences, Ghent University, B-9000 Ghent, Belgium*

ABSTRACT: In order to elucidate the causal relationship between increased intracellular free calcium ($[Ca^{2+}]_i$) and induction of apoptosis, serum-free cultured granulosa cell (GC) explants were subjected to various experimental protocols interfering with normal Ca^{2+} homeostasis. Modulation of apoptotic indices was calculated on DAPI-stained GC explants. In some experiments $[Ca^{2+}]_i$ was measured with the Ca^{2+} probe fura-2 in combination with epifluorescence microscopy. Buffering of $[Ca^{2+}]_i$ with BAPTA-AM resulted in inhibition of apoptosis, while increasing extracellular Ca^{2+} (otherwise called $[Ca^{2+}]_e$ load) resulted in a biphasic response characterized by an initial inhibitory effect on apoptosis followed by a delayed phase of increased apoptosis that became apparent at 4 h after withdrawal of the $[Ca^{2+}]_e$ load. The initial inhibitory effect of the $[Ca^{2+}]_e$ load on apoptosis was dependent on the concentration of the load (range 2–50 mM), was augmented when the $[Ca^{2+}]_e$ load was applied in the presence of the Ca^{2+} channel blocker methoxyverapamil, and was mimicked by applying Mg^{2+} and Gd^{3+}, two Ca^{2+}-receptor agonists. These observations point towards the involvement of an extracellular Ca^{2+}-sensing receptor (CaR). Measurements of $[Ca^{2+}]_i$ showed that the ion was increased just after $[Ca^{2+}]_e$ load, followed by recovery that was complete at 2 h after the load. Collectively these data suggest that a $[Ca^{2+}]_e$ load initiates apoptosis, becoming manifest 4 h later, by the provoked $[Ca^{2+}]_i$ increase, and this effect is preceded by an apoptosis-inhibiting phase presumably involving CaR activation. We conclude that Ca^{2+} may act as a first (extracellular) messenger promoting cell survival and as a second (intracellular) messenger activating the cell death pathway.

KEYWORDS: Apoptosis; Ca^{2+}; Granulosa cells; Ca^{2+}-sensing receptor

INTRODUCTION

Programmed cell death of granulosa cells, the somatic cells surrounding the oocyte, is the primary mover in follicular atresia.[1] Indeed granulosa cell apoptosis is

[b]Address for correspondence: Sylvie Mussche, Department of Anatomy, Embryology, Histology and Medical Physics, Faculty of Medicine and Health Sciences, Ghent University, Godshuizenlaan 4, B-9000 Ghent, Belgium. Voice: +32 (0)9 264.92.25; fax: +32 (0)9 225.94.52.
s.mussche@rug.ac.be

detected in atretic follicles of all vertebrate ovaries analyzed so far, including human ovaries;[1] similarly, in the invertebrate ovary oocyte loss is linked to apoptosis of the surrounding nurse cells.[2] Insights into the regulation of granulosa cell apoptosis will contribute to several clinical applications.[3]

Calcium, one of the most versatile and universal signaling agents involved in cell growth and differentiation, has attracted interest as a potential second messenger in apoptosis since Kaiser and Edelman[4] demonstrated that glucocorticoid-stimulated thymocyte apoptosis is associated with enhanced Ca^{2+} influx.

Calcium coupling to the effector pathway of apoptosis has been documented for a myriad of processes including activation of proteases, transglutaminases, activation of Ca^{2+}-dependent protein kinases or phosphatases leading to alterations in gene transcription, and exposure of phosphatidylserine.[5–7] Moreover, the implication of changes in Ca^{2+} homeostasis during apoptosis was underlined by the finding that Bcl-2 regulates intracellular Ca^{2+} compartmentalization in the endoplasmic reticulum and mitochondria.[8]

In a large number of *in vitro* models a rise in intracellular free calcium triggers the apoptotic cascade.[9] This Ca^{2+} rise can be either transient and physiologically controlled, as is the case for second messengers, or the Ca^{2+} rise can be damage-mediated and hence sustained when for instance mitochondrial Ca^{2+} overload is implicated.[5] The increased $[Ca^{2+}]_i$ can result either from influx from the extracellular space, from mobilization from intracellular stores (endoplasmic reticulum, mitochondria) or from both through the process of calcium-induced calcium release[6] or through store-operated calcium influx.[10,11]

However, several studies challenge the proposed universal role of increased $[Ca^{2+}]_i$ in apoptosis and point to a more complex regulation of survival and apoptosis pathways by Ca^{2+}: in cultured neurons a sustained rise of $[Ca^{2+}]_i$, which is apparently the result of Ca^{2+} influx through voltage activated Ca^{2+}-channels, promotes survival.[12,13] In several model systems apoptosis is accompanied by a decrease in $[Ca^{2+}]_i$ as evidenced by the apoptosis-inducing effect of Ca^{2+} chelators.[8,14,15] Some authors report that mobilization of Ca^{2+} from the intracellular stores *per se* provides the signal for induction of apoptosis.[16]

In previous studies by our group we addressed the question of altered Ca^{2+} homeostasis in apoptosis by measuring fluorometrically $[Ca^{2+}]_i$ in serum-free cultured granulosa explants as compared to their gonadotropin-supplemented controls.[17] Measuring Ca^{2+} in individual cells at various stages of the apoptotic cascade by videomicroscopy in combination with digital imaging and double staining protocols pointed to an early and transient elevation of $[Ca^{2+}]_i$ in our model system.[18] To further elaborate the causal relationship between the measured $[Ca^{2+}]_i$ increase and the onset of apoptosis, we used in the present study a set of experimental tools to interfere with Ca^{2+} homeostasis and evaluated the effect on the apoptotic index. Our results highlight the need for careful evaluation of the consequences of pharmacological treatments interfering with Ca^{2+} homeostasis.

MATERIALS AND METHODS

Animals

Granulosa cell explants were prepared from the largest (F1) preovulatory follicle of the adult Japanese quail (*Coturnix coturnix japonica*). The animals were reared under continuous artificial illumination, with food (fresh lettuce and complete breeding food, Cerafin Dossche, Belgium) and water *ad libitum*. Animal care procedures were conducted in accordance with the guidelines set by the European Community Council Directives 86/6091 EEC.

Isolation and Culture of Granulosa Cell Explants

The monolayered granulosa layer from the largest preovulatory follicle was isolated in Krebs-Ringer solution according to the technique described by Gilbert *et al.*[19] Granulosa cell sheets covering the vegetative pole were divided in parts of circa 4 mm^2 followed by culture in 35 mm petri dishes under serum-free conditions in a 38°C incubator as described previously.[20] Explants were cultured in filter-sterilized M199 (Sigma, cat. no. M-0393, Bornem, Belgium) at pH 7.4, supplemented with 0.1% w/v Bovine Serum Albumin fraction V (Sigma, cat. no. A-4503, Bornem, Belgium), 10 mM HEPES (Acros, Geel, Belgium), 1% v/v Penicillin-Streptomycin (Paisley, Scotland) and 4 mM sodium hydrogen carbonate. In this model apoptosis is elicited by gonadotropin withdrawal; supplementation of follicle-stimulating hormone (FSH; 100 ng/ml) inhibits apoptosis.[17]

Reagents and Experimental Procedures

The BAPTA buffers (AM and cell impermeant), used to control the level of respectively intracellular and extracellular Ca^{2+} and the Ca^{2+} ionophore 4-bromo-A23187 were purchased from Molecular Probes (Leiden, the Netherlands). Both BAPTA buffers were loaded in Normal Buffer Solution (NBS) with 2.5 mM Probenecid, ensuring optimal retention of the chelator.[21] To facilitate cell loading the dispersing agent Pluronic F-127 (Molecular Probes; final concentration < 0.1%) was added to the loading buffer. The complete protocol of the BAPTA-AM treatment consisted of 24 h incubation in serum-free M199 followed by 1 h of BAPTA-AM (10 μM) loading. After rinsing with M199 the GC were incubated in fresh M199 supplemented with 2.5 mM probenecid up to 48 hours.

In initial experiments thapsigargin (TG; Sigma, Bornem, Belgium) treatment consisted of 24 h incubation in serum-free M199 followed by 24 h incubation in TG (1–100 nM) supplemented medium that had a normal or increased $[Ca^{2+}]_e$. The increased $[Ca^{2+}]_e$ was applied to stimulate the capacitative Ca^{2+} influx evoked by TG. Similar experiments were carried out starting the incubation with TG directly after isolation of the explants up to 24 h of serum-free culture in M199 (with normal Ca^{2+} concentrations, 1.25 mM). Adjustments of the protocol to increase the efficiency of store emptying and to activate subsequently capacitative Ca^{2+} influx were carried out in 24 h cultures where TG was first incubated for 30 min in BAPTA cell impermeant (1–10 mM) to buffer $[Ca^{2+}]_e$ followed by 23.5 h TG treatment with readmittance to normal or increased extracellular Ca^{2+} conditions.

The $[Ca^{2+}]_e$ load protocol consisted of 24 h incubation in serum-free M199 followed by 4 h incubation in elevated $[Ca^{2+}]_e$ (5 mM) in the presence or absence of A23187 (100 nM) followed by 20 h in M199 with normal $[Ca^{2+}]_e$. Methoxyverapamil (D600) was purchased from Sigma (Bornem, Belgium) and used during 24 h with or without elevated $[Ca^{2+}]_e$ (5 mM). All stock solutions were made in DMSO (Sigma, Bornem, Belgium). Ca^{2+} ($CaCl_2.2H_2O$), Mg^{2+} ($MgCl_2.6H_2O$) and Gd^{3+} ($GdCl_3.6H_2O$) were purchased from RPL (Leuven, Belgium), Merck (Belgium) and Sigma (Bornem, Belgium), respectively and added to the medium in 24 h incubations. Follicle Stimulating Hormone (FSH) was obtained from Sigma (Bornem, Belgium) and dissolved in 0.9% NaCl.

In control cultures (M199) the medium was rinsed and refreshed with either M199 or buffer as in the treated counterparts. There was no addition of vehicles to controls as these were kept below 1%, which was tested and demonstrated to have no influence on cell death.

Staining and Quantification of Apoptotic Nuclei

GC explants were rinsed in phosphate buffered saline (PBS, 10 mM, pH 7.4), fixed for 20 min in 4% formaldehyde in PBS, again rinsed in PBS and stained by a 1/1000 solution of DAPI (2′,6′-diamidino-2-phenylindole) in PBS. Mounted GC explants were examined on a Leica DM IRB/E inverted microscope, and apoptotic cells were identified by their characteristic fragmented and condensed chromatin masses. Small groups of apoptotic bodies were counted as remnants of one apoptotic cell. Apoptosis was expressed as the number of apoptotic nuclei per number of total nuclei counted in the same microscopic field. This apoptotic index was averaged for 10 fields giving a total number of about 1500 cells counted per treatment.

$[Ca^{2+}]_i$ Measurements

$[Ca^{2+}]_i$ was measured using the ratiometric Ca^{2+} indicator dye fura-2 in combination with epifluorescence microscopy and digital imaging. Phase-contrast images were taken every time to check for integrity of the explant. Granulosa cell explants were loaded with 6.6 μM fura-2 AM as described previously.[18]

Statistical Analysis

Statistical analysis was done using the statistical program SPSS 6.1. (SPSS Inc.). Statistical significance was tested with the Wilcoxon rank sum test, a non-parametric test for paired samples; p-values less than 0.05 were considered as statistically significant. Values represent means ± SEM of at least four independent experiments.

RESULTS

Effect of Chelating $[Ca^{2+}]_i$ with BAPTA-AM

FIGURE 1 indicates that buffering intracellular Ca^{2+} with BAPTA-AM resulted in a significant decrease of apoptotic indices (A.I.) from 26.1 ± 4.6% in serum-free controls to 9.1 ± 1.5% in BAPTA-AM treated explants ($p < 0.05$). Inhibition of apoptosis

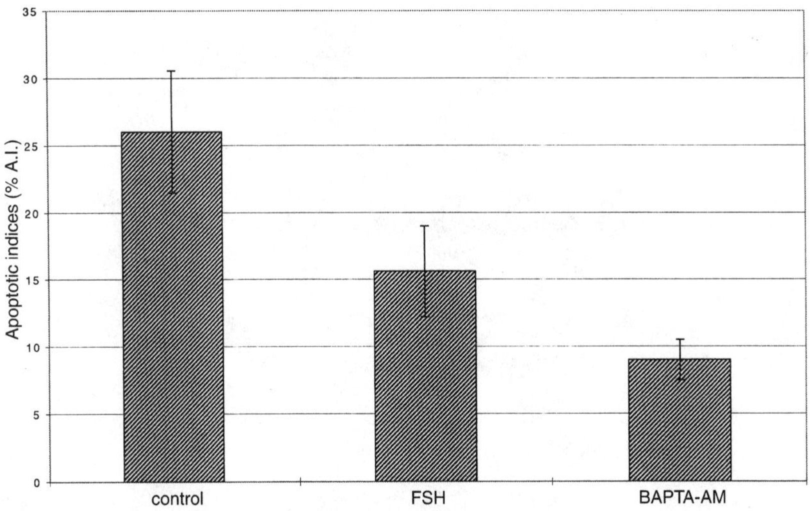

FIGURE 1. Effect of BAPTA-AM (10 μM) on apoptotic indices. GC explants were incubated at 38°C for 24 h in serum-free medium (M199) followed by 1 h BAPTA-AM of intracellular Ca^{2+} buffering in NBS-buffer + 2.5 mM probenecid and evaluated 20 h thereafter by DAPI-staining. From the loading process up to the DAPI staining it was necessary to add the anion transport blocker probenecid to prevent sequestration and secretion of BAPTA-AM.[21] Data represent means ± SEM of at least 4 experiments. Significant differences are observed between all groups ($p < 0.05$).

was more pronounced than the effect of FSH, a known survival factor in our model system.[17] For comparison: BAPTA-AM decreased the A.I. by 65.2% whereas FSH decreased the A.I. by 40.2% ($p < 0.05$).

Effect of Extracellular Ca^{2+} Load on the Apoptotic Index

Granulosa cell explants were incubated for 24 h in M199 followed by 4 h incubation in M199 with $[Ca^{2+}]_e$ elevated 4 times from the normal 1.25 mM to 5 mM and evaluated at 48 h by DAPI-staining (FIG. 2). Results show a significant increase in A.I. from 23.0 ± 3.4% in serum-free controls to 29.3 ± 1.9% in Ca^{2+}-treated explants ($p < 0.01$; FIG. 3). When the Ca^{2+} ionophore A23187 (100 nM) was used in combination with the temporary Ca^{2+}-elevation, an additional significant increase in A.I. is seen from 29.3 ± 1.9% in Ca^{2+}-treated explants to 37.1 ± 2.7% in Ca^{2+} and A23187 treated explants ($p < 0.05$; FIG. 3). The ionophore by itself had no significant influence on the A.I.

Unexpectedly, when apoptosis was evaluated directly after the $[Ca^{2+}]_e$ load a strong inhibition of apoptosis was seen from 7.2 ± 1.6% in serum-free controls to 3.0 ± 1.0% ($p < 0.05$; FIG. 4). FIGURE 4 also shows that when the GC explants were returned under physiological $[Ca^{2+}]_e$ conditions for 2 h, their A.I. are similar to serum-free controls. Four hours after $[Ca^{2+}]_e$ treatment, the A.I. are significantly

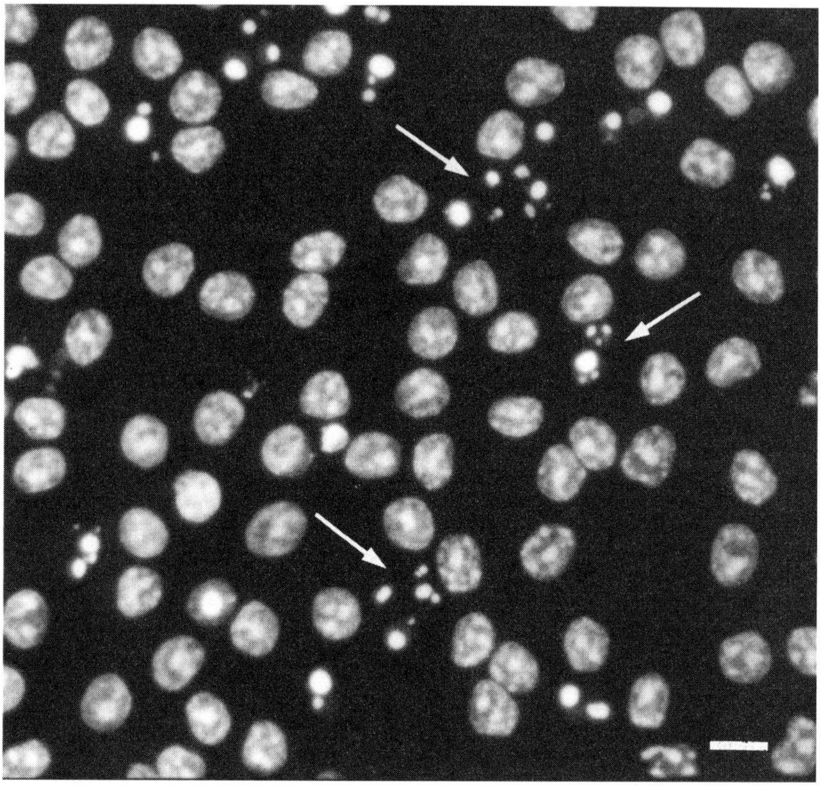

FIGURE 2. DAPI staining of F1 GC explant after 48 h serum-free culture. *Arrows* show condensed nuclear chromatin masses, characteristic for apoptosis. Obj. ×63. Bar = 10 μm.

increased from 8.3 ± 1.3% in serum-free controls to 15.7 ± 3.3% in Ca^{2+}-treated explants ($p < 0.05$).

Effect of Extracellular Ca^{2+} Load on Intracellular Free Calcium Levels

At the various times (0, 2, 4 h) after the $[Ca^{2+}]_e$ load was first applied, $[Ca^{2+}]_i$ was measured using fura-2 and compared to the controls. Our data show a significant $[Ca^{2+}]_i$ increase from 41.0 ± 0.7 nM in controls to 58.6 ± 1.6 nM immediately after the $[Ca^{2+}]_e$ load was applied whereafter $[Ca^{2+}]_i$ is recovering again to control values at 2 and 4 h after the $[Ca^{2+}]_e$ treatment (FIG. 5).

Presence of a Ca^{2+}-Sensing Receptor?

To elucidate the mechanism by which sustained $[Ca^{2+}]_e$ elevation inhibits apoptosis we examined the possibility of the presence of a Ca^{2+}-sensing receptor (CaR)

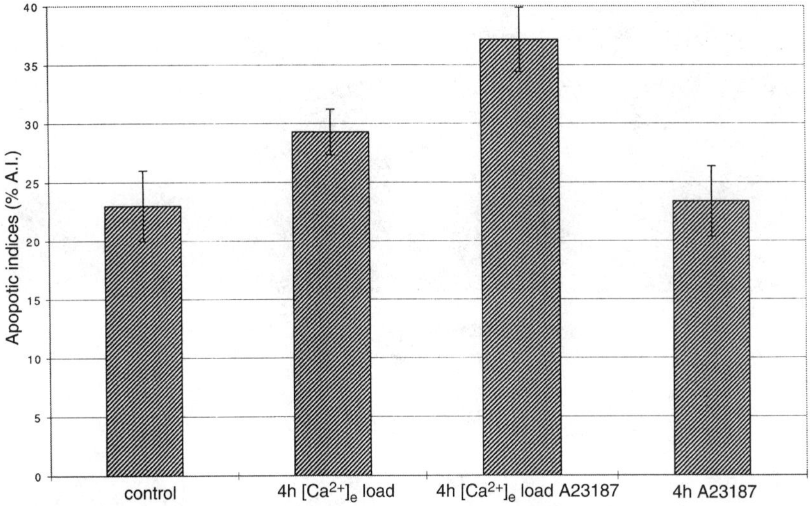

FIGURE 3. Effect of elevated $[Ca^{2+}]_e$—from 1.25 mM to 5 mM, with or without A23187 (100 nM)—on apoptotic indices. GC explants were incubated at 38°C for 24 h in serum-free medium (M199) followed by 4 h treatment and evaluated 20 h thereafter by DAPI-staining. Data represent means ± SEM of at least four experiments. Significant differences are observed between all groups ($p < 0.05$) except between control and A23187.

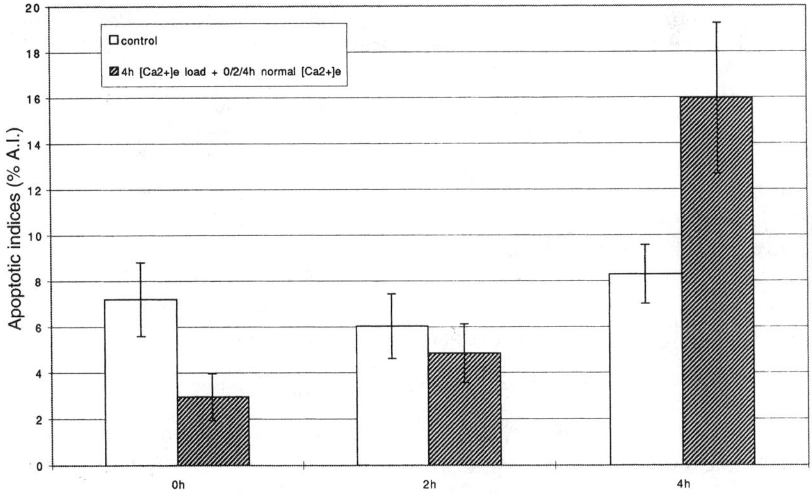

FIGURE 4. Effect of readmission to normal $[Ca^{2+}]_e$ conditions (1.25 mM) after 4 h $[Ca^{2+}]$e load. GC explants were incubated at 38°C for 24 h in serum-free medium (M199) followed by 4 h $[Ca^{2+}]_e$ treatment and evaluated 0, 2 and 4 h thereafter by DAPI-staining. Data represent means ± SEM of at least four experiments. Significant differences are observed between control and treated GC explants at 0 and 4 h after treatment ($p < 0.05$).

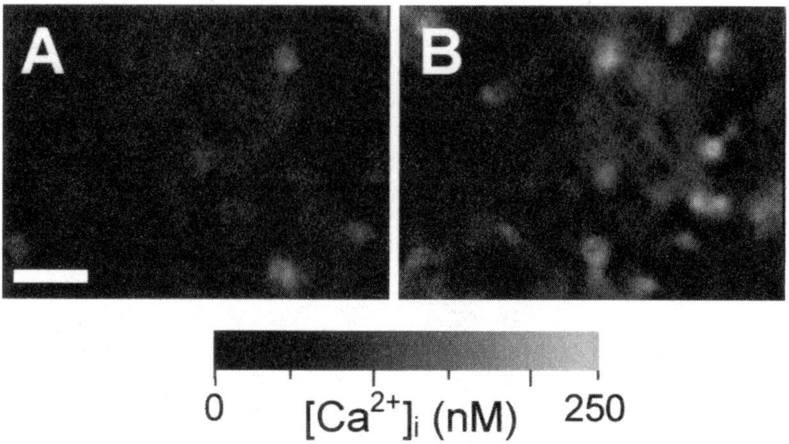

FIGURE 5. Images of $[Ca^{2+}]_i$ measurements in GC explants. GC explants were incubated at 38°C for 24 h in serum-free medium (M199) followed by 4 h treatment. **A** represents control: only a few cells show Ca^{2+} values above physiological level. **B** shows the treated GC explant with more cells with higher Ca^{2+} values according to the calibration bar. *White bar* = 50 μm.

on the plasma membrane of the cells. Elevating $[Ca^{2+}]_e$ during 24 h incubation from 1.25 mM to 3 mM had no significant effect on viability; from 4 to 50 mM apoptosis decreased significantly (A.I. from serum-free controls was normalized at 100%). The normalized A.I. fell to 51.6 ± 10.6%, 39.1 ± 7.2% and 21.3 ± 9.5% at 4, 5 and 10 mM $[Ca^{2+}]_e$ respectively; at higher concentrations the A.I. fell to 0%, ($p < 0.05$); FIG. 6). When the normal medium was supplemented during 24 h with Mg^{2+} (23 mM) or Gd^{3+} (100–500 μM), both acting as agonists of the CaR, significant inhibition of apoptosis was obtained: the A.I. was reduced from 7.1 ± 1.4% in serum-free controls to 1.3 ± 0.6% in Mg^{2+}-treated cells and to 2.8 ± 1.5% in 500 μM Gd^{3+}-treated cells ($p < 0.05$; FIG. 7). When the $[Ca^{2+}]_e$ load was applied in the presence of methoxyverapamil (50 μM), a Ca^{2+} channel blocker, an additional significant decrease of apoptosis (A.I. from 3.0 ± 0.6% in Ca^{2+}-treated cells to 0.7 ± 1.4% in Ca^{2+} + methoxyverapil treated cells, FIG. 7) was obtained. Methoxyverapamil by itself had no influence on viability (data not shown).

Effect of Thapsigargin

In order to investigate the effect of imposing a $[Ca^{2+}]_i$ increase by emptying the endoplasmic reticulum, the GC explants were incubated with thapsigargin, a blocker of sarco/endoplasmatic reticulum Ca^{2+}-ATPase (SERCA). Several concentrations (1 to 100 nM) and protocols (24 h TG after 24 h serum-free culture; 30 min TG with simultaneous buffering of $[Ca^{2+}]_e$ with BAPTA cell impermeant to prevent store refilling, followed by 23.5 h in 1.25 mM or 5 mM $[Ca^{2+}]_e$) were applied at different culture stages but no increases in apoptotic indices were obtained in all these experimental conditions ($p > 0.05$); when TG treatment was performed in combination with a $[Ca^{2+}]_e$ load the A.I. were significantly decreased (FIG. 8 A&B; $p < 0.05$).

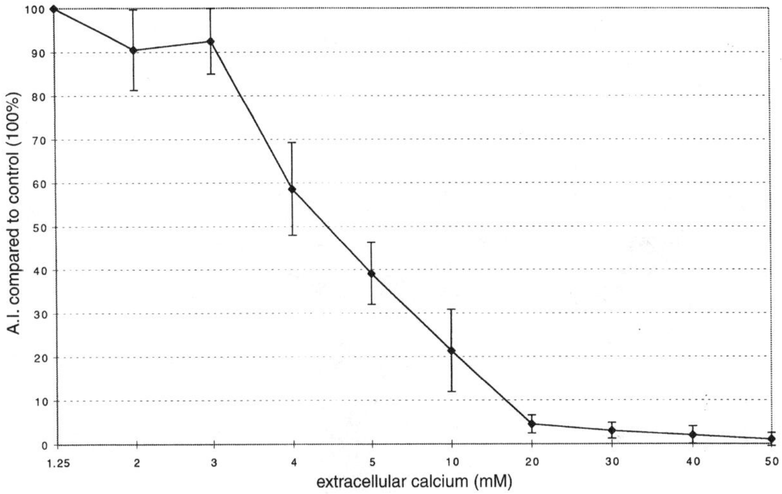

FIGURE 6. Effect of increasing extracellular Ca^{2+} concentrations during 24 h on apoptosis. Control is normalized to 100%. Data represent means ± SEM of at least four experiments. Significant differences are observed between controls and $[Ca^{2+}]_e$ from 4 mM to 50 mM ($p < 0.05$).

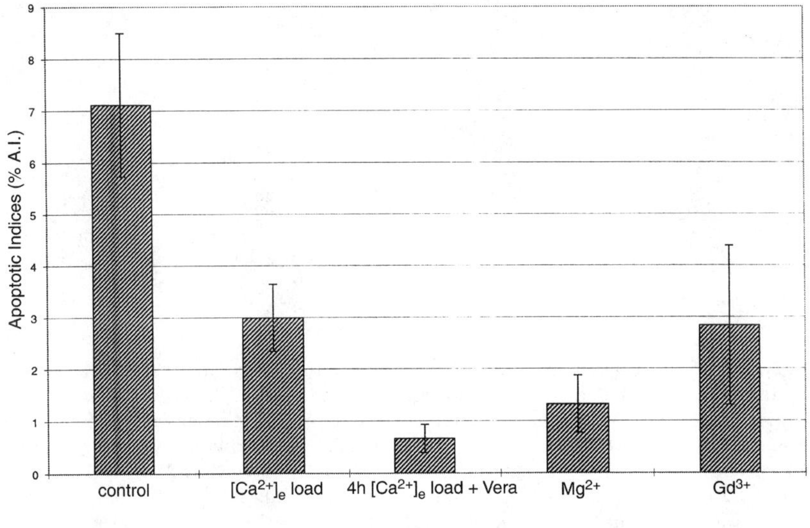

FIGURE 7. Effect of elevated $[Ca^{2+}]_e$ (5 mM) in the absence or presence of the Ca^{2+} channel blocker methoxyverapamil (Vera, 50 μM) and the CaR agonists Mg^{2+} (23 mM) and Gd^{3+} (500 μM) on apoptosis in 24 h serum-free cultures. Data represent means ± SEM of at least four experiments. Significant differences are observed between all groups ($p < 0.05$) except between $[Ca^{2+}]_e$ + Vera and Mg^{2+} and Gd^{3+} mutually.

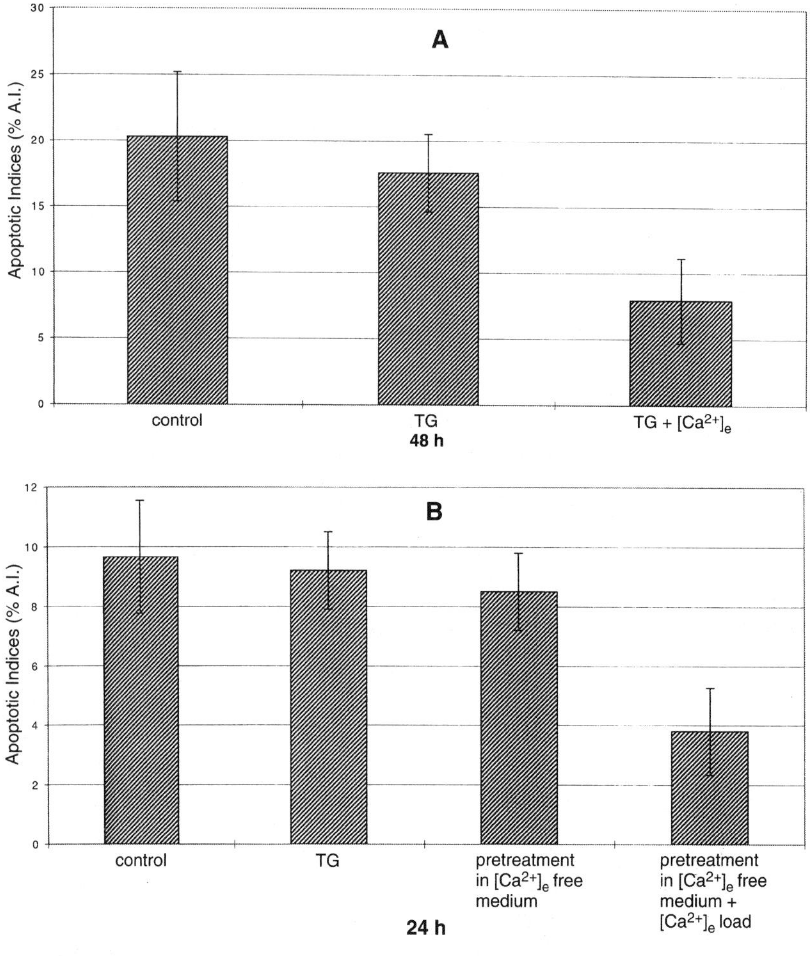

FIGURE 8. Effect of thapsigargin (1–100 nM) on apoptotic indices. **A:** GC explants were incubated at 38°C for 24 h in serum-free medium (M199) followed by 24 h TG treatment in either 1.25 mM or 5 mM calcium conditions. **B:** GC explants were immediately incubated with TG at 38°C during 30 min in nominal Ca^{2+} conditions by intracellular calcium buffering by BAPTA cell impermeant (1, 5, 10 mM) followed by 23.5 h TG treatment in either 1.25 mM or 5 mM calcium conditions. Apoptotic indices were evaluated by DAPI-staining. Data represent means ± SEM of at least four experiments. No significant differences were observed except when TG was carried out in elevated $[Ca^{2+}]_e$ ($p < 0.05$).

DISCUSSION

We previously showed in GC sheets that the initiation phase of apoptosis, as evidenced by phosphatidylserine exposure, is associated with an increase of intracellular calcium ($[Ca^{2+}]_i$). At the final degradation phase of the apoptotic cascade, when nuclear apoptotic morphology is apparent, $[Ca^{2+}]_i$ is recovered and returned to normal resting values.[18]

In order to elucidate the causal relationship between increasing $[Ca^{2+}]_i$ and triggering of apoptosis, serum-free cultured GC explants were subjected to various experimental procedures known to interfere with Ca^{2+} homeostasis. Apoptosis was evaluated quantitatively by DAPI-staining, which clearly shows chromatin condensation and nuclear fragmentation, the morphological features of apoptotic nuclei. In our hands chromatin staining methods (DAPI or Feulgen) are more reliable for quantification of apoptosis than TUNEL staining.[22]

As expected from our working hypothesis, buffering $[Ca^{2+}]_i$ by treatment with an intracellular Ca^{2+} chelator (BAPTA-AM) resulted in significant inhibition of apoptosis. Moreover, the causal role of the $[Ca^{2+}]_i$ increase in triggering of apoptosis is underlined by the fact that the BAPTA effect is more pronounced than the effect of FSH on apoptosis, a known survival factor in our model. We showed previously that FSH also affects $[Ca^{2+}]_i$ levels in our model at the 72 h culture stage.[17] Buffering $[Ca^{2+}]_i$ has been described to prevent or diminish DNA fragmentation and apoptosis in several model systems.[23–26] It is conceivable that BAPTA-AM sequesters cytosolic free Ca^{2+} as well as any Ca^{2+} from intracellular stores that may be released by depleting the cytosolic free Ca^{2+}. Hence the effect of BAPTA-AM on Ca^{2+} homeostasis could involve both buffering of $[Ca^{2+}]_i$ and depletion of the ER stores.

Another frequently used tool to manipulate Ca^{2+} homeostasis is thapsigargin, a compound that acts at nanomolar concentrations to irreversibly inhibit the SERCA Ca^{2+}-ATPases, causing depletion of intracellular stores and, in some cell types, a sustained rise of $[Ca^{2+}]_i$ due to capacitative Ca^{2+} influx through store-operated Ca^{2+} channels (SOCs). We used different concentrations ranging from 1 nM to 100 but none of the protocols resulted in higher apoptotic indices. In contrast, the decreased A.I. obtained from TG treatment with sustained $[Ca^{2+}]_{ex}$ load is probably due to the presence of the CaR (see below). In thymocytes TG was able to induce apoptosis only at low concentrations despite the occurrence of Ca^{2+} influx induced by higher TG concentrations.[27] Although there are several reports describing the $[Ca^{2+}]_i$ increasing and subsequently apoptosis-inducing effects of TG, which can be inhibited by Bcl-2,[28–31] some authors also report an alternative outcome which is possibly valid for our model. In chicken granulosa cells two different response patterns are observed in the same cell type: TG stimulates either a sustained $[Ca^{2+}]_i$ increase or a transient $[Ca^{2+}]_i$ increase.[32] The transient $[Ca^{2+}]_i$ increase in this latter granulosa model does not involve the capacitative model of Ca^{2+} entry. The rapid return to resting $[Ca^{2+}]_i$ after TG stimulation in this chicken model was suggested to be accomplished by sequestration of Ca^{2+} by mitochondria.[32] Other possible explanations for the ineffectiveness of TG in our model could be rapid inactivation of the store-operated Ca^{2+} influx or an increased Ca^{2+} efflux through plasma membrane Ca^{2+}-ATPase, which is insensitive to TG.[33,34] In any case, although TG does probably not provoke the necessary $[Ca^{2+}]_i$ increase for apoptosis induction in our model, the absence of apoptosis induction under TG indicates that depletion of intracellular stores by itself is not a trigger for apoptosis.

As TG was not an effective modulator of the A.I. in our model, we tried to affect $[Ca^{2+}]_i$ by applying a $[Ca^{2+}]_e$ load (a fourfold increase from 1.25 to 5 mM) in the culture medium in the presence or absence of the Ca^{2+} ionophore A23187. Measurements of $[Ca^{2+}]_i$ with fura-2 indicated a significant increase by the applied protocol only immediately after the $[Ca^{2+}]_e$ load (no measurements were done during the $[Ca^{2+}]_e$ load) and a recovery to normal values within 2 hours. Interestingly, experiments with apoptosis evaluation directly after the $[Ca^{2+}]_e$ load showed increasing viability with increasing concentrations of $[Ca^{2+}]_e$, while the $[Ca^{2+}]_e$ load induced apoptosis with a time delay of approximately 4 hours. This latter effect was potentiated by increasing $[Ca^{2+}]_i$ through ionophore (A23187)–mediated Ca^{2+}-influx.

How can one treatment result in a death as well as in a life signal in the same cell? A possible explanation is given by Lin *et al.*,[35] who showed that $[Ca^{2+}]_e$ protection against apoptosis in response to serum deprivation in rat fibroblasts was mediated through Ca^{2+}-sensing receptors (CaR) on the plasma membrane. The only evidence of the CaR in avian species has, to our knowledge, been reported by Diaz *et al.*[36] for chicken. The CaR, a member of the superfamily of G-protein-coupled receptors is found in chicken and other species[35,37,38] in several organs involved in Ca^{2+} homeostasis (parathyroid, kidney, intestine,…) where it senses small changes in $[Ca^{2+}]_e$ and transduces them into the necessary intracellular signals such as cAMP or IP3. But the CaR is not restricted to these organs (reviewed in ref. 39). The interaction of $[Ca^{2+}]_e$ and CaR in tissues not involved in Ca^{2+} regulation suggests that $[Ca^{2+}]_e$ can act as a first messenger to regulate diverse cellular functions in addition to its established role as an intracellular second messenger.[35–37] The dose dependency of the protective effects of $[Ca^{2+}]_e$ followed an S-shaped curve (see FIG. 6) which is typical for ligand-receptor interactions and is less compatible with the hypothesis of changes in the electromotive force of Ca^{2+} across the plasma membrane as the mechanism of signal transduction. Application of a high concentration of Mg^{2+}, a less potent agonist of the CaR as compared to Ca^{2+}, during 24 h, resulted in a decrease of apoptosis. The potent CaR agonist Gd^{3+} already had effect in the μM range. These results indicate that in our model the CaR may be present and linked to a survival pathway.

When in addition to a 24 h $[Ca^{2+}]_e$ load a voltage-sensitive Ca^{2+} channel blocker, methoxyverapamil, was added so as to abrogate Ca^{2+} influx, the survival effect of the $[Ca^{2+}]_e$ load was enhanced. This suggests that a small percentage of the cells are initiated for apoptosis during the $[Ca^{2+}]_e$ load presumably by the Ca^{2+} influx, which is absent in methoxyverapamil-treated cells. From these observations we concluded that the survival effect of $[Ca^{2+}]_e$ is not mediated by a perturbation of $[Ca^{2+}]_i$. As the A.I. in both cases ($[Ca^{2+}]_e$ load and $[Ca^{2+}]_e$ load + methoxyverapamil) are still significantly lower than serum-free controls the effect of the CaR acts in a dominant way on apoptosis induced by $[Ca^{2+}]_i$ increases. The increased A.I. seen after withdrawal of the $[Ca^{2+}]_e$ load is hypothesized to be due to the loss of the CaR-mediated rescuing effect.

In contrast to our findings, Lorget *et al.* describes apoptosis induction under sustained $[Ca^{2+}]_e$ load in osteoclasts due to Ca^{2+} influx as pointed out by use of Ca^{2+}-channel blockers.[40] In their model the calcium-sensing receptors are not coupled to a survival pathway. In conclusion, granulosa cell explants are stimulated for apoptosis by an increase in $[Ca^{2+}]_i$ as evidenced by the higher apoptotic indices resulting from Ca^{2+} influx and decreased apoptotic indices when $[Ca^{2+}]_i$ is buffered.

More importantly, this report suggests a dual role for Ca^{2+}in apoptosis in serum-free cultured granulosa cell explants: it acts as a second messenger to initiate the apoptotic cascade, second as a first messenger to promote survival through calcium-sensing receptors. Further experiments will be directed to obtain more evidence for the presence and functions of the CaR.

ACKNOWLEDGMENTS

We wish to thank Christophe Van Steenkiste, Tom Van Maerken and Jill Vanmassenhove for excellent technical assistance. This study was supported by the Belgian BOZF (no. 01106797 to F. Roels and no. 01115799 to K. D'Herde). We also thank Frank Roels for helpful discussion.

REFERENCES

1. TILLY, J.L. 1997. Apoptosis and the ovary: a fashionable trend or food for thought? Fertil. Steril. **67:** 226–228.
2. CRIEL, G. & K. D'HERDE. 1996. Programmed cell death of the nurse cells at the end of vitellogenesis in the ovary of the brine shrimp *Artemia franciscana.* Proceedings of the Joint Meeting BVM/NVvM, Ghent, 1996, pp. 118–125.
3. TILLY, J.L. & R. ROBLES. 1999. Apoptosis and its impact in clinical reproductive medicine. *In* Molecular Biology in Reproductive Medicine. B.C.J.M. Fauser, A.J. Rutherford, J.F. Strauss III & P. Van Steinteghem, Eds.: 79–101.
4. KAISER, N. & I.S. EDELMAN. 1977. Calcium dependence of glucocorticoid-induced lymphocytolysis. Proc. Natl. Acad. Sci. USA **74:** 638–642.
5. MCCONKEY, D.J. & S. ORRENIUS. 1997. Breakthroughs and views. The role of calcium in the regulation of apoptosis. Biochem. Biophys. Res. Commun. **239:** 357–366.
6. TRUMP, B.F. & I.K. BEREZESKY. 1996. The role of altered $[Ca^{2+}]_i$ regulation in apoptosis, oncosis, and necrosis. Biochem. Biophys. Acta **1313:** 173–178.
7. KREBS, J. 1998. The role of calcium in apoptosis. BioMetals **11:** 375–382.
8. BAFFY, G., T. MIYASHITA, J.R. WILLIAMSON & J.C. REED. 1993. Apoptosis induced by withdrawal of interleukin-3 (IL-3) from an IL-3-dependent hematopoietic cell line is associated with repartitioning of intracellular calcium and is blocked by enforced Bcl-2 oncoprotein production. J. Biol. Chem. **268:** 6511–6519.
9. MCCONKEY, D.J. & S. ORRENIUS. 1996. The role of calcium in the regulation of apoptosis. J. Leukoc. Biol. **59:** 775–783.
10. PAREKH, A.B. & R. PENNER. 1997. Store depletion and calcium influx. Physiol. Rev. **77:** 901–930.
11. BERRIDGE J.M., M.D. BOOTMAN & P. LIPP. 1998. Calcium—a life and death signal. Nature **395:** 645–648.
12. FRANKLIN, J.L. & E.M. JOHNSON, JR. 1992. Suppression of programmed cell death by sustained elevation of cytoplasmic calcium. TINS **15:** 501–508.
13. LAMPE, P.A., E.B. CORNBROOKS, A. JUHASZ, *et al.* 1995. Suppression of programmed cell death by a thapsigargin-induced Ca^{2+} influx. J. Neurobiol. **26:** 205–212.
14. BANSAL, N., A.G. HOULE & G. MELNYKOVYCH. 1990. Dexamethasone-induced killing of neoplastic cells of lymphoid derivation: lack of early calcium involvement. J. Cell. Physiol. **143:** 105–109.

15. KLUCK, R.M., C.A. MCDOUGALL, B.V. HARMON & J.W. HALLIDAY. 1994. Calcium chelators induce apoptosis—evidence that raised intracellular ionised calcium is not essential for apoptosis. Biochem. Biophys. Acta **1223:** 247–254.
16. ZHU, W.H. & T.T. LOH. 1995. Roles of calcium in the regulation of apoptosis in HL-60 promyelocytic leukemia cells. Life Sci. **57:** 2091–2099.
17. D'HERDE, K. & L. LEYBAERT. 1997. Intracellular free calcium related to apoptotic cell death in quail granulosa sheets kept in serum-free culture. Cell Death Differ. **4:** 59–65.
18. D'HERDE, K. & L. LEYBAERT. 1998. Apoptotic granulosa cells have moderately increased intracellular free calcium during phosphatedidylserine exposure and a normal resting calcium level during DNA fragmentation. Apoptosis **3:** 337–343.
19. GILBERT, A.B., A.J. EVANS, M.M. PERRY & M.H. DAVIDSON. 1977. A method for separating the granulosa cells, the basal lamina and the theca of the preovulatory ovarian follicle of the domestic fowl (*Gallus domesticus*). J. Reprod. Fertil. **50:** 79–81.
20. FARIOLI-VECCHIOLI, S., S. RAES, M. ESPEEL, *et al.* 2000. Inversed expression of peroxisomes and connexin-43 in the granulosa cells of the quail follicle. J. Histochem. Cytochem. **48:** 167–177.
21. DIVIRGILIO, F., T.H. STEINBERG, J.A. SWANSON & S.C. SILVERSTEIN. 1988. Fura-2 secretin and sequestration in macrophages. A blocker of organic anion transport reveals that these processes occur via a membrane transport system for organic anions. J. Immunol. **140:** 915–920.
22. D'HERDE, K., G. DE PESTEL & F. ROELS. 1994. In situ end labeling of fragmented DNA in induced ovarian atresia. Biochem. Cell. Biol. **72:** 573–579.
23. MCCONKEY, D.J., P. NICOTERA, P. HARTZELL, *et al.* 1989. Glucocorticoids activate a suicide process in thymocytes through an elevation of cytosolic Ca^{2+} concentration. Arch. Biochem. Biophys. **269:** 365–370.
24. STORY, M.D., L.C. STEPHENS, S.P. TOMASOVIC & R.E. MEYN. 1992. A role for calcium in regulating apoptosis in rat thymocytes irradiated *in vitro*. Int. J. Rad. Biol. **61:** 243–251.
25. BELLOMO, G., M. PEROTTI, F. TADDEI, *et al.* 1992. Tumor necrosis factor alpha induces apoptosis in mammary adenocarcinoma cells by an increase in intracellular free Ca^{2+} concentration and DNA fragmentation. Cancer Res. **52:** 1342–1346.
26. PEROTTI, M., F. TADDEI, F. MIRABELLI, *et al.* 1990. Calcium-dependent DNA fragmentation in human synovial cells exposed to cold shock. FEBS Lett. **59:** 331–334.
27. BEAVER, J.P. & P. WARING. 1994. Lack of correlation between early intracellular calcium ion rises and the onset of apoptosis in thymocytes. Immunol. Cell. Biol. **72:** 489–499.
28. GULATI, R. & J.J. PELUSO. 1997. Opposing actions of hepatocyte growth factor and basic fibroblast growth factor on cell contact, intracellular free calcium levels, and rat ovarian surface epithelial cell viability. Endocrinology **138:** 1847–1856.
29. HE, H., M. LAM, T.S. MCCORMICK & C.W. DISTELHORST. 1997. Maintenance of calcium homeostasis in the endoplasmic reticulum by Bcl-2. J. Cell. Biol. **138:** 1219–1228.
30. MARIN, M.C., A. FERNANDEZ, R.J. BICK, *et al.* Apoptosis suppression by bcl-2 is correlated with the regulation of nuclear and cytosolic Ca^{2+}. Oncogene **12:** 2259–2266.
31. DISTELHORST, C.W. & T.S. MCCORMICK. 1996. Bcl-2 acts subsequent to and independent of Ca^{2+} fluxes to inhibit apoptosis in thapsigargin- and glucocorticoid-treated mouse lymphoma cells. Cell Calcium **19:** 473–483.
32. MORLEY, P., B.K. TSANG, J.F. WHITFIELD & J.-L. SCHWARTZ. 1992. Thapsigargin increases cytoplasmic free Ca^{2+} without influencing steroidogenesis in chicken granulosa cells. Cell Calcium **13:** 263–271.
33. TAKEMURA, H., O. THASTRUP & J.W. PUTNEY, JR. 1990. Calcium efflux across the plasma membrane of rat parotid acinar cells is unaffected by receptor activation or by the microsomal calcium ATPase inhibitor, thapsigarin. Cell Calcium **11:** 11–17.
34. THASTRUP, O., P.J. CULLEN, B.K. DROBAK, *et al.* 1990. Thapsigargin, a tumor promoter, discharges intracellular Ca^{2+} stores by specific inhibition of the endoplasmic reticulum Ca^{2+}-ATPase. Proc. Natl. Sci. USA **87:** 2466–2470.

35. LIN, K-I., N. CHATTOPADHYAY, M. BAI, *et al.* 1998. Elevated extracellular calcium can prevent apoptosis via the calcium-sensing receptor. Biochem. Biophys. Res. Commun. **249:** 325-331.
36. DIAZ, R., S. HURWITZ, N. CHATTOPADHYAY, *et al.* 1997. Cloning, expression, and tissue localization of the calcium-sensing receptor in chicken (*Gallus domesticus*). Am. J. Physiol. **273:** R1008–R1016.
37. BROWN, E.M., M. POLLAK, Y.-H. CHOU, *et al.* 1995. Cloning and functional characterization of extracellular Ca^{2+}-sensing receptors from parathyroid and kidney. Bone **17:** 7S–11S.
38. BROWN, E.M. 1999. Physiology and pathophysiology of the extracellular calcium-sensing receptor. Am. J. Med. **106:** 238–253.
39. CHATTOPADHYAY, N., P.M. VASSILEV & E.M. BROWN. 1997. Calcium-sensing receptor: roles in and beyond systemic calcium homeostasis. Biol. Chem. **378:** 759–768.
40. LORGET, F., S. KAMEL, A.R.E. MENTAVERRI, A. WATTEL, M. NAASSILA, M. MAAMER & M. BRAZIER. 2000. High extracellular calcium concentrations directly stimulate osteoclast apoptosis. Biochem. Biophys. Res. Commun. **268:** 899–903.

Primary Human Fibroblasts from a Maple Syrup Urine Disease Patient Undergo Apoptosis following Exposure to Physiological Concentrations of Branched Chain Amino Acids

PHILIPPE JOUVET, MARY KOZMA, AND HUSEYIN MEHMET[a]

Weston Laboratory, Division of Paediatrics, Obstetrics and Gynaecology, Imperial College of Science, Technology and Medicine, Hammersmith Hospital, London W12 0NN, United Kingdom

ABSTRACT: Maple syrup urine disease (MSUD) is an inborn error of metabolism caused by a deficiency in branched chain α-ketoacid dehydrogenase. We have recently found that MSUD neurodegeneration may result, at least in part, from apoptosis triggered by branched chain amino acids and their α-ketoacid derivatives. In the present study, we investigated the sensitivity of MSUD fibroblasts to defined mixtures of MSUD metabolites. Defined combinations of MSUD metabolites, at levels comparable to those in MSUD patients, triggered cell death in skin fibroblasts from a MSUD patient, while control fibroblasts were resistant. The mechanism of cell death was confirmed as apoptosis by *in situ* end labeling.

KEYWORDS: Apoptosis; α-ketoacids; Maple syrup urine disease

INTRODUCTION

Maple syrup urine disease (MSUD) is an inborn error of metabolism caused by a deficiency in branched chain α-ketoacid dehydrogenase, leading to the accumulation of branched chain amino-acids (BCAA: leucine, valine, isoleucine) and a corresponding increase in their α-ketoacid derivatives (BCKA: α-keto isocaproic acid, α-keto valeric acid, α-keto-β-methyl-n-valeric acid) levels.[1] Acute neurological deterioration in children is often associated with increased plasma and cerebrospinal fluid (CSF) concentrations of BCAA and BCKA.[2,3] We have recently shown that MSUD neurodegeneration may result, at least in part, from an accumulation of branched chain amino acids and their α-ketoacid derivatives that trigger apoptosis through a cytochrome *c*-independent pathway.[4] In the present study, we compared

[a]Address for correspondence: Dr. Huseyin Mehmet, Weston Laboratory, Division of Paediatrics, Obstetrics and Gynaecology, Imperial College School of Science Technology and Medicine, Hammersmith Hospital, DuCane Road, London W12 0NN, U.K.Voice: 44 020 8383 3358; fax: 44 020 8383 2474.
h.mehmet@ic.ac.uk

the toxic effects of BCAA's and BCKA's on human skin fibroblasts obtained from a MSUD patient to those from a healthy age-matched control.

MATERIALS AND METHODS

Materials

Glucose-rich (4.5 g/L) Dulbecco's modified Eagle's medium (DMEM), leucine, valine, isoleucine, α-keto isocaproic acid, α-keto valeric acid, α-keto-β-methyl-*n*-valeric acid, MTT (3,(4,5-dimethylthiazol-2-yl)2,5-diphenyl-tetrazolium bromide), staurosporine and cycloheximide were obtained from Sigma (Poole, UK). Fetal calf serum (FCS) and tissue culture plastics were obtained from Life Technologies (Paisley, UK).

Cell Culture

Human skin fibroblasts were isolated using standard techniques[5] and cultured in 10 cm tissue culture dishes, in DMEM containing 10% heat inactivated FCS, supplemented with penicillin and streptomycin. Cells were seeded at 2×10^5 per dish, subcultured twice weekly and incubated at 37°C in a humidified atmosphere of 10% CO_2/90% air. Individual cultures were maintained for no more than 4 weeks.

Survival Assay by the Tetrazolium Salt Method (MTT Assay)

The tetrazolium salt assay relies on the conversion of MTT to colored formazan by succinate dehydrogenase in metabolically active cells and provides an indirect measurement of cell viability.[6] For the present series of experiments, we used our modification[4] of this method, where the metabolic activity of cells is directly related to the amount of MTT cleaved to produce formazan. Results were expressed as percent viability according to the equation:

$$\%\ \text{viability} = \frac{[\text{OD of treated cells}] - [\text{OD of DMEM/0.5\% FCS without cells}]}{[\text{OD of healthy cells (DMEM/0,5\% FCS}] - [\text{OD of DMEM/0.5\% FCS without cells}]}.$$

To exclude the possibility of false absorbance readings (e.g., increased metabolism in surviving cells) cell death was also quantified on the basis of apoptotic morphology combined with the absence of MTT metabolism. Both methods yielded comparable results. For all cell survival experiments, either the protein synthesis inhibitor cycloheximide or the protein kinase blocker staurosporine were used as positive controls for apoptosis.

In Situ *End-Labeling*

Human fibroblasts were cultured in 8-well chamber slides at a density of 10^5/ml (300 µl / well) in DMEM containing 0.5% FCS alone or in the presence of defined concentrations of BCAA, BCKA, SSP or cycloheximide. Following a 20 h incubation at 37°C, *in situ* end-labelling was performed as described previously[7] with minor modifications.[8]

RESULTS

We have recently demonstrated that BCAA and BCKA can trigger apoptosis in murine neurons and glia both in culture and *in vivo*.[4] Two questions arose from our observations in these studies. First, are the relatively high concentrations of MSUD metabolites required for an apoptotic effect physiologically relevant and, second, can differences be observed between healthy and MSUD-derived human cells? We therefore used a defined mixture of MSUD metabolites (MSUD mixture), at clinically relevant concentrations derived from our previous observations of three neonates with severe acute onset MSUD.[9] The concentrations of the corresponding keto acids were calculated using the molar ratio between BCAA's and their ketoacid derivatives determined in plasma by Snyderman *et al.*[10] These MSUD mixtures are defined in TABLE 1.

Quantitative MTT assays indicated a reduced viability in MSUD fibroblasts following exposure to the MSUD mixture: 14% of MSUD fibroblasts failed to reduce MTT in the presence of MSUD mixture compared to only 2% of cells treated with 0.5% FCS alone. In contrast, normal fibroblasts showed no significant reduction in viability following treatment with the MSUD mixture (FIG. 1C).

ISEL revealed that fibroblasts obtained from healthy donors contained healthy nuclei that were not labeled by Klenow (FIG. 1A). In contrast, following exposure to the physiological MSUD mixture, the majority of MSUD fibroblasts had a reduced

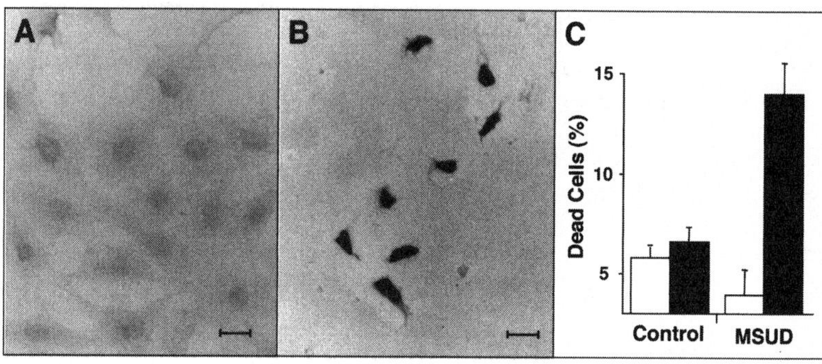

FIGURE 1. A and **B:** *In situ* end labeling of primary human MSUD fibroblasts following a 20 hour exposure to DMEM/0.5% FCS alone **(A)** or in the presence of MSUD mixture **(B)**. No ISEL staining is detectable in control cultures. However, after treatment with the MSUD mixture, numerous ISEL positive cells are seen. These cells also display morphological characteristics of apoptotsis including cytoplasmic shrinkage and nuclear pyknosis. (Bar = 10 μm). **C:** Comparative effects of MSUD metabolites on MTT metabolism in human skin fibroblasts from a healthy donor or from a MSUD patient. Human fibroblasts were treated for 24 hours with either DMEM/0.5% FCS alone or containing MSUD mixture (defined in TABLE 1). Cells obtained from healthy donor or a MSUD patient were treated with DMEM/0.5% FCS alone (□), or in the presence of MSUD mixture (■) for 20 hours. C6 cells were treated in parallel for comparison and cell death was then assessed by MTT assay using microscopic quantitation of live (reducing) and dead (non-reducing) cells. Per cent cell death was calculated and the data expressed as mean (± SEM) of five independent experiments.

cytoplasmic volume and distinct nuclear pyknosis. These nuclei were labeled dark brown by ISEL following treatment with the MSUD mixture, consistent with apoptosis (FIG. 1B).

DISCUSSION

Toxic effects of the metabolites that accumulate in MSUD have been previously demonstrated in the rat cerebellum, where myelination was severely impaired,[11] in lymphoblastoid cell lines from MSUD patients, where significant growth inhibition was observed[12] and in C6 cells, where the results were interpreted as an increase in cell cycle time, although morphological changes indicative of apoptosis were present but not appreciated.[13] In the present study, we demonstrate that MSUD metabolites are also toxic to primary human fibroblasts in culture. Moreover, cell death occurs by apoptosis, verified by ISEL of apoptotic nuclei.

These results are consistent with those of Bissel *et al.*[14] who observed that the replication of mouse fibroblasts was inhibited by the three BCKA's (KICA, α-keto valeric acid and KILE) and those of Patel *et al.*[15] who found that all three BCKA's inhibited the mitochondrial BCKA dehydrogenase complex in the developing rat brain. These findings emphasize the importance of this synergy: leucine at high concentrations is only slightly toxic to neural cells but at lower concentrations acts synergistically with BCKA to trigger apoptosis.[4] This effect may be particularly important in the brain where the high aminotransferase activity rapidly metabolizes leucine to KICA.[16] We found that the two other BCKA's that accumulate in MSUD are also toxic to C6 rat glioma cells[4] and may therefore contribute to neurological damage in MSUD patients although their concentration in plasma is relatively low (see TABLE 1).

The increased toxic effect of MSUD metabolites on human MSUD fibroblasts suggests that the cells of these patients are predisposed to apoptosis. This could be because they lack the enzyme branched chain α-ketoacid dehydrogenase and so cannot metabolize BCKA's. Alternatively, they may already contain a pool of BCAA's and BCKA's and so exposure to further doses may result in an increased intracellular concentration.

Our observations also provide useful information for patient management. Plasma leucine levels are routinely used to monitor the treatment of MSUD and plasma

TABLE 1. Definition of maple syrup urine disease (MSUD) mixture

BCAA/BCKA	Concentration in MSUD mixture (μM)
Leucine/α-keto-isocaproic acid	4000/4000
Isoleucine/α-keto-β-methyl-*n*-valeric acid	800/200
Valine/α–keto-isovaleric acid	1000/700

NOTE: The branched chain amino acid (BCAA) concentrations were derived using the mean plasma concentrations of three neonates with classic MSUD.[9] The corresponding branched chain keto-acid (BCKA) concentrations were calculated using the molar ratio of ketoacid analogues to the parent amino acids that occur in plasma.[10]

concentrations of BCAA's are a good indicator of the risk of neurological injury.[2] Consistent with a direct role for BCKA in neurodegeneration associated with MSUD, intracerebral injection of KICA triggers neuronal apoptosis in the hippocampus in 14-day-old rat pups.[4] The apoptotic effects are dose and time dependent and may explain, at least in part, the clinical neurological sequelae associated with high plasma concentrations of MSUD. If such a mechanism is confirmed in brain specimens from MSUD patients, it will have important implications for the design of therapeutic strategies to prevent or limit acute cerebral injury that occurs during acute onset MSUD. Such a neuronal rescue therapy might be combined with conventional treatment of MSUD. Furthermore, our observations provocatively suggest that other conditions associated with ketoacid accumulation (e.g. diabetic acidosis) may also have an apoptotic component in the neurological deficit.

ACKNOWLEDGMENT

We would like to thank the Weston Foundation for their continued financial support.

REFERENCES

1. Snyderman, S. 1988. Treatment outcome of maple syrup urine disease. Acta Paediatr. Jpn. **30:** 417–424.
2. Riviello, J., I. Rezvani, A. DiGeorge & C. Foley. 1991. Cerebral edema causing death in children with maple syrup urine disease. J. Pediatr. **119:** 42–45.
3. Levin, M., A. Scheinmann, R. Lewis & A. Beaudet. 1993. Cerebral edema in maple syrup urine disease. J. Pediatr. **122** (1):167–168.
4. Jouvet, P., P. Rustin, D.L. Taylor, J.M. Pocock, U. Felderhoff-Mueser, N.D. Mazarakis, C. Sarraf, U. Joashi, M. Kozma, K. Greenwood, A.D. Edwards & H. Mehmet. 2000. Branched chain amino acids induce apoptosis in neural cells without mitochondrial membrane depolarization or cytochrome c release: implications for neurological impairment associated with maple syrup urine disease. Mol. Biol. Cell **11** (5):1919–1932.
5. Freshney, R.I. 1987. Culture of Animal Cells: A Manual of Basic Technique. Wiley-Liss. New York, pp. 107–126.
6. Hansen, M., S. Nielsen & K. Berg. 1989. Re-examination and further development of a precise and rapid dye method for measuring cell growth/cell kill. J. Immunol. Methods **119:** 203–210.
7. Ansari, B., P. Coates, B. Greenstein & P. Hall. 1993. In situ end-labelling detects DNA strand breaks in apoptosis and other physiological and pathological states. J. Pathol. **170:** 1–8.
8. Joashi, U.C., K. Greenwood, D. Taylor, M. Kozma, N.D. Mazarakis, A.D. Edwards & H. Mehmet. 1999. Poly(ADP ribose) polymerase cleavage precedes neuronal death in the hippocampus and cerebellum following injury to the developing rat forebrain. Eur. J. Neurosci. **11:** 91–100.
9. Jouvet, P., F. Poggi, D. Rabier, J. Michel, P. Hubert, M. Sposito, J. Saudubray & N. Man. 1997. Continuous venovenous hemodiafiltration in the acute phase of neonatal maple syrup urine disease. J. Inherited Metab. Dis. **20:** 463–472.
10. Snyderman, S., F. Goldstein, C. Sansaricq & P. Norton. 1984. The relationship between the branched chain amino acids and their α-ketoacids in maple syrup urine disease. Pediatr. Res. **18:** 851–853.
11. Silberberg, D. 1969. Maple syrup urine disease metabolites studied in cerebellum cultures. J. Neurochem. **16:** 1141–1146.

12. SKAPER, S., D. MOLDEN & J. SEEGMILLER. 1976. Maple syrup urine disease: branched-chain amino acid concentrations and metabolism in cultured human lymphoblasts. Biochem. Genet. **14:** 527.
13. LIAO, C., M. HERMAN & K. BENSCH. 1978. Prolongation of G1 and S phase in C-6 glioma cells treated with maple syrup urine disease metabolites. Morphologic and cell cycle studies. Lab. Invest. **38:** 122–133.
14. BISSEL, M. *et al.* 1974. Effects of maple syrup urine disease metabolites on mouse L-fibroblasts in vitro: a fine structural and biochemical study. J. Neurochem. **22:** 957–964.
15. PATEL, M. 1974. Inhibition by the branched chain 2-oxoacids of the 2-oxoglutarate dehydrogenase complex in developing rat and human brain. Biochem. J. **144:** 91–97.
16. BRAND, K., S. HAUSCHILDT & J. LÜTHJE. 1984. Effect of diets on the activity of enzymes involved in branched chain α-keto acid metabolism. *In* Branched Chain Amino and Keto Acids in Health and Disease, Vol. 1. S. A. Adibi, W. Fekl, U. Langenbeck & P. Schauder, Eds. Karger. Basel, pp. 100–111.

Heat Shock Protein 70 Is Required for the Survival of Cancer Cells

JESPER NYLANDSTED,[a] KARSTEN BRAND,[b] AND MARJA JÄÄTTELÄ[a,c]

[a]*Apoptosis Laboratory, Institute of Cancer Biology, Danish Cancer Society, DK-2100 Copenhagen Ø, Denmark*

[b]*Max-Planck-Gesellschaft, Humbolt University, Max Delbrück Centre for Molecular Medicine, D-13122 Berlin-Buch, Germany*

ABSTRACT: The major stress-inducible heat shock protein, Hsp70, is a chaperone protein abundantly and preferentially expressed in human tumors and tumor cell lines. Owing to the ability of Hsp70 to protect cells from a wide range of apoptotic and necrotic stimuli, it has been assumed that Hsp70 may confer survival advantage to tumor cells. To investigate this hypothesis in human tumor cell lines, we generated an adenovirus expressing antisense Hsp70 (Ad.asHsp70). The effective and specific depletion of Hsp70 by Ad.asHsp70 resulted in massive cell death of all tumorigenic cell lines tested (carcinomas of breast, colon, prostate and liver as well as glioblastoma). In spite of an effective depletion of Hsp70, Ad.asHsp70 had no effect on the survival or growth of fetal fibroblasts or non-tumorigenic epithelial cells of breast or prostate. Anti-apoptotic proteins Bcl-2, Bcl-X_L and CrmA as well as peptide-inhibitors of caspases, DEVD-CHO and zVAD-FMK, failed to rescue tumor cells from Ad.asHsp70-induced cell death. These results indicate that the high expression of Hsp70 is a prerequisite for the survival of human cancer cells of various origins and reveal Hsp70 as the only protein described so far whose expression is specifically needed for the survival of tumorigenic cells.

KEYWORDS: Apoptosis; Hsp70; Cancer therapy

Resistance of tumor cells to apoptosis enhances the spontaneous growth of tumors and renders them resistant to host defense mechanisms as well as to various forms of therapy.[1,2] Hsp70 is an anti-apoptotic protein abundantly and preferentially expressed in malignant human tumors.[1,3] The role of Hsp70 in tumorigenesis is supported by data showing that it enhances the tumorigenic potential of rodent cells,[4–6] and that its expression is essential for the growth of a human T lymphoblast cell line.[7] Furthermore, the expression of Hsp70 correlates with increased cell proliferation, poor differentiation, lymph node metastases and poor therapeutic outcome in human breast cancer.[8–11] Thus, the neutralization of the function of Hsp70 or inhibition of its expression may inhibit tumor growth and/or sensitize tumor cells to chemotherapeutic agents without affecting normal cells with low or absent Hsp70 expression.

[c]Author for correspondence: Dr. Marja Jäättelä, Apoptosis Laboratory, Institute of Cancer Biology, Danish Cancer Society, Strandboulevarden 49, DK-2100 Copenhagen Ø, Denmark. Voice: +45-35257500; fax +45-35257721.
mhj@biobase.dk

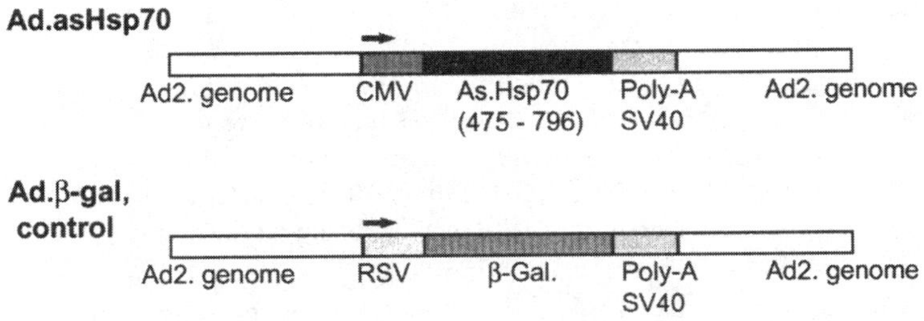

FIGURE 1. Structure of the antisense-Hsp70 adenovirus (Ad.asHsp70) and β-gal control virus (Ad.β-gal.). An expression unit containing the CMV immediate early promoter, a sequence corresponding to bases 475–796 of the published Hsp70 sequence[14] in antisense orientation (spanning the ATG codon) and a SV40 polyadenylation signal was inserted into the human adenovirus type 2 genome. An adenovirus harboring the β-galactosidase gene under the Rous sarcoma virus (RSV) promoter was used as control.[15]

This study was designed to explore the possible usefulness of Hsp70 depletion in cancer therapy. For this purpose we generated an adenoviral vector carrying a fragment of human Hsp70 cDNA in antisense orientation capable of inhibiting the expression of Hsp70 without depleting the closely related heat shock cognate protein, Hsc70 (FIG. 1). A panel of human tumor cell lines was then infected with

TABLE 1. Depletion of Hsp70 preferentially kills human tumorigenic cells of various origins

Cell line	Origin of the cell line	Tumorgenicity	Response to Ad.asHsp70	Response to Ad.β-gal	MOI[a]
MCF-7	Breast cancer	+	Death	None	1,000
MDA-MB-468	Breast cancer	+	Death	None	100
BT-549	Breast cancer	+	Death	None	500
SK-BR-3	Breast cancer	+	Death	None	500
LOVO-36	Colon cancer	+	Death	None	100
HUH-7	Hepatocellular cancer	+	Death	None	100
LN-18	Glioblastoma	+	Death	None	1,000
PC-3	Prostate cancer	+	Death	None	200
DU145	Prostate cancer	+	Death	None	200
PNT1A	Prostate epithelia	–	None	None	1,000
HBL-100	Breast epithelia	–	None	None	1,000
MCF-10A	Breast epithelia	–	None	None	250
WI-38	Fetal lung fibroblast	–	None	None	1,000

[a]MOI; Multiplicity of infection resulting in a 100% infection and used in the experiment.

Ad.asHsp70 at a level needed to achieve 100% infection (TABLE 1). Surprisingly, the majority of Ad.asHsp70-infected tumor cells died concomitant with decreasing Hsp70 protein levels, approximately three to five days after the infection. To study whether Hsp70 expression was also required for the survival of non-tumorigenic cells, we infected normal fetal fibroblasts and immortalized but non-tumorigenic epithelial cells with Ad.asHsp70. Interestingly, the Ad.asHsp70-induced reduction in Hsp70 expression had no effect on the survival of any of the non-tumorigenic cells tested (TABLE 1).

Tumor cells infected with Ad.asHsp70 presented with several typical signs of apoptosis. First, 95% of the dying cells contained fragmented DNA visualized by TdT-mediated dUTP nick end labeling (TUNEL). Second, analysis by light microscopy revealed loss of adherence, condensation of the cytoplasm and formation of apoptotic bodies in dying cells prior to the breakage of the plasma membrane. Third, electron microscopy of Ad.asHsp70-infected MCF-7, MDA-MB-468 and BT-549 cells revealed chromatin condensed in lumps at the inner side of the nuclear envelope. In spite of these classical signs of apoptosis, lysates prepared from Ad.asHsp70-infected cells did not contain increased activity of caspase-3-like proteases as measured by DEVDase enzyme assay. In concordance with the absence of caspase activation, neither inhibition of caspases by overexpression of CrmA nor treatment with caspase inhibitors zVAD-FMK and DEVD-CHO had an effect on Ad.asHsp70-induced cell death. Furthermore, potent anti-apoptotic proteins Bcl-2 and Bcl-X_L failed to rescue cells from apoptosis induced by reduced expression of Hsp70.

The precise nature of the death pathway silenced by Hsp70 remains to be studied. In spite of the apoptosis-like morphological changes, depletion of Hsp70 induces a caspase-independent death pathway. Furthermore, Ad.Hsp70-induced apoptosis differs also from most other caspase-independent apoptotic pathways by being insensitive to protection conferred by anti-apoptotic proteins of Bcl-2 family. Thus, it is interesting to note that Hsp70-mediated protection from apoptosis induced by, for example, TNF occurs without the inhibition of either the release of cytochrome *c* from mitochondria or the activation of effector caspases.[12] Thus, it is tempting to speculate that Hsp70 inhibits another as yet unknown caspase-independent death signaling cascade. One clue to the nature of the death induced by Hsp70 depletion comes from the results showing that Ad.asHsp70 preferentially kills tumorigenic cells. Although proliferation and death appear opposing and mutually contradictory, substantial evidence now indicates that several growth-inducing oncoproteins, Myc and Ras, for example, indeed promote cell death.[13] The coupling of cell division to cell death can thus function as an effective barrier that must be circumvented for cancer to occur. Our results suggest that in the case of Hsp70, this barrier may form the Achilles heel of the cancer and offer a novel and specific therapeutic target for future cancer therapy.

ACKNOWLEDGMENTS

We thank Birgit Poulsen and Klaus Kokholm for excellent technical assistance, and Lone Bastholm and Folmer Elling for performing electron microscopy.

This work was supported by Danish Medical Research Council and Danish Cancer Society.

REFERENCES

1. JÄÄTTELÄ, M. 1999. Escaping cell death: survival proteins in cancer. Exp. Cell Res. **248:** 30–43.
2. REED, J.C. 1999. Mechanisms of apoptosis avoidance in cancer. Curr. Opin. Oncol. **11:** 68–75.
3. JÄÄTTELÄ, M. 1999. Heat shock proteins as cellular lifeguards. Ann. Med. **31:** 261–271.
4. JÄÄTTELÄ, M. 1995. Overexpression of hsp70 confers tumorigenicity to mouse fibrosarcoma cells. Int. J. Cancer **60:** 689–693.
5. SEO, J.S. *et al.* 1996. T cell lymphoma in transgenic mice expressing the human hsp70 gene. Biochem. Biophys. Res. Commun. **218:** 582–587.
6. VOLLOCH, V.Z. & M.Y. SHERMAN. 1999. Oncogenic potential of Hsp72. Oncogene **18:** 3648–3651.
7. WEI, Y. *et al.* 1995. Inhibition of proliferation and induction of apoptosis by abrogation of heat-shock protein (HSP) expression in tumor cells. Cancer Immunol. Immunother. **40:** 73–78.
8. CIOCCA, D.R. *et al.* 1993. Heat shock protein hsp70 in patients with axillary lymph node-negative breast cancer: prognostic implications. J. Natl. Cancer Inst. **85:** 570–574.
9. LAZARIS, A.C. *et al.* 1997. Proliferating cell nuclear antigen and heat shock protein 70 immunolocalization in invasive ductal breast cancer not otherwise specified. Breast Cancer Res. Treat. **43:** 43–51.
10. VARGAS-ROIG, L.M. *et al.* 1997. Heat shock proteins and cell proliferation in human breast cancer biopsy samples. Cancer Detect. Prev. **21:** 441–451.
11. VARGAS-ROIG, L.M. *et al.* 1998. Heat shock protein expression and drug resistance in breast cancer patients treated with induction chemotherapy. Int. J. Cancer **73:** 468–475.
12. JÄÄTTELÄ, M. *et al.* 1998. Hsp70 exerts its anti-apoptotic function downstream of caspase-3-like proteases. EMBO J. **17:** 6124–6134.
13. HANAHAN, D. & R.A. WEINBERG. 2000. The hallmarks of cancer. [In Process Citation]. Cell **100:** 57–70.
14. MILARSKI, K.L. & R.I. MORIMOTO. 1986. Expression of human HSP70 during synthetic phase of the cell cycle. Proc. Natl. Acad. Sci. USA **83:** 9517–9521.
15. STRATFORD-PERRICAUDET, L.D. *et al.* 1992. Widespread long-term gene transfer to mouse skeletal muscles and heart. J. Clin. Invest. **90:** 626–630.

The Interaction of HSP27 with Daxx Identifies a Potential Regulatory Role of HSP27 in Fas-Induced Apoptosis

STEVE J. CHARETTE AND JACQUES LANDRY[a]

Centre de recherche en cancérologie de l'Université Laval, L'Hôtel-Dieu de Québec, Québec, Canada G1R 2J6

ABSTRACT: The heat shock protein HSP27 protects cells against a wide variety of toxic treatments and blocks apoptosis induced by exposures to anticancer drugs and activation of the death receptor Fas. The molecular mechanisms of protection are unknown but appear to be regulated by phosphorylation of HSP27. Two apoptotic pathways can be activated downstream of Fas. The Fas-adaptor FADD mediates a caspase-dependent pathway. Fas also activates a caspase-independent pathway which correlates with Fas-induced translocation of Daxx from the nucleus to the cytoplasm and involves the interaction of Daxx with Fas and Ask1. We found that phosphorylated dimers of HSP27 interact with Daxx, preventing its interaction with Ask1 and Fas and blocking Daxx-mediated apoptosis. Expression of HSP27 also prevents the translocation of Daxx from the nucleus to the cytoplasm which is induced upon expression of Ask1 or stimulation of Fas. The observations reveal a new level of regulation of the Fas pathway. Whereas the FADD axis can be modulated by expression of FLIP, a natural inhibitor of FADD, our results show that HSP27 can accomplish a similar function for the Daxx axis.

KEYWORDS: Apoptosis; Fas; Heat shock protein; Daxx; HSP27; Ask1

INTRODUCTION

The Fas (CD95/Apo-1) receptor belongs to the tumor necrosis factor (TNF) receptor superfamily. When activated by its ligand Fas-L,[1] it triggers an apoptotic signal, which, in many physiological processes such as the regulation of the lymphocyte population, is essential for the elimination of specific cells. A number of pathological conditions result from inappropriate activation or blockage of the Fas signaling pathway. After multimerization by its ligand, the Fas receptor initiates two distinct pathways depending on the adapter proteins that interact with its death domain (FIG. 1). The interaction with Fas-associated-death-domain (FADD) protein leads to the recruitment and activation of caspase-8, and subsequently to the activation of caspase-3 which may (type 2) or may not (type 1) involve the mitochondria depending on caspase-8 activation levels. Caspase activation leads to the cleavage of

[a]Address for correspondence: Jacques Landry, Ph.D., Centre de recherche en cancérologie de l'Université Laval, L'Hôtel-Dieu de Québec, Centre hospitalier universitaire de Québec, 9, rue McMahon, Québec, Canada G1R 2J6. Voice: 418-525-4444 (ext. 5555); fax: 418-691-5439. jacques.landry@med.ulaval.ca

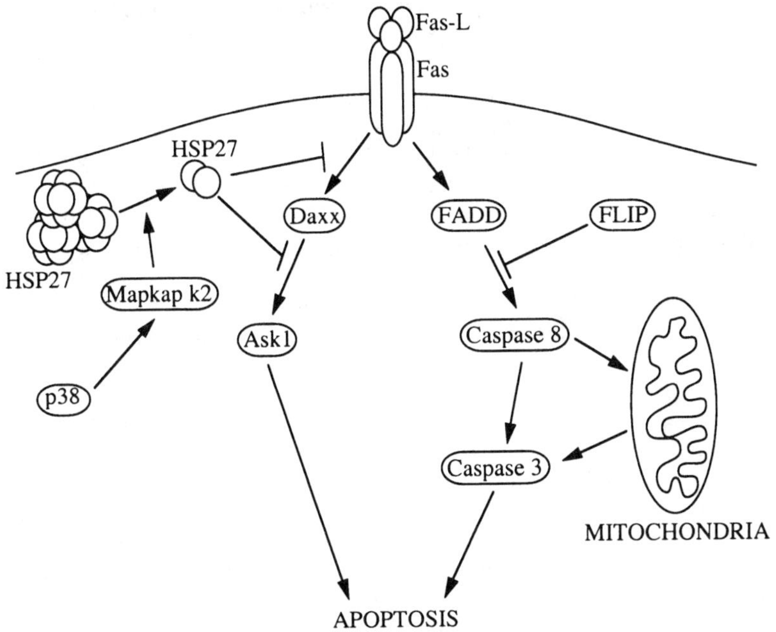

FIGURE 1. Implication of HSP27 in Daxx-dependent apoptosis induced downstream of Fas.

essential proteins and to cell death.[2] Another adapter protein of Fas is Daxx. The mechanism of Daxx-induced apoptosis is less characterized. Daxx interacts with Fas through its Fas-binding domain (FBD) and induces apoptosis though its apoptosis-activation domain (AAD). Daxx-mediated apoptosis is caspase-independent and involves the recruitment of the apoptosis-signal-regulating-kinase-1 (Ask1), a MAP kinase kinase kinase that activates c-Jun N-terminal kinase (JNK).[3,4,4a]

The Fas signaling pathway is tightly regulated in such a way that not all cells expressing the Fas receptor will undergo apoptosis in the presence of Fas-L. For example, some cells at particular stages of differentiation express high level of FLIP, a protein that acts as a dominant negative inhibitor of caspase-8 and specifically blocks caspase-8 activation by FADD.[5] Virus-infected cells can also be blocked in apoptosis signaling downstream of FADD owing to the expression of a viral form of FLIP or else the caspase inhibitor CrmA. Blocking the FADD-dependent arm of the Fas pathway may not, however, be sufficient to totally protect from Fas activation since, at least in some cell lines, activation of the Daxx-dependent arm is sufficient to induce cell death although at a reduced rate. Indeed, Daxx-dependent cell death was observed upon activation of Fas in cells that express a Fas receptor mutant that could not bind FADD or in cells that expressed a dominant negative form of FADD.[4a,6] Therefore, a total inhibition of death signaling downstream of Fas also implies the existence of a mechanism to inhibit the Daxx pathway. Such a role has recently been ascribed to Heat Shock Protein 27.

PHOSPHORYLATED HSP27 ASSOCIATES WITH DAXX AND BLOCKS FAS-INDUCED APOPTOSIS DOWNSTREAM OF DAXX

Heat Shock Protein 27 (HSP27) is a member of the small Heat Shock Protein (sHSP) family that includes many phylogenically conserved homologues from bacteria, yeast, plants and mammals. Like many proteins of this family, HSP27 is up-regulated at the transcriptional level and accumulates to high levels in cells during stress. HSP27 is also regulated at the post-translational level. A wide array of stimuli like heat shock, oxidative stress, growth factors and cytokines lead to phosphorylation of HSP27 catalyzed by MAPKAP kinase-2, a serine kinase activated by the MAP kinase SAKP2/p38.[7–10] Phosphorylation is associated with a modification of the quaternary structure of HSP27. Unphosphorylated HSP27 forms in the cells large homo-oligomers of up to 700-kD that dissociate into dimers and monomers upon phosphorylation.[11,12]

HSP27 has well-documented protective activities against a variety of toxic agents, including heat shock, oxidative stress and various toxic anticancer drugs, and also against death receptor agonists such as TNF or Fas. Although a function as a regulator of actin filament dynamics or as a chaperon protein have been suggested to play a role in resistance, recent studies suggested a more direct function in blocking apoptosis. HSP27 expression was implicated in regulation of apoptosis associated with cessation of growth and differentiation.[13–16] HSP27 also specifically blocks apoptosis induced by some chemicals such as staurosporin, doxorubicin and etoposide.[17–19]

HSP27 interacts both *in vivo* and *in vitro* with Daxx, blocks its interaction with both Ask1 and Fas and, as a result, totally blocks apoptosis induced by co-expression of Ask1 and Daxx. Consequently, HSP27 can also block Fas-induced apoptosis, at least under conditions where signaling downstream of FADD is blocked. This activity of HSP27 is modulated by its phosphorylation/oligomerization status (FIG. 1). Only dimeric phosphorylated HSP27, not unphosphorylated oligomers, can bind Daxx and block apoptosis.[4a]

THE RELOCALIZATION OF DAXX FROM THE NUCLEUS TO THE CYTOPLASM CORRELATES WITH DAXX-INDUCED APOPTOSIS AND IS BLOCKED BY HSP27

The mechanisms by which Daxx mediates apoptosis and how HSP27 blocks this pathway are unknown. Daxx has been described as a nuclear protein and thus it is unclear how it can interact with the Fas receptor at the membrane and mediate downstream signaling through cytoplasmic Ask1. To better understand the mechanisms that control Daxx-Ask1–induced apoptosis, we performed immunolocalization studies with Daxx and Ask1 in 293 cells. We first confirmed that after transfection most newly synthesized Daxx accumulated in the nucleus as concentrated points (FIG. 2A, *panel 1*) previously identified as the centrosome and a sub-class of nuclear bodies involving promyelocytic leukemia protein (PML).[20,21]Transfected Ask1 showed in contrast an exclusive cytoplasmic staining (FIG. 2A, *panel 2*). At concentrations that caused apoptosis, the co-expression of Ask1 with Daxx induced a major intracellular redistribution of Daxx. The percentage of cells with staining for Daxx in the cytoplasm increased from 15 to 50% upon co-expression of Ask1 (FIG. 2B). In most of the cells, Daxx then co-localized with Ask1 at the cell periphery (FIG. 2A, *panel 3*). We also found that Daxx relocalizes to the cytoplasm upon activation of the Fas receptor.[4a] Consistent with their protective function, phosphorylated dimers of HSP27 (HSP27-EE) efficiently restored the nuclear localization of Daxx co-expressed with Ask1 (FIG. 2A, *panel 4* and FIG. 2B) whereas the multimers (HSP27-AA) that are inefficient in blocking apoptosis also produced no effect on Daxx localization (FIG. 2B).

These results suggest that Daxx cytotoxicity is linked with its translocation from the nucleus to the cytoplasm and that HSP27 protection results from an inhibition of this relocalization. The fact that Daxx-induced apoptosis and translocation depend on Ask1 may suggest that Ask1 is involved in maintaining Daxx in the cytoplasm.

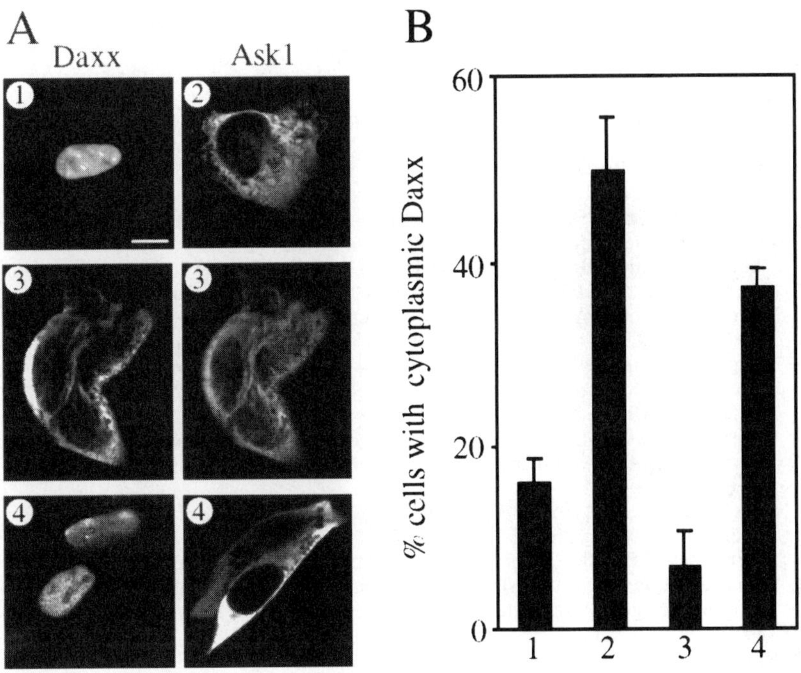

FIGURE 2. Ask1-induced nuclear to cytoplasmic translocation of Daxx is blocked by HSP27. **A.** 293 cells were transfected either with Daxx (1), Ha-tagged-Ask1 (2), Daxx and Ha-Ask1 (3), or Daxx, HA-Ask1 and HSP27EE (4). Daxx was immunolocalized using an anti-Daxx antibody (EL-68); Ask1 was detected using anti-HA (12CA5). Scale bar = 10 μm. *Panel 1*, one cell with Daxx localized in the nucleus; *panel 2*, one cell with Ask1 localized in the cytoplasm; *panel 3*, double immunofluorescence micrograph of 3 cells showing a co-localization of Daxx and Ask1 in the cytoplasm; *panel 4*, double immunofluorescence micrograph of 2 cells showing nuclear Daxx and cytoplasmic Ask1. **B.** Proportion of cells displaying a cytoplasmic staining of Daxx in 293 cells transfected with Daxx alone (1), both Daxx and Ha-Ask1 (2), or Daxx and Ask1 with either HSP27EE (3) or HSP27AA (4). Each value corresponds to the average of at least three different experiments.

Daxx has two possible nuclear localization signal sequences. A direct interaction between Daxx and Ask1 may mask the active nuclear relocalization signal and inhibit Daxx active nuclear transport. By blocking Daxx-Ask1 interaction,[4a] HSP27 could then restore or prevent the inhibition of the active nuclear transport of Daxx.

Because of the high capacity of HSP27 to maintain Daxx in the nucleus and prevent Daxx-mediated apoptosis, and because HSP27 level of expression is extremely variable between different cell lines, it will become important that the level of HSP27 be determined in studies examining mechanisms of Fas and Daxx-mediated apoptosis.

REFERENCES

1. Nagata, S. 1997. Apoptosis by death factor. Cell **88:** 355–365.
2. Green R.D. 1998. Apoptotic pathways: the roads to ruin. Cell **94:** 695–698.
3. Yang, X. *et al.* 1997. Daxx, a novel Fas-binding protein that activates JNK and apoptosis. Cell **89:** 1067–1076.
4. Chang, H.Y. *et al.* 1998. Activation of apoptosis signal-regulating kinase 1 (ASK1) by the adapter protein Daxx. Science **281:** 1860–1863.

4a. Charette, S.J. *et al.* 2000. Inhibition of daxx-mediated apoptosis by heat shock protein 27. Mol. Cell Biol. **20:** 7602–7612.

5. Tschopp J., M. Irmler & M. Thome. 1998. Inhibition of fas death signals by FLIPs. Curr. Opin. Imm. **10:** 552–558.
6. Chang, H.Y. *et al.* 1999. Dissecting Fas signaling with an altered-specificity death-domain mutant: requirement of FADD binding for apoptosis but not Jun N- terminal kinase activation. Proc. Natl. Acad. Sci. USA **96:** 1252–1256.
7. Huot, J. *et al.* 1995. Characterization of 45-kDa/54-kDa HSP27 kinase, a stress-sensitive kinase which may activate the phosphorylation-dependent protective function of mammalian 27-kDa heat-shock protein HSP27. Eur. J. Biochem. **227:** 416–427.
8. Landry, J. *et al.* 1992. Human HSP27 is phosphorylated at serines 78 and 82 by heat shock and mitogen-activated kinases that recognize the same amino acid motif as S6 kinase II. J. Biol. Chem. **267:** 794–803.
9. Rouse, J. *et al.* 1994. A novel kinase cascade triggered by stress and heat shock that stimulates MAPKAP kinase-2 and phosphorylation of the small heat shock proteins. Cell **78:** 1027–1037.
10. Stokoe, D. *et al.* 1992. Identification of MAPKAP kinase 2 as a major enzyme responsible for the phosphorylation of the small mammalian heat shock proteins. FEBS Lett. **313:** 307–313.
11. Lambert, H. *et al.* 1999. HSP27 multimerization mediated by phosphorylation-sensitive intermolecular interactions at the amino terminus. J. Biol. Chem. **274:** 9378–9385.
12. Rogalla, T. *et al.* 1999. Regulation of Hsp27 oligomerization, chaperone function, and protective activity against oxidative stress/tumor necrosis factor alpha by phosphorylation. J. Biol. Chem. **274:** 18947–18956.
13. Mehlen, P. *et al.* 1999. Small stress protein Hsp27 accumulation during dopamine-mediated differentiation of rat olfactory neurons counteracts apoptosis. Cell Death Differ. **6:** 227–233.
14. Mehlen, P. *et al.* 1997. hsp27 as a switch between differentiation and apoptosis in murine embryonic stem cells. J. Biol. Chem. **272:** 31657–31665.
15. Spector, N. L. *et al.* 1992. Growth arrest of human B lymphocytes is accompanied by induction of the low molecular weight mammalian heat shock protein (Hsp28). J. Immunol. **148:** 1668–1673.
16. Stahl, J. *et al.* 1992.The small heat shock protein hsp25 is accumulated in P19 embryonal carcinoma cells and embryonic stem cells of line BLC6 during differentiation. Differentiation **51:** 33–37.

17. MEHLEN, P., K. SCHULZE-OSTHOFF & A.P. ARRIGO. 1996 Small stress proteins as novel regulators of apoptosis. Heat shock protein 27 blocks Fas/APO-1- and staurosporine-induced cell death. J. Biol. Chem. **271:** 16510–16514.
18. GARRIDO, C. *et al.* 1999. HSP27 inhibits cytochrome c-dependent activation of procaspase-9. FASEB J. **13:** 2061–2070.
19. HUOT, J. *et al.* 1991. Increased survival after treatments with anticancer agents of Chinese hamster cells expressing the human Mr 27,000 heat shock protein. Cancer Res. **51:** 5245–5252.
20. EVERETT, R.D. *et al.* 1999. A dynamic connection between centromeres and ND10 proteins. J. Cell Sci. **112:** 3443–3454.
21. PLUTA, A.F. *et al.* 1998. Interphase-specific association of intrinsic centromere protein CENP-C with HDaxx, a death domain-binding protein implicated in Fas-mediated cell death. J. Cell Sci. **111:** 2029–2041.

A Single Cell Analysis of Apoptosis

Ordering the Apoptotic Phenotype

JOSHUA C. GOLDSTEIN,[a,b] RUTH M. KLUCK,[a] AND DOUGLAS R. GREEN[a,c]

[a]*Division of Cellular Immunology, La Jolla Institute for Allergy and Immunology, San Diego, California 92121, USA*

[b]*Biology Department, University of California San Diego, La Jolla, California 92093, USA*

ABSTRACT: Microcinematography was applied to the analysis of the kinetics of apoptotic cell death. Apoptosis was found to be a process that proceeds in different cells at different times after an initial stress, and therefore kinetic studies of apoptotic events in bulk cultures can be problematic. Using single cell analysis we found that stronger apoptotic stimuli induce an earlier onset of apoptosis, but that there is no relationship between time of onset and duration of the apoptotic process. That is, cells that initiate apoptosis shortly after induction do not proceed more rapidly through the process. This suggests an all-or-none mechanism that is supported by some models of the biochemical pathways of apoptosis.

KEYWORDS: Apoptosis; Annexin V; Caspases; Apoptotic interval

INTRODUCTION

Apoptosis, or programmed cell death is defined by a set of morphological changes that occurs in cells as they die.[1–3] Apoptotic morphology includes changes in cell shape, (rounding, blebbing), membrane dynamics (phosphatidylserine, PS, exposure[4]) and organelle dysfunction (mitochondria[5]). Some apoptotic changes signal phagocytosis and aid in cell degradation. Specifically, phosphatidylserine externalization from the inner to the outer leaflet of the cell membrane signals phagocytosis by macrophages.[6] Although there are exceptions in cell lines that do not show a specific phenotype, apoptotic morphology is exhibited by a wide variety of animal cells induced to die by a myriad of stimuli. These morphological changes distinguish apoptosis from other kinds of cell death such as necrosis.[1–3] Genetic and biochemical approaches have characterized many of the molecular pathways underlying the apoptotic phenotype such as DNA fragmentation, chromatin condensation, and blebbing.[7–11]

Apoptotic morphology is largely the result of the activation of a family of cysteine proteinases called caspases.[12] Caspase activation occurs through at least two types of complexes, called apoptosomes. The first type of apoptosome forms through

[c]Address for correspondence: Dr. Douglas R. Green, Division of Cellular Immunology, La Jolla Institute for Allergy and Immunology, 10355 Science Center Dr., San Diego, CA 92121, USA. Voice: 858-558-3500.
dgreen5240@aol.com

the ligation of death receptors and subsequent recruitment and activation of apical caspases such as caspases 8 and 10, which can process downstream caspases, such as caspases 3 and 7, in a proteolytic cascade.[13] Another type of apoptosome is involved in apoptotic pathways involving mitochondrial damage. In this scenario, cellular stress induces the release of cytochrome *c* from the intermembrane space of mitochondria. Binding of cytochrome *c* to the adapter molecule Apaf-1 initiates its oligomerization and formation of an apoptosome. This apoptosome recruits and activates caspase 9, and subsequently other downstream caspases.[14–19] The importance of the mitochondrial pathway in whole cells has recently been demonstrated by the marked reduction of stress-induced apoptosis in cytochrome *c*–deficient cells.[20]

The timing of different apoptotic events has remained less defined, in part because studies of cell populations provide findings based on the population averages. Here we analyze single cells over time using time-lapse microscopy to (a) place in order several morphological and biochemical apoptotic changes, and (b) to characterize the time intervals that separate these events. Single cell analysis also allowed us to consider the behavior of cells that deviate from that of the bulk of the population. These findings have led to a better understanding of the apoptotic process.

RESULTS AND DISCUSSION

Using time-lapse microscopy the fate of a field (~100 cells) of HeLa cells was observed at 15 min intervals after exposure to UV light. The field was monitored over the course of 18 h and the time at which each cell began to round up was recorded (FIG. 1, *top*). For 5 h few cells rounded up, after which several additional rounded cells accumulated each hour until around 13 hours. As rounding is the first morphological sign of apoptosis, this profile of cell rounding could be interpreted as the majority of stimulated cells undergoing an initial 5 h lag or judgment phase followed by apoptosis.

The same experiment was performed at two different UV doses to determine the dose effect on the kinetics of UV-induced apoptosis. At higher UV doses the onset of cell rounding (and thus apoptosis) was faster. For example, at 8 h approximately 35% versus 70% of cells had rounded after 90 and 180 mJ/cm^2 UV, respectively (FIG. 1). At low UV doses, the time at which cells began to undergo apoptosis followed a roughly Gaussian distribution. At higher UV doses, the distribution was similar, with a skewing to the left. These data indicate that the initiation of apoptosis follows an approximate time course that can be accelerated with stronger apoptotic stimuli.

We next examined whether the length of the apoptotic process itself changed with apoptotic stimuli of different strengths. Detectable beginning and end points of apoptosis were first established. In microscopy studies, cell rounding represents the first apoptotic shape change, but what is the last? FIGURE 2 shows time-lapse images of HeLa cells after treatment with UV as described in FIGURE 1. In this epi-fluorescent time-lapse, loss of plasma membrane integrity was followed by measuring the uptake of the DNA-binding dye propidium iodide (PI). This dye binds to DNA only if it can enter cells and thus provides a convenient measure of the loss of plasma membrane integrity characteristic of late apoptosis. Using markers of cell shape and membrane integrity, FIGURE 2 shows the order of a few apoptotic events. At 6 h and 15 min after a low dose of UV, the cells appeared normal and there was no PI staining. An hour and 15 minutes later, one of the cells began to round and subsequently blebs formed. Much later (15 h and 15 min) high contrast bright field images reveal the existence of large blebs. Importantly, these large blebs, or "airbags" as they are affectionately called, correlate with loss of plasma membrane integrity. Thus we define an "apoptotic interval" as the time from which cells undergo rounding to when they become PI-positive. This simple system can be used to ask two fundamental

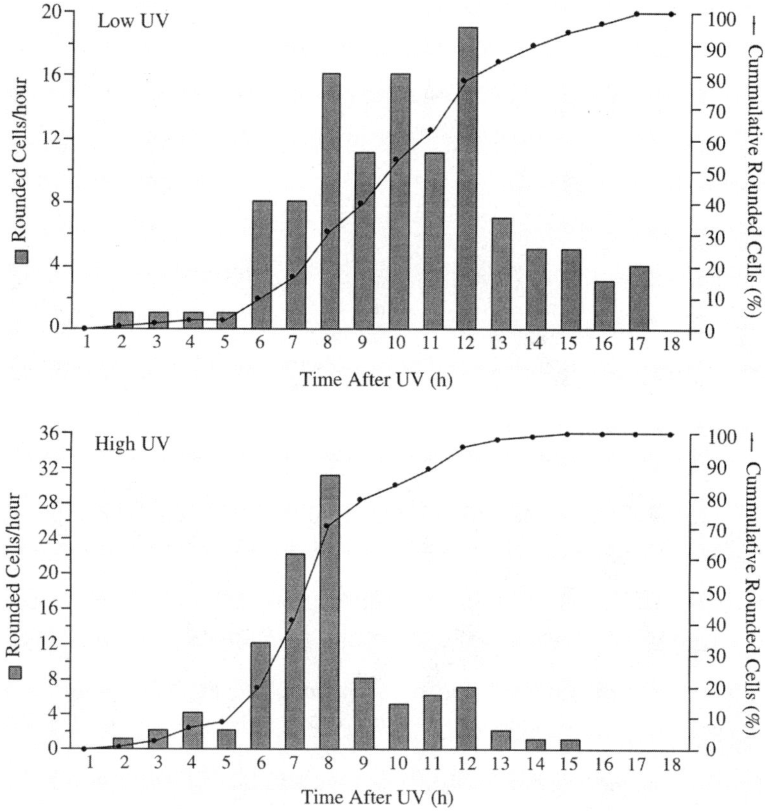

FIGURE 1. Stronger apoptotic stimulus induces a faster onset of apoptosis. HeLa cells treated with a low UV dose (90 mJ/cm^2, *Top*) undergo apoptosis more slowly than after receiving a high dose (180 mJ/cm^2, *Bottom*). A field of cells was exposed to UV light and images were acquired at 15 min intervals. The left *y*-axis (*bar graph*) represents the number of cells that rounded during each hour, while the *y*-axis on the right (*line graph*) shows cumulative cell rounding.

questions about the process of apoptosis. How long is the apoptotic process? And does this length change?

Apoptotic interval measurements were made on HeLa cells treated with either high or low dose UV (FIG. 3). The average apoptotic interval in cells treated with a low dose of UV was 2.99 ± 1.5 hours, with some ranging from 0.5 to 9 hours. The distribution is again reminiscent of a bell-shaped curve. In cells treated with high dose UV the distribution was similar with an average apoptotic interval of 2.8 ± 1.3 hours. In population studies where apoptosis proceeds faster as doses get higher, one might conclude that the entire process, not just the initiation of apoptosis, is also faster. On the basis of FIGURE 3, we found that the duration of apoptosis was not affected by the strength of the apoptotic stimulus. Although there was variation between individual cells, the distribution was not significantly different at high or low UV doses. These findings suggest that the process of apoptosis is an all or nothing response, and once initiated, proceeds down a pre-determined path.

Analysis of individual cells gives the opportunity to examine cells whose behavior is not average for the population. For example, the times at which cells begin to round up follows a roughly bell shaped distribution, as does the apoptotic interval, raising the question of whether there is any relation between when cells begin to round and the length of the apoptotic interval. Do cells that round up earlier have greater apoptotic capacity (such as caspase activity) and thus shorter apoptotic intervals? This is consistent with the ability of caspase inhibitors to abrogate most apoptotic morphological changes (ref. 21; unpublished observations). Perhaps all the cells receive (and process) the apoptotic signal at the same time, but cells with less caspases take longer to round and then longer to complete the apoptotic interval. By this rationale, cells that round at 2 h would be expected to have a shorter apoptotic interval than cells that round up 14 h after induction of apoptosis. However, this does not seem to be the case. There was no significant correlation between the time at which individual cells began to round and the length of time required for them to become PI-positive (FIG. 4). Thus, in each cell the threshold for responding to an apoptotic stimulus appears to be independent of the speed/efficiency of the apoptotic machinery itself.

Single cell analysis may be the best way to determine the order of apoptotic events. Other methods such as Western blotting (even in conjunction with flow cytometry) are notoriously hard to quantify. This problem is exacerbated when trying to resolve events that are relatively fast,[22] such that two events separated by a short amount of time require frequent sampling. Furthermore, events that occur many hours after the apoptotic stimulus necessitate the processing of large numbers of samples. FIGURE 5 illustrates the problem by presenting the same data either as single cell data or as population data. As previously published,[23] when single cells undergoing apoptosis are observed over time they proceed through cytochrome *c* release, annexin V binding, and then PI positivity in that order (FIG. 5, *top*). By this analysis the order of events is clear, limited of course by the relative sensitivity of the detection methods for each event. However, when the same data are presented as a percentage of the total (FIG. 5, *bottom*) incorrect interpretations might be drawn. Although cytochrome *c* is released before annexin V binding, at some time points (>16 h) there seem to be more annexin V-positive cells than cells that have released cytochrome *c*. This issue would be further confused if cytochrome *c* release were

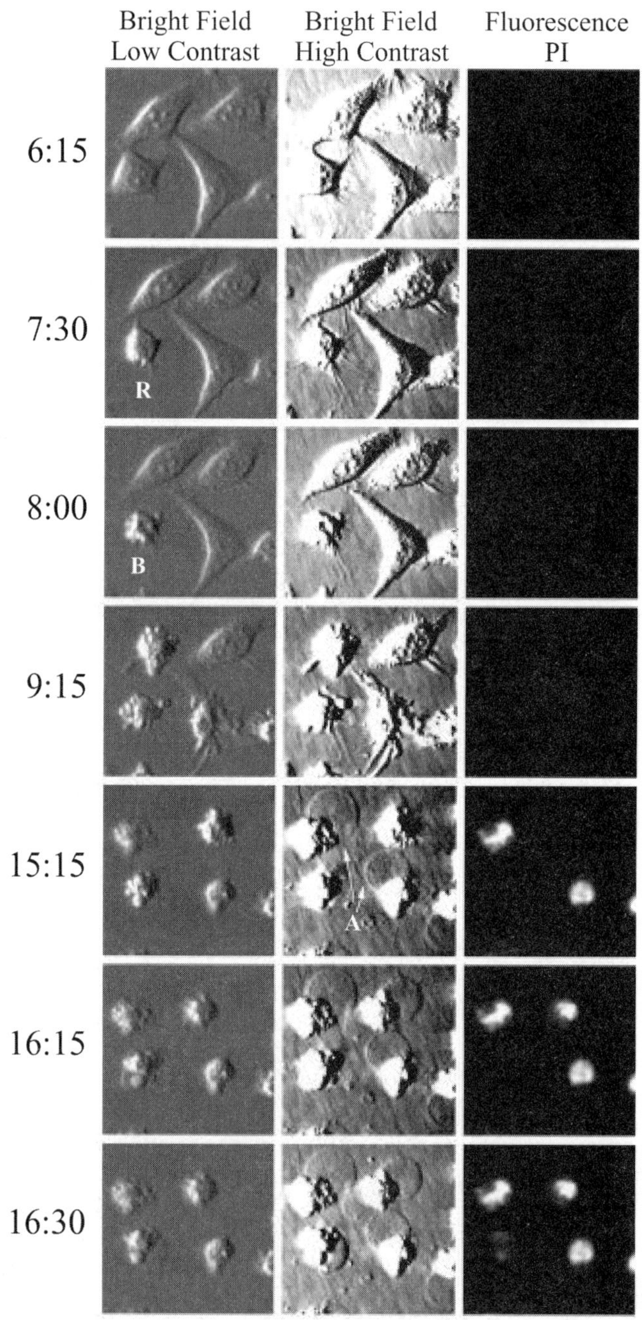
Bright Field
Low Contrast
Bright Field
High Contrast
Fluorescence
PI
6:15
7:30
8:00
9:15
15:15
16:15
16:30
R
B
A

FIGURE 2. Ordering morphological changes in apoptotic cells. Selected images from epi-fluorescence time-lapse microscopy of HeLa cells following exposure to UV light (90 mJ/cm^2). The contrast of each bright field image was adjusted post acquisition to accentuate features that are best visualized by low (*left column*) or high (*middle column*) contrast settings. *Right column* shows propidium iodide fluorescence. Time (h) after exposure is indicated to the left of each row. The lower contrast bright field setting shows cells rounding (7:30 labeled "R") and blebbing (8:00 labeled "B").The high contrast version (*middle column*) shows terminal "airbag" blebs (15:15 labeled "A") that correspond to a breakdown in plasma membrane integrity as visualized by propidium iodide staining (*right column*). Images were acquired with a 40× objective.

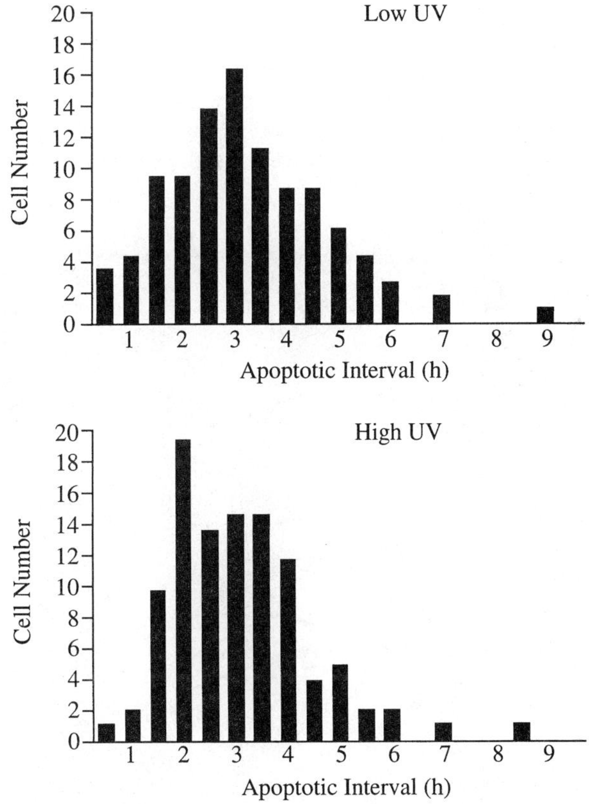

FIGURE 3. Stronger apoptotic stimulus does not change the duration of apoptosis. HeLa cells were treated with UV light (*Top*, low dose [90 mJ/cm^2]; *Bottom*, high dose [180 mJ/cm^2]) and followed individually by time-lapse. The apoptotic interval is defined as time from when a cell begins to round to when it becomes PI-positive. A histogram where the x-axis represents the apoptotic interval (see text) rounded to the nearest half hour and the y-axis shows the corresponding number of cells for each interval. The mean apoptotic interval after both high and low UV doses was three hours.

assayed by Western blot, where 10–20% differences in intensity are difficult to distinguish. This may lead an investigator to conclude that cytochrome *c* release is concurrent with annexin V binding, or may be an event that is downstream of caspase activation rather than upstream. Population analyses are by definition an average where outliers blur the distinction between two events that are temporal neighbors.

The order of apoptotic events can be used to determine the time intervals between them (FIG. 6). The average time between the apoptotic insult and mitochondrial cytochrome *c* release was about 13 hours. This interval corresponds to a judgment phase where individual cells assess damage and perhaps make decisions regarding cell cycle progression, arrest, or apoptosis. Mitochondrial cytochrome *c* release is thought to be the end of this phase in many apoptotic pathways.[23] Once cytochrome *c* is released, it takes about 2 hours for annexin V binding (phosphatidylserine exposure) to be detected. If apoptosis is a process designed to cleanly remove dead cells (and thus avoid inflammation) then phagocytosis may be the most important event that occurs in dying cells. In this model, cells must be phagocytosed (annexin V-positive) before loss of plasma membrane integrity. UV-treated cells were annexin V-positive for about 5 h before they lost plasma membrane integrity. As macrophages can phagocytose 4.4 apoptotic thymocytes per hour,[24] the 5 h window observed here is easily sufficient for removal of cells before loss of the plasma membrane integrity.

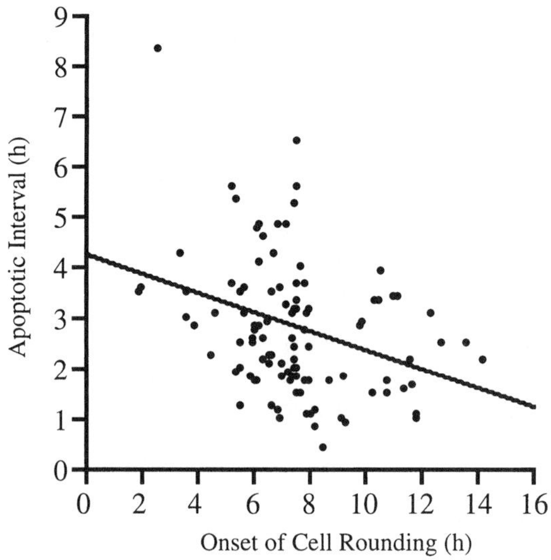

FIGURE 4. The onset of apoptosis is not correlated to the apoptotic interval. Cells were treated with a high dose of UV (180 mJ/cm^2) and the time at which each cell began to round up was plotted against the length of the apoptotic interval. The R^2 for the best-fit line is 0.11, indicating a poor correlation.

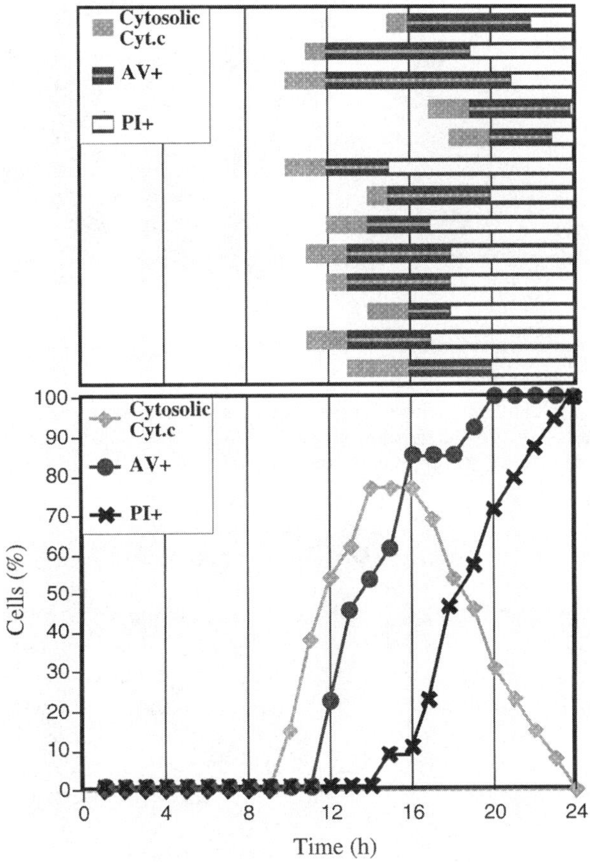

FIGURE 5. Rather than cell population studies, single cell analysis provides a clear assessment of the order of apoptotic events. HeLa cells expressing cytochrome c-GFP were treated with UV (90 mJ/cm^2) and analyzed by confocal time-lapse microscopy. The apoptotic events examined were cytochrome c-GFP release, annexin V binding and PI staining. The same data are presented in two ways, with the *top panel* representing the data from each cell separately and the *bottom panel* showing an average of the data from all cells (simulating a population measurement). The *top panel* shows 13 cells, represented by horizontal bars over a period of 24 hours. Cytosolic cytochrome c-GFP is represented by light gray. Annexin V-positive cells are dark gray with a light gray interior, as these cells presumably still have cytochrome c in the cytoplasm. PI-positive cells are also Annexin v-positive and are represented by dark gray with a white interior. At this stage, cytosolic cytochrome *c* can diffuse out of the cell. The *bottom panel* shows the same data graphed as a population study. Light gray diamonds represent cytosolic cytochrome *c*–GFP; dark gray circles show annexin V binding; and black X's show PI-positivity.

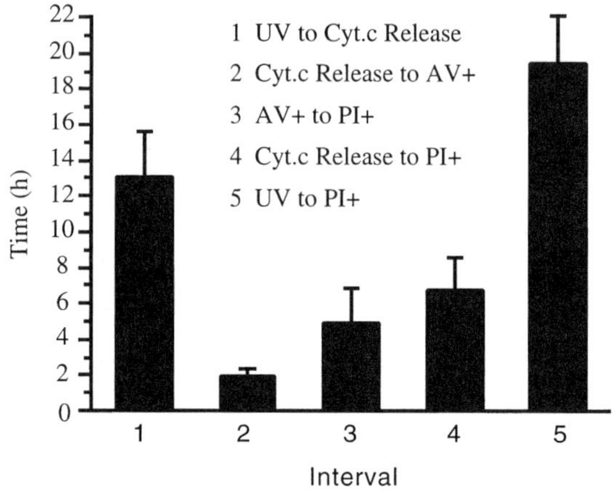

FIGURE 6. Time intervals between different apoptotic events. This bar graph shows the mean time between different apoptotic events in the UV-initiated apoptotic pathway (same experiment as FIG. 5). Error bars represent standard deviation.

CONCLUSIONS

In this report we utilized single cell analysis to examine the kinetics and ordering of several apoptotic events in HeLa cells exposed to UV. Cells that initiate apoptosis (cell rounding) earlier than other cells do not go through the apoptotic process itself more quickly. The length of the apoptotic process was also constant with different doses of UV. We conclude that once initiated, apoptosis proceeds along a general program, which depends more on the original status of the individual cell than on the strength of the apoptotic stimulus. Strengthening this view is establishment of the order and the time interval between of apoptotic events. We find that the length of time that cells exhibit one phagocytosis signal (PS externalization) is far greater than the time required for macrophages to engulf them.

METHODS

Epifluorescent time-lapse microscopy was performed much as reported.[23] Briefly, HeLa cells (approx. 200,000) that had been plated in 3.5 cm dishes the previous day in phenol red free Dulbecco's modified Eagle medium (Gibco) (supplemented with 10% FCS L-glutamine, 200 μg/ml penicillin and 100 μg/ml streptomycin sulfate), were covered with mineral oil (Sav-On, Buena Park, CA) and maintained at 37°C by an MS-C Temp Controller (Narishige, New York, NY) on a Nikon Diaphot inverted microscope. Propidium iodide (5 μg/ml) was added to detect plasma membrane integrity and 4 mM glutathione, 1mM ascorbic acid and 0.5 mM dithiothreitol were added to mitigate light toxicity. Fluorescent and bright field (Hoffman Optics) images were acquired with a Sensys Digital (Photometrics, Tucson, AZ) using 2×2

binning (Fluorescent images) at 15 min intervals using MetaMorph 3.5 software (Universal Imaging, West Chester, PA).

Confocal time-lapse was performed as described.[23]

REFERENCES

1. Walker, N.I., B.V. Harmon, G.C. Gobe & J.F. Kerr. 1988. Methods Achiev. Exp. Pathol. **13:** 18–54.
2. Robertson, A.M., C.C. Bird, A.W. Waddell & A.R. Currie. 1978. J. Pathol. **126:** 181–187.
3. Wyllie, A.H., J.F. Kerr & A.R. Currie. 1980. Int. Rev. Cytol. **68:** 251–306.
4. Martin, S.J., C.P. Reutelingsperger, A.J. McGahon, J.A. Rader, R.C. van Schie, D.M. LaFace & D.R. Green. 1995. J. Exp. Med. **182:** 1545–1556.
5. Marchetti, P., M. Castedo, S.A. Susin, N. Zamzami, T. Hirsch, A. Macho, A. Haeffner, F. Hirsch, M. Geuskens & G. Kroemer. 1996. J. Exp. Med. **184:** 1155–1160.
6. Fadok, V.A., D.L. Bratton, D.M. Rose, A. Pearson, R.A. Ezekewitz & P.M. Henson. 2000. Nature **405:** 85–90.
7. Enari, M., H. Sakahira, H. Yokoyama, K. Okawa, A. Iwamatsu & S. Nagata. 1998. Nature **391:** 43–50.
8. Liu, X., H. Zou, C. Slaughter & X. Wang. 1997. Cell **89:** 175–184.
9. Wolf, B.B., M. Schuler, F. Echeverri & D.R. Green. 1999. J. Biol. Chem. **274:** 30651-30656.
10. Sahara, S., M. Aoto, Y. Eguchi, N. Imamoto, Y. Yoneda & Y. Tsujimoto. 1999. Nature **401:** 168–173.
11. Rudel, T. & G.M. Bokoch. 1997. Science **276:** 1571-1574.
12. Wolf, B.B. *et al.* 1999. J. Biol. Chem. **274:** 20049–20052.
13. Ashkenazi, A. & V.M. Dixit. 1998. Science **281:** 1305–1308.
14. Green, D.R. & J.C. Reed. 1998. Science **281:** 1309–1312.
15. Liu, X., C.N. Kim, J. Yang, R. Jemmerson & X. Wang. 1996. Cell **86:** 147–157.
16. Li, P., D. Nijhawan, I. Budihardjo, S.M. Srinivasula, M. Ahmad, E.S. Alnemri & X. Wang. 1997. Cell **91:** 479–489.
17. Saleh, A., S.M. Srinivasula, S. Acharya, R. Fishel & E.S. Alnemri. 1999. J. Biol. Chem. **274:** 17941–17945.
18. Zou, H., W.J. Henzel, X. Liu, A. Lutschg & X. Wang. 1997. Cell **90:** 405–413.
19. Zou, H., Y. Li, X. Liu & X. Wang. 1999. J. Biol. Chem. **274:** 11549–11556.
20. Li, K., Y. Li, J.M. Shelton, J.A. Richardson, E. Spencer, Z.J. Chen, X. Wang & R.S. Williams. 2000. Cell **101:** 389–399.
21. Martin, S.J., D.M. Finucane, G.P. Amarante-Mendes, G.A. O'Brien & D.R. Green. 1996. J. Biol. Chem. **271:** 28753–28756.
22. Chen, Q., B. Gong & A. Almasan. 2000. Cell Death Differ. **7:** 227–233.
23. Goldstein, J.C., N.J. Waterhouse, P. Juin, G.I. Evan & D.R. Green. 2000. Nat. Cell Biol. **2:** 156–162.
24. Licht, R., C.W. Jacobs, W.J. Tax & J.H. Berden. 1999. J. Immunol. Methods **223:** 237–248.

Molecular Mapping of the Physiological Cell Death Process

Mitochondrial Events May Be Disordered

DANIEL FLORYK AND DAVID S. UCKER[a]

Department of Microbiology and Immunology, University of Illinois College of Medicine, Chicago, Illinois 60612, USA

ABSTRACT: The mitochondrion plays a central role in Bcl-2-inhibitable physiological cell deaths. The detailed order of mitochondrial and other events during cell death *in vivo* remains ambiguous, however. As part of an effort to explore this issue, we have asked whether mitochondrial dissolution during physiological cell death occurs in an orderly and concerted process. Here, we describe the characterization of two elements of mitochondrial disintegration on the level of individual cells. Using a novel cytofluorimetric approach, we have assessed simultaneously the release of cytochrome *c* (specifically a fluorescently tagged transfected construct) from mitochondria and the dissipation of mitochondrial membrane potential. Our results indicate that mitochondrial disintegration does not follow a strictly ordered process and is not concerted. We are extending these studies to further characterize mitochondrial events in the context of Bcl-2 family members and place them definitively within the context of the caspase cascade.

KEYWORDS: Bcl-2; Caspases; Cytochrome *c*; Membrane potential; Transient transfection

INTRODUCTION

Physiological cell deaths follow a thematically conserved and ordered pathway. In general terms, idiosyncratic signaling inputs act afferently to trigger an amplifying proteolytic cascade of caspases; the downstream elements of this cascade exert profound changes in cellular structure, at the level of the cytoskeleton, plasma membrane, and intracellular organelles, and cause the activation of molecules that lead directly to the demise of the cell, including the activation of effector cyclin dependent kinase complexes.[1]

A central role for the mitochondrion in Bcl-2-inhibitable (especially death receptor–independent) deaths has been implicated by studies demonstrating the localization of Bcl-2 and other Bcl-2 family members to that organelle and by findings that mitochondrial disruption triggered by disparate suicidal stimuli is blocked when death is spared by Bcl-2. Cytochrome *c*, which is released specifically from mitochondria during

[a]Address for correspondence: Dr. David S. Ucker, Department of Microbiology and Immunology, University of Illinois College of Medicine, Rm. E803 (M/C 790), 835 South Wolcott, Chicago, IL 60612. Voice: 312-413-1102; fax: 312-996-6415.
DUCK@UIC.EDU

death,[2,3] can complex with APAF-1; this appears sufficient to activate apical caspase 9 (and initiate the caspase cascade) *in vitro.*[4] Consistent with this view, the absence of either caspase 9 or APAF-1 abrogates death severely and pleiotropically in mutant mice.[5–8]

The detailed order of events *in vivo* remains ambiguous, however. In particular, the notion that mitochondrial disruption is necessary for activation of the caspase cascade implies that all caspase activity occurs subsequent to Bcl-2-inhibitable mitochondrial events. However, we have identified apical caspase activity that is upstream of the action of Bcl-2.[9]

The precise mechanism of cytochrome c release also has not been ascertained. Distinct modes of cytochrome c release, through selective pores or following mitochondrial membrane rupture, have been proposed. Compromise of the outer mitochondrial membrane may result from opening of the permeability-transition pore (PTP) or following loss of mitochondrial membrane potential ($\Delta\Psi_m$) that results from the interaction of pro-death Bcl-2 family members with the Adenine Nucleotide Transporter (ANT; ref. 10) or Voltage-Dependent Anion Channel (VDAC; ref. 11), two components of the PTP. Alternatively, some researchers have suggested that Bcl-2 family members form pores in the intact outer mitochondrial membrane through which cytochrome *c* can pass directly. Recent data have suggested that mitochondrial depolarization is neither required for, nor is it a direct consequence of, cytochrome *c* release and that the rapid loss of $\Delta\Psi_m$ is dependent on caspase activity.[12]

As part of the effort to order death-specific mitochondrial events affirmatively, we have characterized the loss of mitochondrial membrane potential and the release of cytochrome *c* kinetically during cell death and on the level of individual cell.

EXPERIMENTAL METHODS

Cell Culture, Transfection, and Death Induction

Freshly cloned populations of HeLa cells were grown in DMEM medium (Mediatech, Herndon, VA) supplemented with heat inactivated fetal calf serum (10% v/v; HyClone Laboratories, Logan, UT) and 2 mM L-glutamine (Mediatech). Six μg of pCyt *c*-GFP[13] was co-transfected into 10^5 HeLa cells seeded the night before in six well plates utilizing a standard calcium phosphate precipitation method, as described.[1] Forty eight hours after transfection, cells were induced to die with 1 μM staurosporine (Sigma Chemical Co.).

Cytofluorimetric Assessment of Mitochondrial Membrane Potential and Cytochrome c Release

Mitochondrial membrane potential was assessed with tetramethyl rhodamine methyl ester (TMRM; Molecular Probes, Eugene, OR), a fluorescent dye rapidly responsive to mitochondrial membrane potential.[14] TMRM was stored as a stock solution at 25 mg/ml in DMSO; a fresh working solution (2 μM) was prepared in Mitochondria Buffer (80 mM KCl; 10 mM Tris-HCl, pH 7.2; 3 mM $MgCl_2$; 1 mM Na_2EDTA; 5 mM KH_2PO_4; and 10 mM K_2succinate as a mitochondrial substrate). Cells were harvested, resuspended in Mitochondria Buffer, and permeabilized by treatment with digitonin (at a digitonin:protein mass ratio of 0.12) for 5 min.[15] Permeabilized cells were stained with 20 nM TMRM (15 min, 25°C) and analyzed immediately. TMRM (Ex_λ = 488 nm, Em_λ = 580 nm) and cytochrome *c*-GFP (Ex_λ = 488 nm, Em_λ = 525 nm) signals were collected (FACSCaliber instrument and CellQuest software; Beckton Dickinson, San Jose, CA). Data were processed with WinMDI software (Joe Trotter, Scripps Research Institute; La Jolla, CA).

RESULTS AND DISCUSSION

Different methodological approaches have been employed to examine the release of cytochrome c from mitochondria, and to quantify the loss of mitochondrial membrane potential. The detection of post-mitochondrial cytochrome c most typically is accomplished by immunofluorescence microscopic examination of fixed cells or by Western blot analysis of the cytosolic fractions of lysed cell extracts. Neither of these techniques permits the simultaneous mutiparameter analyses of cell death events in individual cells.

In contrast, a variety of mitochondria-specific potentiometric dyes have permitted the examination of mitochondrial membrane potential ($\Delta\Psi_m$) in intact cells, although complications limit the utility of some of these probes. The behavior of rhodamine 123, for example, is not strictly membrane-potential dependent. In addition to $\Delta\Psi_m$-dependent accumulation in the mitochondrial matrix, rhodamine 123 binds to the inner mitochondrial membrane in a non-Nernstian manner[16] and further associates with mitochondrial ATP-synthase.[17] Another dye, $DiOC_6(3)$, was shown not responsive to several agents that depolarize mitochondria, including valinomycin, suggesting that $DiOC_6(3)$ may be more reliable for estimation of plasma membrane potential rather than mitochondrial membrane potential.[18] Finally, while JC-1[19] is a reliable reporter of $\Delta\Psi_{m,}$[20] its utility in multiparameter cytofluorimetric analyses is limited by the bimodal emission spectrum of JC-1 monomers and aggregates, and uncertainties associated with the redistribution of these alternative conformational states.

We have employed a fluorescently tagged cytochrome c variant, a fusion of cytochrome c with green fluorescent protein (GFP), to examine the release of this protein from mitochondria. The cytochrome *c*-GFP construct was shown to be suitable for the kinetic studies of cytochrome *c* release by confocal microscopy.[13] We have found that selective disruption of the plasma membrane with digitonin permits a convenient assessment of cytochrome *c* release from the mitochondrion. When it is contained within the mitochondrion, the tagged cytochrome *c* molecule is not solubilized by digitonin treatment; in contrast, upon release from the mitochondrion, the now-cytosolic fluorescent molecule is liberated entirely from the cell. Cytofluorimetrically, this reflects a greater than 10-fold diminution of its fluorescent signal.

Simultaneously, digitonin is of great analytical value in the assessment of $\Delta\Psi_{m.}$[21] We have described previously the utility of tetramethyl rhodamine esters (tetramethyl rhodamine ethyl ester [TMRE] and tetramethyl rhodamine methyl ester [TMRM]; ref. 22) in the evaluation of $\Delta\Psi_m$ in multiparameter cytofluorimetric analyses.[1] These membrane-permeable dyes distribute in cells strictly according to the Nernstian potential and provide a direct measure of membrane potential.[23] However, the

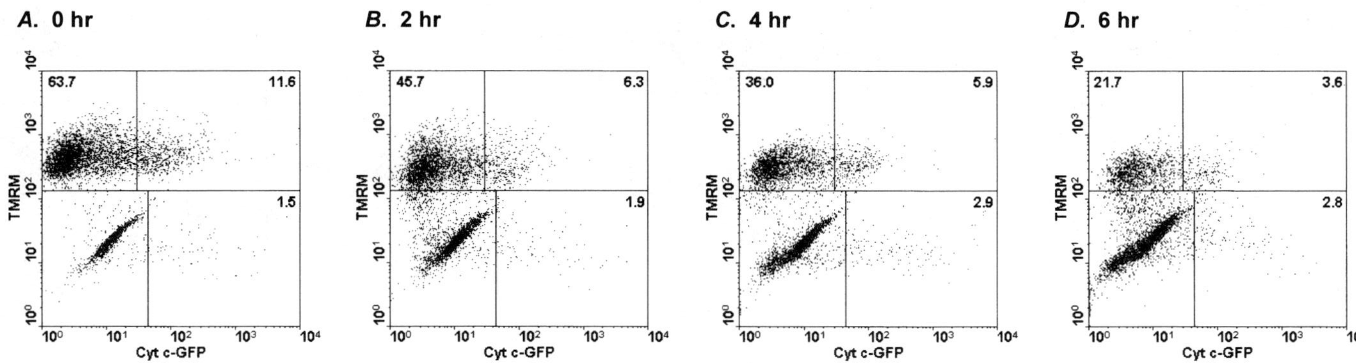

FIGURE 1. Multiparameter assessment of mitochondrial disintegration during staurosporine-induced cell death.The loss of mitochondrial membrane potential ($\Delta\Psi_m$) and release of cytochrome *c*-GFP (Cyt *c*-GFP) were assessed kinetically in transiently transfected HeLa cells following the induction of cell death with staurosporine. $\Delta\Psi_m$ was assessed by staining with tetramethyl rhodamine methyl ester (TMRM); cells were permeabilized with digitonin prior to analysis.

signals obtained in unmanipulated cells are not exclusively mitochondrial; plasma membrane potential also can contribute. In cells in which a plasma membrane potential is retained although $\Delta\Psi_m$ has been dissipated, a TMRM signal that is intermediate between the fully polarized and depolarized extremes is obtained. This plasma membrane effect is eliminated by digitonin treatment.[15]

HeLa cells, previously transfected with the cytochrome *c*-GFP construct, were induced to undergo physiological cell death by treatment with staurosporine; this death response involves mitochondrial events and is inhibitable by Bcl-2.[1] As revealed in FIGURE 1, cytochrome *c*-GFP that is present in the mitochondria of viable cells is not solubilized by digitonin treatment (panel A), but cytochrome *c*-GFP is released during the death process; in the presence of digitonin, this protein is effectively solubilized and lost from the cell (see panels B–D). The kinetics of staurosporine-induced cytochrome *c* release are quantified in FIGURE 2.

The same population of cells was stained with TMRM, and the loss of mitochondrial membrane potential ($\Delta\Psi_m$) was assessed following staurosporine treatment. Again, highly polarized mitochondria in viable cells (panel A) lose their potential rapidly following the induction of death with staurosporine (panels B–D). The kinetics of this aspect of mitochondrial disintegration are tabulated in FIGURE 2. In contrast, HeLa cells dying in response to treatment with TNF-α (a process dependent on the TNF-RI death receptor and uninhibitable by Bcl-2, ref. 1) did not suffer either of these manifestations of mitochondrial disintegration at comparably early times (data not shown).

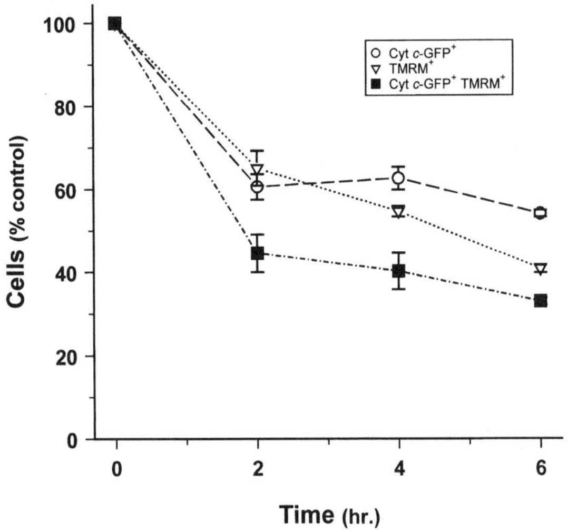

FIGURE 2. Death-associated mitochondrial events are disordered.The kinetics of loss of mitochondrial membrane potential and release of cytochrome *c*-GFP were determined as in FIGURE 1. Three populations were quantified: cells that had not lost mitochondrial membrane potential ($TMRM^+$), those that retained cytochrome *c*-GFP (Cyt *c*-GFP^+), and cells that preserved both markers of intact mitochondria ($TMRM^+$ Cyt *c*-GFP^+). The results are presented as the mean (± SEM) of four independent experiments.

The critical question here is whether the disintegration of the mitochondrion during cell death occurs in an orderly fashion, for example with cytochrome *c* release preceding depolarization, or whether the loss of mitochondrial integrity is cataclysmic. The two markers we have employed to monitor mitochondrial dissolution allow us to address this question. Because the loss of cytochrome *c*-GFP and TMRM signals can be monitored independently, we can ask whether one event precedes the other. Specifically, we could ask whether cells that become cytochrome *c*-GFP$^-$, for example, do so when they are still TMRM$^+$. Technically, because our transiently transfected population is heterogeneous, and always contains cytochrome *c*-GFP$^-$ cells, we have instead formulated the question in terms of the disappearance of the cytochrome *c*-GFP$^+$ cells. If disintegration occurs in a concerted order, the disappearance of the cytochrome *c*-GFP$^+$ TMRM$^+$ (double positive) population should occur at a rate identical with the loss of the first (rate-limiting) marker.

As seen in FIGURE 2, the data do not support such a model. The loss of the double positive population occurs more rapidly than the loss of either marker alone. This suggests that mitochondrial disintegration does not follow a strictly ordered process and is not concerted. We are continuing to refine this analysis, especially utilizing stable (and homogenous) clones expressing cytochrome *c*-GFP. The additional delineation on the single cell level of the activation of apical caspase 9 relative to these mitochondrial events and within the context of the action of Bcl-2 family members will provide essential and definitive resolution in understanding the molecular choreography of the cell death pathway.

ACKNOWLEDGMENTS

We are grateful to Kaisa Heiskanen and Anna-Liisa Nieminen for generously providing pCyt *c*-GFP. This work was supported by grants to D.S.U. from the National Institutes of Health.

REFERENCES

1. HARVEY, K.J., D. LUKOVIC & D.S. UCKER. 2000. Caspase-dependent Cdk activity is a requisite effector of apoptotic death events. J. Cell Biol. **148:** 59–72.
2. KLUCK, R.M., E. BOSSY-WETZEL, D.R. GREEN & D.D. NEWMEYER. 1997. The release of cytochrome c from mitochondria: a primary site for Bcl-2 regulation of apoptosis. Science **275:** 1132–1136.
3. YANG, J., X. LIU, K. BHALLA, C.N. KIM, A.M. IBRADO, J. CAI, T.-I. PENG, D.P. JONES & X. WANG. 1997. Prevention of apoptosis by Bcl-2: release of cytochrome c from mitochondria blocked. Science **275:** 1129–1132.
4. LI, P., D. NIJHAWAN, I. BUDIHARDJO, S. M. SRINIVASULA, M. AHMAD, E. S. ALNEMRI & X. WANG. 1997. Cytochrome c and dATP-dependent formation of Apaf-1/caspase-9 complex initiates an apoptotic protease cascade. Cell **91:** 479–489.
5. CECCONI, F., G. ALVAREZ-BOLADO, B.I. MEYER, K.A. ROTH & P. GRUSS. 1998. Apaf1 (CED-4 homolog) regulates programmed cell death in mammalian development. Cell **94:** 727–737.
6. HAKEM, R., A. HAKEM, G.S. DUNCAN, J.T. HENDERSON, M. WOO, M.S. SOENGAS, A. ELIA, J.L. DE LA POMPA, D. KÄGI, W. KHOO, J. POTTER, R. YOSHIDA, S.A. KAUFMAN, S.W. LOWE, J.M. PENNINGER & T.W. MAK. 1998. Differential requirement for caspase 9 in apoptotic pathways in vivo. Cell **94:** 339–352.

7. KUIDA, K., T.F. HAYDAR, C.Y. KUAN, Y. GU, C. TAYA, H. KARASUYAMA, M.S. SU, P. RAKIC & R.A. FLAVELL. 1998. Reduced apoptosis and cytochrome c-mediated caspase activation in mice lacking caspase 9. Cell **94:** 325–337.
8. YOSHIDA, H., Y.-Y. KONG, R. YOSHIDA, A.J. ELIA, A. HAKEM, R. HAKEM, J.M. PENNINGER & T.W. MAK. 1998. Apaf1 is required for mitochondrial pathways of apoptosis and brain development. Cell **94:** 739–750.
9. HARVEY, K.J., J.F. BLOMQUIST & D.S. UCKER. 1998. Commitment and effector phases of the physiological cell death pathway elucidated with respect to Bcl-2, caspase, and cyclin-dependent kinase activities. Mol. Cell. Biol. **18:** 2912–2922.
10. MARZO, I., C. BRENNER, N. ZAMZAMI, J.M. JÜRGENSMEIER, S.A. SUSIN, H.L. VIEIRA, M.-C. PRÉVOST, Z. XIE, S. MATSUYAMA, J.C. REED & G. KROEMER. 1998. Bax and adenine nucleotide translocator cooperate in the mitochondrial control of apoptosis. Science **281:** 2027–2031.
11. SHIMIZU, S., M. NARITA & Y. TSUJIMOTO. 1999. Bcl-2 family proteins regulate the release of apoptogenic cytochrome *c* by the mitochondrial channel VDAC. Nature **399:** 483–487.
12. GOLDSTEIN, J.C., N.J. WATERHOUSE, P. JUIN, G.I. EVAN & D.R. GREEN. 2000. The coordinate release of cytochrome c during apoptosis is rapid, complete and kinetically invariant. Nat. Cell Biol. **2:** 156–162.
13. HEISKANEN, K.M., M.B. BHAT, H.-W. WANG, J. MA & A.-L. NIEMINEN. 1999. Mitochondrial depolarization accompanies cytochrome *c* release during apoptosis in PC6 cells. J. Biol. Chem. **274:** 5654–5658.
14. FARKAS, D.L., M.-D. WEI, P. FEBBRORIELLO, J.H. CARSON & L.M. LOEW. 1989. Simultaneous imaging of cell and mitochondrial membrane potentials. Biophys. J. **56:** 1053–1069.
15. FLORYK, D. & J. HOUSTEK. 1999. Tetramethyl rhodamine methyl ester (TMRM) is suitable for cytofluorometric measurements of mitochondrial membrane potential in cells treated with digitonin. Biosci. Rep. **19:** 27–34.
16. SALVIOLI, S., A. ARDIZZONI, C. FRANCESCHI & A. COSSARIZZA. 1997. JC-1, but not $DiOC_6(3)$ or rhodamine 123, is a reliable fluorescent probe to assess $\Delta\Psi$ changes in intact cells: implications for studies on mitochondrial functionality during apoptosis. FEBS Lett. **411:** 77–82.
17. O'CONNOR, J.E., J.L. VARGAS, B.F. KIMLER, J. HERNANDEZ-YAGO & S. GRISOLIA. 1988. Use of rhodamine 123 to investigate alterations in mitochondrial activity in isolated mouse liver mitochondria. Biochem. Biophys. Res. Commun. **151:** 568–573.
18. TERASAKI, M., J. SONG, J.R. WONG & L.B. CHEN. 1984. Localization of endoplasmic reticulum in living and glutaraldehyde-fixed cells with fluorescent dyes. Cell **38:** 101–108.
19. REERS, M., T.W. SMITH & L.B. CHEN. 1991. J-aggregate formation of a carbocyanine as a quantitative fluorescent indicator of membrane potential. Biochemistry **30:** 4480-4486.
20. COSSARIZZA, A., M. BACCARANI-CONTRI, G. KALASHNIKOVA & C. FRANCESCHI. 1993. A new method for the cytofluorimetric analysis of mitochondrial membrane potential using the J-aggregate forming lipophilic cation 5,5′,6,6′-tetrachloro-1,1′,3,3′-tetraethylbenzimidazolcarbocyanine iodide (JC-1). Biochem. Biophys. Res. Commun. **197:** 40–45.
21. SCHONFELD, P., P. JEZEK, E.A. BELYAEVA, J. BORECKY, V.S. SLYSHENKOV, M.R. WIECKOWSKI & L. WOJTCZAK. 1996. Photomodification of mitochondrial proteins by azido fatty acids and its effect on mitochondrial energetics: further evidence for the role of the ADP/ATP carrier in fatty-acid-mediated uncoupling. Eur. J. Biochem. **240:** 387–393.
22. EHRENBERG, B., V. MONTANA, M.-D. WEI, J.P. WUSKELL & L.M. LOEW. 1988. Membrane potential can be determined in individual cells from the Nernstian distribution of cationic dyes. Biophys. J. **53:** 785–794.
23. LOEW, L.M., R.A. TUFT, W. CARRINGTON & F.S. FAY. 1993. Imaging in five dimensions: time-dependent membrane potentials in individual mitochondria. Biophys. J. **65:** 2396-2407.

Mitochondrial Control of Cell Death Induced by HIV-1-Encoded Proteins

KARINE F. FERRI,[a,b] ETIENNE JACOTOT,[a,b] JULIÀ BLANCO,[c] JOSÉ A. ESTÉ,[c] AND GUIDO KROEMER[a,d]

[a]*Centre National de la Recherche Scientifique, UMR1599, Institut Gustave Roussy, 39 rue Camille-Desmoulins, F-94805 Villejuif, France*

[c]*Institut de Recerca de la SIDA-Caixa, Laboratori de Retrovirologia, Hospital Universitari Germans Trias i Pujol, Ctra. Canyet s/n, 08916 Badalona, Catalonia, Spain*

ABSTRACT: In most examples of physiological or pathological cell death, mitochondrial membrane permeabilization (MMP) constitutes an early critical event of the lethal process. Signs of MMP that precede nuclear apoptosis include the translocation of cytochrome *c* and apoptosis-inducing factor (AIF) from mitochondria to an extra-mitochondrial localization, as well as the dissipation of the mitochondrial transmembrane potential. MMP also occurs in HIV-1-induced apoptosis. Different HIV-1 encoded proteins (Env, Vpr, Tat, PR) can directly or indirectly trigger MMP, thereby causing cell death. The gp120/gp41 Env complex constitutes an example for an indirect MMP inducer. Env expressed on the plasma membrane of HIV-1 infected (or Env-transfected) cells mediates cell fusion with CD4/CXCR4-expressing uninfected cells. After a cell type-dependent latency period, syncytia then undergo MMP and apoptosis. Vpr exemplifies a direct MMP inducer. Vpr binds to the adenine nucleotide translocator (ANT), a mitochondrial inner membrane protein which also interacts with apoptosis-regulatory proteins from the Bcl-2/Bax family. Binding of Vpr to ANT favors formation of a non-specific pore leading to MMP. The structural motifs of the Vpr protein involved in MMP are conserved among most pathogenic HIV-1 isolates and determine the cytotoxic effect of Vpr. These data suggest the possibility that viruses employ multiple strategies to regulate host cell apoptosis by targeting mitochondria.

KEYWORDS: HIV; AIF, Bcl-2; Caspases; Apoptosis; Cytochrome *c*; gp120; Vpr

INTRODUCTION
MITOCHONDRIAL CONTROL OF APOPTOSIS

Apoptosis research has recently experienced a change from a paradigm in which the nucleus determined the apoptotic process[1] to a paradigm in which caspases[2] and, more recently, mitochondria constitute the center of death control.[3–5] Today, it is almost generally accepted that the mitochondrial membrane constitutes (one of) the battle ground(s) on which opposing molecules determine whether cell death

[b]K.F.F. and E.J. contributed equally to this paper.

[d]Address for correspondence: Dr. Guido Kroemer, 39 rue Camille-Desmoulins, F-94805 Villejuif, France. Voice: 33-1-42 11 60 46; fax: 33-1-42 11 60 47.
kroemer@igr.fr

ensues.[5–10] Several observations support the hypothesis that mitochondria control cell death.

- First, mitochondrial membrane permeabilization (MMP[e]) mostly precedes the signs of advanced apoptosis or necrosis, irrespective of the cell type or the death-inducing stimulus. Mitochondria are organelles with two well-defined compartments, the matrix surrounded by the inner membrane (IM) and the intermembrane space surrounded by the outer membrane (OM). Apoptotic OM permeabilization involves the release of proteins which are normally confined to the intermembrane space of these organelles.[11–17] As compared to the OM, apoptotic IM permeabilization is selective, not resulting in a massive leakage of matrix proteins. IM permeabilization may occur in a step-wise and reversible fashion, with increasing permeability to solutes up to ~1500 D, culminating in the dissipation of the proton gradient responsible for the transmembrane potential ($\Delta\Psi_m$).[18–20]
- Second, MMP has a better predictive value for cell death than other parameters including caspase activation. As an example, crosslinking of the CXCR4 surface receptor or transfection-enforced overexpression of the tumor suppressor gene PML induce apoptosis with MMP but without detectable caspase activation.[21,22] Even if inhibition of caspases blocks some of the morphological and biochemical manifestations of apoptosis, it fails to prevent MMP and cell death induced via a variety of receptors (e.g., glucocorticoid receptor, CD2, CD4, CD45, CD47, CD99, MHC class II), anti-oncogenes (e.g., *Bax, Bak*), chemotherapeutic agents, irradiation, and pro-apoptotic second messengers (e.g., reactive oxygen species, nitric oxide, ceramide, ganglioside GD3) (reviewed in refs. 8, 23). Caspase activation can even be required for the transmission of mitogenic signals[24,25] and thus may be uncoupled from cell death.
- Third, an increasing number of pro-apoptotic effectors directly induce MMP when added to purified mitochondria *in vitro*. This applies to pro-apoptotic members of the Bcl-2 family such as Bax,[26] pro-apoptotic signal transducing molecules such as Ca^{2+}, ceramide, ganglioside GD3, fatty acids and their oxidation products, reactive oxygen species, nitric oxide, and cytotoxic agents including arsenite, betulinic acid, CD437, lonidamine, and photosensitizing porphyrin derivatives (reviewed in refs. 5, 23).
- Fourth, anti-apoptotic members of the Bcl-2 family interact with mitochondrial membranes or membrane proteins and inhibit cell death by virtue of their capacity to prevent MMP.[11,27] Similarly, inhibition of mitochondrial membrane permeabilization by specific pharmacological interventions (that is by agents acting on mitochondrial proteins) prevents or retards cell death.

[e]**ABBREVIATIONS:** AIF, apoptosis inducing factor; ANT, adenine nucleotide translocator; COX, cytochrome *c* oxidase; Cyt.c, cytochrome *c*; $\Delta\Psi_m$, mitochondrial transmembrane potential; Env, envelope glycoprotein complex; HIV, human immunodeficiency virus; MMP, mitochondrial membrane permeabilization; PBL, peripheral blood lymphocytes; Z-VAD.fmk, N-benzyloxycarbonyl-Val-Ala-Asp-fluoromethylketone.

This applies to N-methyl-4-valine-cyclosporin A (a ligand of the mitochondrial cyclophilin D),[28] bongkrekic acid (a ligand of the adenine nucleotide translocator),[29] and, in some experimental systems, oligomycin (an inhibitor of the F1 ATPase).[30]

- Fifth, cell-free systems have identified soluble intermembrane proteins (SIMPs) from mitochondria that are rate-limiting for the activation of catabolic hydrolases, mainly caspases and nucleases. The heme protein cytochrome *c*, for example, triggers the assembly of a caspase-9/caspase-3 activation complex, the apoptosome, in the cytosol.[31] Released mitochondrial hsp10 and hsp60 may also facilitate caspase activation.[32] Certain caspases can be found sequestered within mitochondria and, when liberated, may contribute to setting off caspase activation cascades.[14,15] Some SIMPs stimulate caspase-independent catabolic events. Once released from mitochondria, for instance, "apoptosis-inducing factor" (AIF), a flavoprotein oxidoreductase, is imported into nuclei where it stimulates large scale DNA fragmentation to ~50 kbp and peripheral chromatin condensation.[13,17] Additional catabolic enzymes released from mitochondria include arginase, sulfite oxidase, glycine cleavage system h protein, and soluble epoxide hydrolase.[33]

Altogether, these observations suggest a three-step model of apoptosis: a premitochondrial phase during which signal transduction cascades or damage pathways are activated (initiation phase); a mitochondrial phase, during which mitochondrial membranes are permeabilized (decision/effector phase); and a post-mitochondrial phase during which proteins released from mitochondria cause the activation of catabolic enzymes (degradation phase). This scheme, initially developed by our group,[3,4,34] has now been adopted by many of the leading scientists in the field.[7,9,10,35]

IMPACT OF HIV-1 INFECTION ON MITOCHONDRIA *IN VIVO* AND *IN CELLULA*

Peripheral blood lymphocytes (PBL) from HIV-1 infected patients are notoriously prone to spontaneously undergo apoptosis upon *in vitro* culture.[36] Apoptosis mainly affects cells in which the endogenous MMP inhibitor Bcl-2 has been down-regulated.[36] Interleukin-2 and interleukin-15, which up-regulate Bcl-2, inhibit spontaneous apoptosis of PBL.[37,38] Both $CD4^+$ and $CD8^+$ cells from the peripheral blood of HIV-1 infected individuals exhibit an increased susceptibility to lose the $\Delta\Psi_m$ *ex vivo*, before they manifest signs of nuclear apoptosis.[39,40] This latter result has been confirmed for peripheral blood lymphocytes from patients shortly after primary infection.[41] PBL from chronically HIV-1 infected donors also have an enhanced capacity to produce superoxide anion[39,40,42] in a reaction that is suppressed by the respiratory chain inhibitor rotenone,[40] suggesting that superoxide anion is generated in mitochondria. Moreover, in freshly isolated lymph node cells from HIV-1-infected donors, mitochondrial swelling has been detected by electron microscopy *ex vivo*.[43] These results suggest that cell death induced by HIV-1 is associated with MMP.

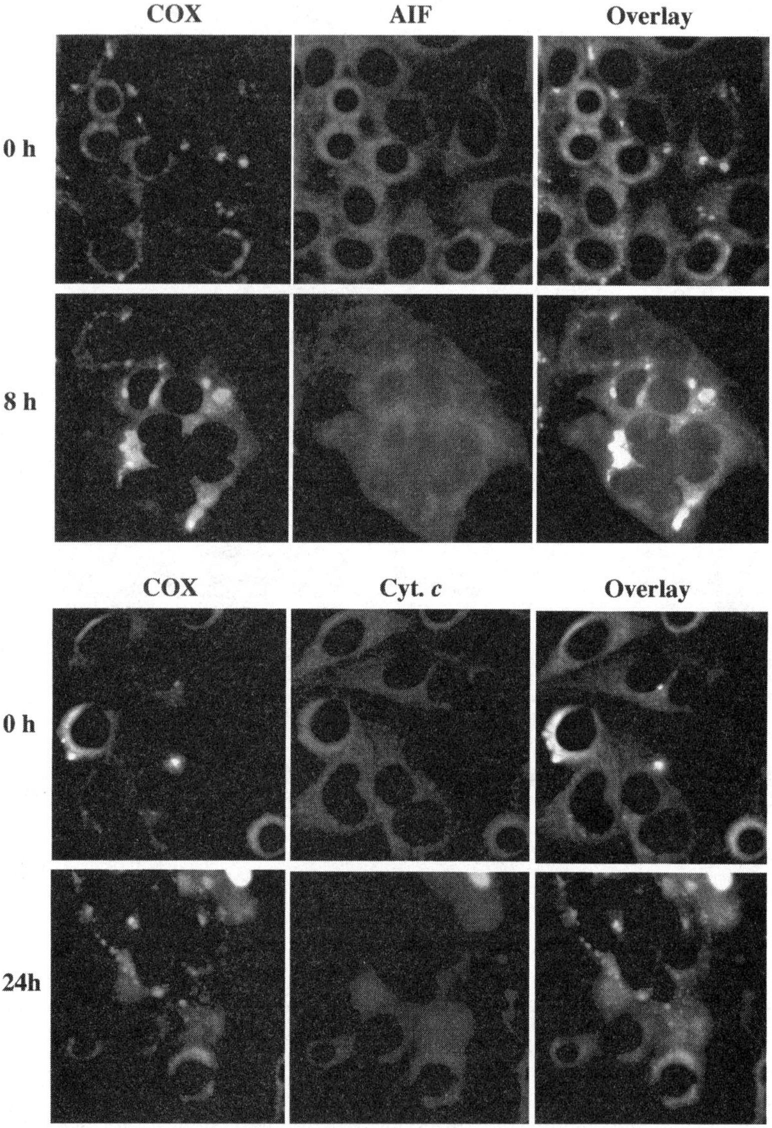

FIGURE 1. Translocation of AIF translocation and cytochrome *c* in HIV-1 infected cells. CD4-expressing HeLa cells were either cultured alone (Co.) or co-cultured with chronically HIV-1-infected H9/IIIB cells at a 2:1 ratio. At the indicated time, cells were fixed and permeabilized to assess the subcellular localization of AIF (*upper panels*) or that of cytochrome *c* (*lower panels*), detected by fluoresceine isothiocyanate-labeled antibodies, in relationship to mitochondria (*both upper and lower panels*), detected with phycoerythrin-conjugated antibodies directed against cytochrome *c* oxidase, COX, a sessile inner membrane protein. Note the punctate staining pattern for AIF and cytochrome *c* in controls (overlapping with the staining pattern of COX) and the more diffuse pattern in HIV-1–induced syncytia. Note also the nuclear AIF staining pattern of syncytia.

A further hint to a general mitochondrial effect of HIV-1 has been obtained by treating transgenic mice expressing replication-incompetent HIV-1 (NL4-3 Δ gag/pol) with 3′-azido-2′,3′-deoxythymidine (AZT).[44] AZT, as well as analogous anti-retroviral nucleosides, negatively affect mitochondrial DNA replication and can provoke a persistent mitochondrial myopathy in AIDS patients, especially in infants.[45] In mice, transgenic HIV sensitizes to AZT-induced myocardial mitochondriopathy,[44] suggesting direct or indirect HIV-mediated effects on mitochondria. A similar result was obtained when AZT was administered to mice expressing the HIV-1 one-exon-encoded 72 amino acid Tat protein and full-length 86 amino acid Tat proteins. In Tat-expressing mice, AZT-treatment caused a greatly enhanced suppression of the mitochondrial isoform of superoxide dismutase (Mn-SOD or SOD-2), when compared to control mice. In addition, Tat exacerbated signs of AZT-induced (presumably mitochondrion-generated) oxidative stress.[46] Thus, HIV-1 and specifically Tat can affect mitochondrial metabolism *in vivo*, in the intact organism.

When T cells are infected with HIV-1 *in vitro*, they manifest mitochondrial swelling[47,48] and lose the expression of several nucleus-encoded mRNA species coding for mitochondrial proteins.[49,50] In addition, HIV-1 downregulates the cytochrome *b* mRNA from the mitochondrial genome.[50] Downregulation of mitochondrial DNA transcription is also a distinctive feature of apoptosis.[51] Cytopathogenic HIV-1 infected cells overproduce superoxide anion in mitochondria[52] and manifest a decline in the $\Delta\Psi_m$.[53] CD4-expressing HeLa cells co-cultured with a lymphoid cell lines chronically infected with a syncytium-inducing HIV-1 isolate exhibit the release of AIF and cytochrome *c* from mitochondria clearly before nuclear apoptosis occurs.[54] (FIG. 1). In U937 or Jurkat cells, overexpression of Bcl-2 or Bcl-X_L decreases HIV-1-induced apoptosis.[55,56] Bcl-2 overexpression has also been reported to suppress syncytial apoptosis in HIV-1 infected CD4 T cell lines.[57] These data suggest that HIV-1 affects mitochondrial function and causes a type of apoptosis which, at least in some cell types, depends on the (Bcl-2/Bcl-X_L inhibitable) MMP.

EFFECT OF INDIVIDUAL HIV-1-ENCODED PROTEINS ON MITOCHONDRIA

HIV-1 infection induces apoptosis via a cornucopia of mechanisms, and there is considerable debate about which viral protein is the dominant inducer, either in HIV-1 infected cells or in non-infected "bystander" cells. Extracellular addition of several HIV-1 encoded proteins to cultured cells, as well as transient transfection of cells with the corresponding genes, induces apoptosis. This applies to Tat, Vpr, Nef, and Env. In contrast, cell lines stably transfected with either Vpr or Tat may become resistant to exogenous apoptosis inducers, a fact that may be ascribed to an adaptive response and/or selection of resistant clones. In this section, we will briefly discuss how Env, Vpr, Tat, and PR induce MMP and cell death.

Env

Env and its soluble product gp120 have been thought for long to be the principal HIV-1 encoded killer protein. Three mechanisms have been proposed for the action of Env. First, Env-mediated interaction of virions with the cell surface has been suggested to perturb ion fluxes of the plasma membrane, leading to ballooning of the cell and its necrotic demise.[58] Second, interaction of the gp120/gp41 envelope complex, present on the membrane of HIV-infected cells, with the CD4/CXCR4 complex of uninfected cells, could mediate a cell type-dependent pro-apoptotic signal transduction process. Thus, cross-linking of CD4 or CXCR4 causes a rapid $\Delta\Psi_m$ disruption and apoptosis in lymphocytes.[21] gp120-mediated cross-linking of CD4 causes a decrease in Bcl-2 expression in T cells.[59] Soluble gp120 also triggers calcium mobilization, ATP depletion, decline in $\Delta\Psi_m$, and neurotoxicity in fetal monolayer cultures from a number of brain regions.[60] Third, interaction of the gp120/gp41 complex present on HIV-infected cells with the CD4/CXCR4 complex of uninfected cells can result in cell-to-cell fusion.[61] Such syncytia then undergo apoptosis after a lag phase which depends on the cell type involved in the fusion process.[54] It remains an open possibility that the nonphysiological formation of syncytia by itself, rather than acute signals mediated via gp120/gp41 in interaction, accounts for cell death. Syncytia arising from the fusion of HeLa cells expressing the HIV-1 encoded Env with cells expressing the CD4/CXCR4 complex spontaneously undergo apoptosis in a process which is accompanied by MMP.[54] Caspase inhibition does not suppress the AIF- and cytochrome *c*–translocation, yet prevents all signs of nuclear apoptosis. Bcl-2-mediated inhibition of MMP also prevents the subsequent nuclear chromatin condensation and DNA fragmentation. This suggests that MMP occurs upstream of caspase activation in syncytial apoptosis. Moreover, we have found that the release of AIF occurs before that of cytochrome *c* and before caspase activation.[54] Microinjection of AIF into syncytia suffices to trigger rapid, caspase-independent cytochrome *c* release. Neutralization of enogenous AIF by injection of an AIF-specific antibody prevents all signs of spontaneous apoptosis occurring in syncytia, including the mitochondrial cytochrome *c* release and nuclear apoptosis. In contrast, cytochrome *c* neutralization prevents only nuclear apoptosis and does not affect the release of AIF. Taken together, our results establish that the following molecular sequence governs apoptosis of Env-induced syncytia: Bax-mediated/Bcl-2-inhibited MMP → AIF release → cytochrome *c* release → caspase activation → nuclear apoptosis.[54]

Vpr

If T cells are infected with vesicular stomatitis virus envelope G (VSV-G) glycoprotein-pseudotyped HIV-1 *in vitro*, Vpr becomes rate-limiting for cell killing.[62–64] The 14 kD Vpr protein (96 amino acids) kills lymphocytes,[63] monocytes,[62] and neurons,[65] either upon infection with *vpr* positive HIV-1 isolates[62–64] or upon extracellular addition of the Vpr protein.[65,66] Virion-associated Vpr may cause apoptosis in the absence of viral replication.[67]

We have discovered recently[68] that synthetic Vpr, when added to intact cells or to purified mitochondria, causes a rapid dissipation of the mitochondrial transmembrane potential ($\Delta\Psi_m$), as well as the mitochondrial release of cytochrome *c* and AIF. This effect is mimicked by the C-terminal moiety of the molecule (Vpr52–96), not by its N-terminal moiety (Vpr1-51), and is inhibited by overexpression of Bcl-2. Vpr52–96 also causes a rapid $\Delta\Psi_m$ loss when added to intact cells, and this effect precedes the induction of apoptotic nuclear chromatin condensation (FIG. 2). If add-

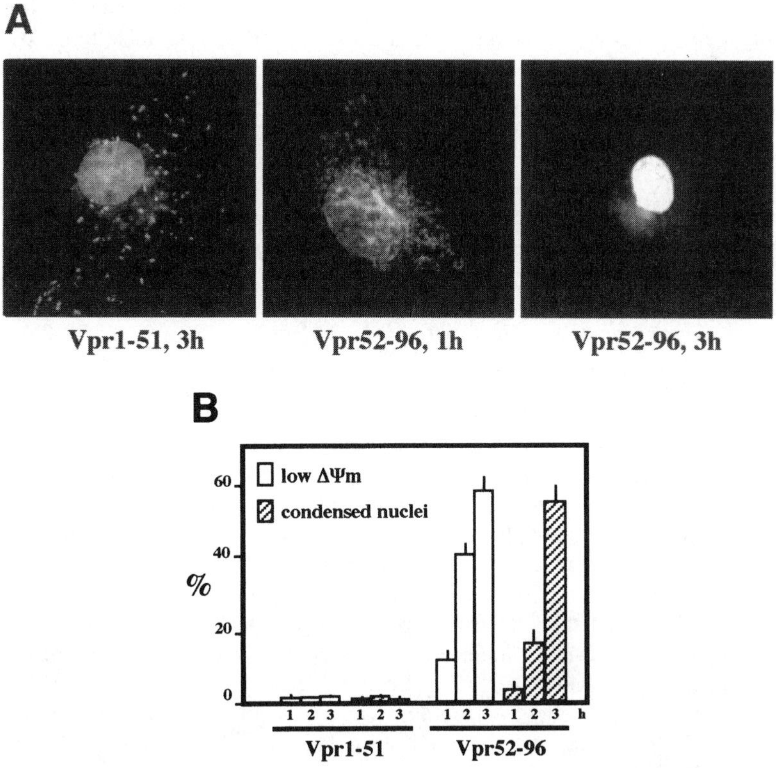

FIGURE 2. Induction of mitochondrial and nuclear signs of apoptosis by Vpr. COS cells were cultured in the presence of 1 μM Vpr52–96 (or as a negative control Vpr1-51) for 1 or 3 h and stained during the last 30 min of culture with Hoechst 33324 (which detects nuclear chromatin condensation: blue fluorescence) and the $\Delta\Psi_m$ sensitive dye JC-1 (which marks mitochondria with a high $\Delta\Psi_m$ in red, with Vpr1-51 at 3 h, and mitochondria with a low $\Delta\Psi_m$ in green, with Vpr52–96 at 1 h). Note that the dissipation of the $\Delta\Psi_m$ is detectable well before nuclear apoptosis becomes visible (at 3 h with Vpr52–96). Representative images are shown in **A**, and the kinetics of the mitochondrial and nuclear signs of apoptosis (X ± SEM, n = 3) are shown in **B**.

ed to lymphocytes or COS cells, a substantial fraction of FITC-labeled Vpr52–96 binds to mitochondria.[68] A biotinylated Vpr-derived peptide (Vpr52–96) may be employed as a bait to specifically purify a mitochondrial molecular complex containing ANT. Surface plasmon resonance indicates that the Vpr C-terminus binds purified ANT with an affinity of $7.4 \times 10^8\ M^{-1}$. We have found that Vpr favors the permeabilization of proteoliposomes containing ANT. This effect is enhanced by addition of recombinant Bax and prevented by recombinant Bcl-2. In addition, *Saccharomyces cerevisiae* strains lacking ANT are less susceptible to Vpr52–96-induced killing than control cells, yet recover Vpr52–96 sensitivity when retransfected with ANT. Thus, Vpr kills cells via a direct effect on the mitochondrial membranes and in particular on ANT.[68] The same structural motifs relevant for cell killing are responsible for the mitochondriotoxic effects of Vpr. Substitution of critical Arg residues (R73, R80) in the mitochondriotoxic domain (aa 71–82) strongly diminishes both the mitochondrial and the cytotoxic effects of Vpr.[68] These observations correlate with the facts that R80 mutations reduce cell killing by VSV-G pseudotyped HIV-1 *in vitro*[63] and that R73 and R80 are extremely conserved among pathogenic HIV-1 isolates. Intriguingly, the same domain that we have implied in the apoptogenic activity of Vpr is also critical for the G2 arrest of the cell cycle induced by Vpr.[63] Since G2 arrest and apoptosis induction by Vpr have similar structural requirements, it is tempting to speculate that both Vpr effects (cytotoxicity and cell cycle arrest) may be functionally interconnected at the level of mitochondria.

Tat

Cells stably transfected with Tat are more susceptible than control cells to undergo apoptosis upon serum withdrawal. According to one study, Tat causes a decrease in Bcl-2 expression and an increase in Bax expression.[69] Another study reports that Tat translocates to mitochondria concomitantly with the $\Delta\Psi_m$ loss.[70] Addition of extracellular Tat causes apoptosis, at least in some cell types.[71] Tat down-regulates the mitochondrial isoform of superoxide dismutase (SOD2), either at the post-transcriptional[72] or transcriptional levels.[73] Down-regulation of SOD2 is thought to sensitize cells to the lethal effects of reactive oxygen species produced in mitochondria. Exogenous Tat can also induce the expression of pro-apoptotic proteins. Such Tat-induced proteins include CD95 (in T cells)[74,75] and Par-4 (in cultured hippocampal neurons), which induces apoptosis via a mitochondrion-dependent mechanism.[76,77] These results suggest multiple connections between Tat-induced cell death and MMP.

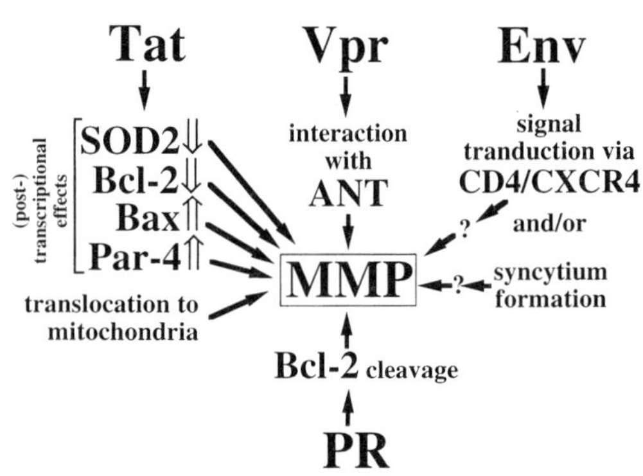

FIGURE 3. Indirect and direct effects of HIV-1 encoded proteins on mitochondria. Tat, Vpr, Env, and PR may induce mitochondrial membrane permeabilization (MMP) via a variety of different mechanisms. For details and references consult main text.

PR

Transfection-enforced overexpression of HIV-1-encoded protease (PR), a cysteine protease essential for viral replication, can induce cell death.[78] According to one report, cell death is preceded by PR-mediated cleavage of the anti-apoptotic, MMP-inhibitory protein Bcl-2 between Phe112 and Ala113.[79] Clipping of the BH4 domain may be expected to neutralize Bcl-2 or to transform it into a pro-apoptotic, MMP-inducing protein, as this has been demonstrated for a cleavage event caused by caspase-3.[80,81]

Based on the results summarized above, it appears that HIV-1 uses several independent strategies to induce MMP and apoptosis (FIG. 3). However, it remains an ongoing conundrum whether these manifold strategies are designed to cooperate among each other in an additive or synergistic fashion, in the same cell.

VIRAL PROTEINS ACTING ON MITOCHONDRIA: A GENERAL STRATEGY OF APOPTOSIS MODULATION?

Why, in teleological terms, do viruses regulate apoptosis? It appears obvious that, especially at the beginning of the viral life cycle, before lysis, the virus must aim at preventing its host cells from dying. However, the virus may trigger apoptosis at late stages of the viral life cycle, either to facilitate viral spreading to adjacent phagocytic cells (which recognize and engulf the dying cell) or to hijack the phagocytic system (which may be overwhelmed by apoptotic bodies). In the particular case that the infected host cell belongs to the non-specific defense system (e.g., phagocytes) or to the immune system, the virus may employ a strategy to induce rapid cell death shortly after infection. In other words, the virus could kill antiviral cells and productively infect (or persist in) other cells, not involved in antiviral defense. Here, we will briefly discuss the current literature on mitochondrial effects mediated by viruses other than HIV-1.

Viral Apoptosis Inhibitors Acting on Mitochondria

Recent studies have unravelled the existence of several viral apoptosis inhibitors acting on mitochondria. Several pathogenic viruses produce apoptosis inhibitory (and presumably MMP-inhibitory) Bcl-2 analogues (reviewed in ref. 34): adenovirus (protein: E1B19K), african swine fever virus (5-HL/A179L), herpesvirus saimiri (HVS-Bcl-2), Kaposi sarcoma–associated herpes virus 8 (Ksbcl-2), and murine gammaherpesvirus-68 (M11). Epstein-Barr virus even encodes two structural and function homologues of Bcl-2: BHRF1 and BALF1. Among these proteins, HVS-Bcl-2 and 5-HL/A179L have been described to inhibit MMP, much as Bcl-2 does. According to our working hypothesis, Bcl-2 prevents MMP via its interaction with the adenine nucleotide translocator (ANT).[82] It is tempting to speculate that viral Bcl-2 analogues inhibit MMP via a similar mechanism. A cytomegalovirus inhibitor of apoptosis (CIA), without any obvious sequence homology to Bcl-2, has been found to bind to ANT,[83] suggesting that it acts as a functional Bcl-2 homologue. Another mechanism of apoptosis inhibition may be the neutralization of pro-apoptotic proteins from the Bcl-2 family. As an example, the large T antigen from simian virus 40 (SV40) binds a pro-apoptotic member of the Bcl-2 family, p193.[84]

Viral Apoptosis Inducers Acting on Mitochondria

An influenza virus B protein (PB2) has been the first viral protein to be shown to interact with mitochondria via its N-terminus.[85] This is an interesting finding because influenza virus B is one of major causative agents of Reye's syndrome, an

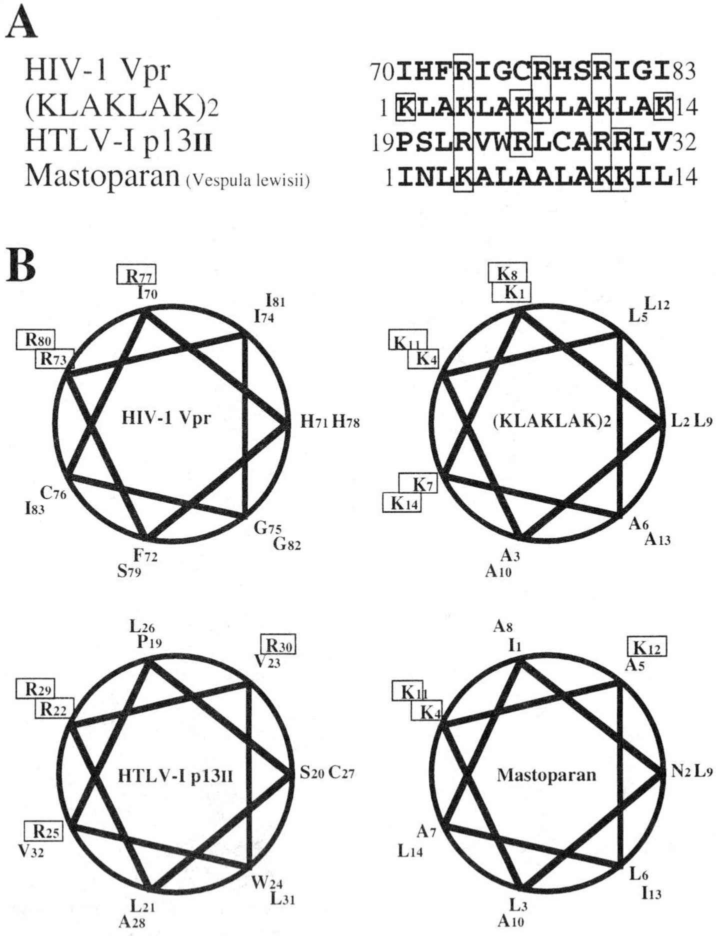

FIGURE 4. Comparison of the mitochondriotoxic domains of four proteins or peptides known to induce MMP. The primary structure is shown in **A** and the helical wheel presentation in **B**. Positively charged residues are boxed. Numbers refer to the position of each amino acid in the entire protein or peptide. Note the common amphipathic distribution of residues in each of the α-helices.

acute mitochondriopathy associated with massive cell death.[19] However, the exact effects of this influenza virus protein have not been determined in molecular terms. More recently, it has been found that the p13 (II) protein derived from the X-II ORF of HTLV-1 is targeted to mitochondria, where it induces a dissipation of the $\Delta\Psi_m$ and matrix swelling.[86] Mitochondrial targeting of this protein has been mapped to a decapeptide sequence which contains several Arg residues that are asymmetrically distributed in an α-helix and which resemble the mitochondriotoxic motif of HIV-1–encoded Vpr (FIG. 4). In contrast with Vpr, however, Arg→Ala substitutions did not abolish the mitochondrial targeting of p13.[86] Hepatitis B virus X protein (HBV-X) is a potent apoptosis inducer whose effect is neutralized by Bcl-2 overexpression.[87] HBV-X protein colocalizes to mitochondria with a human voltage-dependent anion channel (VDAC) isoform, HVDAC3, and alters its transmembrane potential.[88] Thus, HBV-X may act in a similar fashion as it has been suggested for Bax, a Bcl-2 antagonist that may cause VDAC to form a large protein-permeant channel.[89] Intriguingly VDAC can interact with ANT,[90] suggesting that HBV-X targets the same molecular complex as Vpr, CIA, and viral Bcl-2 analogues.

The above examples clearly illustrate that viruses may target different mitochondrial proteins to induce or prevent apoptotic MMP. Clearly, our current information on the submolecular mode of action of these apoptosis regulators is scarce. Future investigation will identify additional virus-encoded MMP regulators and determine to which extent mitochondrial targeting of viral proteins impinges on the viral life cycle. Irrespective of these general considerations, it becomes clear, however, that mitochondria may constitute a specific target of HIV-1 mediated cytopathogenicity.

ACKNOWLEDGMENTS

Work reported here was supported by a special grant by the Ligue Nationale contre le Cancer, as well as by grants from ANRS, FRM, European Commission, the Picasso Program (to G.K), Fundació IrsiCaixa (to J.A.E), and FIS 00/0893 (J.B.). K.F.F. receives a fellowship from the French Ministry of Science; E.J. receives an ANRS fellowship. J.B. is a researcher from the "Fundació per a la Recerca Biomèdica Germans Trias i Pujol".

REFERENCES

1. ANRENDS, M.J., R.G. MORRIS & A.H. WYLLIE. 1990. Apoptosis: the role of the endonuclease. Am. J. Pathol. **136:** 593–608.
2. MARTIN, S.J. & D.R. GREEN. 1995. Protease activation during apoptosis: death by a thousand cuts? Cell **82:** 349–352.
3. KROEMER, G., P.X. PETIT, N. ZAMZAMI, J.-L. VAYSSIÈRE & B. MIGNOTTE. 1995. The biochemistry of apoptosis. FASEB J. **9:** 1277–1287.
4. KROEMER, G., N. ZAMZAMI & S.A. SUSIN. 1997. Mitochondrial control of apoptosis. Immunol. Today **18:** 44–51.
5. KROEMER, G. & J.C. REED. 2000. Mitochondrial control of cell death. Nat. Med. **6:** 513–519.
6. KROEMER, G., B. DALLAPORTA & M. RESCHE-RIGON. 1998. The mitochondrial death/life regulator in apoptosis and necrosis. Annu. Rev. Physiol. **60:** 619–642.

7. Green, D.R. & J.C. Reed. 1998. Mitochondria and apoptosis. Science **281:** 1309–1312.
8. Jacotot, E., P. Costantini, E. Labourau, N. Zamzami, S.A. Susin & G. Kroemer. 1999. Mitochondrial membrane permeabilization during the apoptotic process. Ann. N.Y. Acad. Sci. **887:** 18–30.
9. Gross, A., J.M. McDonnell & S.J. Korsmeyer. 1999. Bcl-2 family members and the mitochondria in apoptosis. Genes Devel. **13:** 1988–1911.
10. Vander Heiden, M.G. & C.B. Thompson. 1999. Bcl-2 proteins: Inhibitors of apoptosis or regulators of mitochondrial homeostasis? Nat. Cell Biol. **1:** E209–E216.
11. Susin, S.A., N. Zamzami, M. Castedo, T. Hirsch, P. Marchetti, A. Macho, E. Daugas, M. Geuskens & G. Kroemer. 1996. Bcl-2 inhibits the mitochondrial release of an apoptogenic protease. J. Exp. Med. **184:** 1331–1342.
12. Liu, X.S., C.N. Kim, J. Yang, R. Jemmerson & X. Wang. 1996. Induction of apoptotic program in cell-free extracts: requirement for dATP and cytochrome C. Cell **86:** 147–157.
13. Susin, S.A., H.K. Lorenzo, N. Zamzami, I. Marzo, B.E. Snow, G.M. Brothers, J. Mangion, E. Jacotot, P. Costantini, M. Loeffler, N. Larochette, D.R. Goodlett, R. Aebersold, D.P. Siderovski, J.M. Penninger & G. Kroemer. 1999. Molecular characterization of mitochondrial apoptosis-inducing factor. Nature **397:** 441–446.
14. Krajewski, S., M. Krajewska, L.M. Ellerby, K. Welsh, Z.H. Xie, Q. Deveraux, G.S. Salvesen, D.E. Bredesen, R.E. Rosenthal, G. Fiskum & J.C. Reed. 1999. Release of caspase-9 from mitochondria during neuronal apoptosis and cerebral ischemia. Proc. Natl. Acad. Sci. USA **96:** 5752–5757.
15. Susin, S.A., H.K. Lorenzo, N. Zamzami, I. Marzo, N. Larochette, P.M. Alzari & G. Kroemer. 1999. Mitochondrial release of caspases-2 and -9 during the apoptotic process. J. Exp. Med. **189:** 381–394.
16. Goldstein, J.C., N.J. Waterhouse, P. Juin, G.I. Evan & D.R. Green. 2000. The coordinate release of cytochrome c is rapid, complete and kinetically invariant. Nat. Cell Biol. **2:** 156–162.
17. Daugas, E., S.A. Susin, N. Zamzami, K. Ferri, T. Irinopoulos, N. Larochette, M.C. Prevost, B. Leber, D. Andrews, J. Penninger & G. Kroemer. 2000. Mitochondrio-nuclear redistribution of AIF in apoptosis and necrosis. FASEB J. **14:** 729–739.
18. Zamzami, N., P. Marchetti, M. Castedo, C. Zanin, J.-L. Vayssière, P.X. Petit & G. Kroemer. 1995. Reduction in mitochondrial potential constitutes an early irreversible step of programmed lymphocyte death in vivo. J. Exp. Med. **181:** 1661–1672.
19. Lemasters, J.J., A.-L. Nieminen, T. Qjan, L.C. Trost, S.P. Elmore, Y. Nishimura, R.A. Crowe, W.E. Cascio, C.A. Bradham, D.A. Brenner & B. Herman. 1998. The mitochondrial permeability transition in cell death: a common mechanism in necrosis, apoptosis and autophagy. Biochim. Biophys. Acta **1366:** 177–196.
20. Bernardi, P., L. Scorrano, R. Colonna, V. Petronilli & F. Di Lisa. 1999. Mitochondria and cell death—mechanistic aspects and methodological issues. Eur. J. Biochem. **264:** 687–701.
21. Berndt, C., B. Möpps, S. Angermüller, P. Gierschik & P.H. Krammer. 1998. CXCR4 and CD4 mediate a rapid CD95-independent cell death in $CD4^+$ cells. Proc. Natl. Acad. Sci. USA **95:** 12556–12561.
22. Quignon, F., F. DeBels, M. Koken, J. Feunteun, J.C. Ameisen & H. de Thé. 1998. PML induces a novel caspase-independent death process. Nat. Gen. **20:** 259–265.
23. Costantini, P., E. Jacotot, D. Decaudin & G. Kroemer. 2000. The mitochondrion as a novel target of anti-cancer chemotherapy. J. Natl. Cancer Inst. **92:** 1042–1052.
24. Alam, A., L.Y. Cohen, S. Aouad & R.P. Sekaly. 1999. Early activation of caspases during T lymphocyte stimulation results in selective substrate cleavage in nonapoptotic cells. J. Exp. Med. **190:** 1879–1890.
25. Kennedy, N.J., T. Kataoka, J. Tschopp & R.C. Budd. 1999. Caspase activation is required for T cell proliferation. J. Exp. Med. **190:** 1891–1895.

26. MARZO, I., C. BRENNER, N. ZAMZAMI, J. JÜRGENSMEIER, S.A. SUSIN, H.L.A. VIEIRA, M.-C. PRÉVOST, Z. XIE, S. MUTSIYAMA, J.C. REED & G. KROEMER. 1998. Bax and adenine nucleotide translocator cooperate in the mitochondrial control of apoptosis. Science **281:** 2027–2031.

27. KLUCK, R.M., E. BOSSY-WETZEL, D.R. GREEN & D.D. NEWMEYER. 1997. The release of cytochrome c from mitochondria: a primary site for Bcl-2 regulation of apoptosis. Science **275:** 1132–1136.

28. ZAMZAMI, N., P. MARCHETTI, M. CASTEDO, D. DECAUDIN, A. MACHO, T. HIRSCH, S.A. SUSIN, P.X. PETIT, B. MIGNOTTE & G. KROEMER. 1995. Sequential reduction of mitochondrial transmembrane potential and generation of reactive oxygen species in early programmed cell death. J. Exp. Med. **182:** 367–377.

29. MARCHETTI, P., M. CASTEDO, S.A. SUSIN, N. ZAMZAMI, T. HIRSCH, A. HAEFFNER, F. HIRSCH, M. GEUSKENS & G. KROEMER. 1996. Mitochondrial permeability transition is a central coordinating event of apoptosis. J. Exp. Med. **184:** 1155–1160.

30. MATSUYAMA, S., Q. XU, J. VELOURS & J.C. REED. 1998. Mitochondrial F0F1-ATPase proton pump is required for function of proapoptotic protein Bax in yeast and mammalian cells. Mol. Cell **1:** 327–336.

31. BUDIJARDJO, I., H. OLIVER, M. LUTTER, X. LUO & X. WANG. 1999. Biochemical pathways of caspase activation during apoptosis. Annu. Rev. Cell Dev. Biol. **15:** 269–290.

32. SAMALI, A., J. CAI, B. ZHIVOTOVSKY, D.P. JONES & S. ORRENIUS. 1999. Presence of a pre-apoptotic complex of pro-caspase-3, hsp60, and hsp10 in the mitochondrial fraction of Jurkat cells. EMBO J. **18:** 2040–2048.

33. PATTERSON, S., C.S. SPAHR, E. DAUGAS, S.A. SUSIN, T. IRINOPOULOS, C. KOEHLER & G. KROEMER. 2000. Mass spectrometric identification of proteins released from mitochondria undergoing permeability transition. Cell Death Differ. **7:** 137–144.

34. KROEMER, G. 1997. The proto-oncogene Bcl-2 and its role in regulating apoptosis. Nature Med. **3:** 614–620.

35. PETER, M.E. & P.H. KRAMMER. 1998. Mechanisms of CD95 (APO-1/Fas)-mediated apoptosis. Curr. Op. Immunol. **10:** 545–551.

36. GOUGEON, M.L. & L. MONTAGNIER. 1999. Programmed cell death as a mechanism of CD4 and CD8 T cell depletion in AIDS—molecular control and effect of highly active anti-retroviral therapy. Ann. N.Y. Acad. Sci. **887:** 199–212.

37. REGAMEY, N., T. HARR, M. BATTEGAY & P. ERB. 1999. Downregulation of Bcl-2, but not of Bax or Bcl-x, is associated with T lymphocyte apoptosis in HIV infection and restored by antiretroviral therapy or by interleukin 2. AIDS Res. Hum. Retroviruses **15:** 803–810.

38. NAORA, H. & M.L. GOUGEON. 1999. Interleukin-15 is a potent survival factor in the prevention of spontaneous but not CD95-induced apoptosis in CD4 and CD8 T lymphocytes of HIV-infected individuals. Correlation with its ability to increase Bcl-2 expression. Cell Death Differ. **6:** 1002–1011.

39. CASTEDO, M., A. MACHO, N. ZAMZAMI, T. HIRSCH, P. MARCHETTI, J. URIEL & G. KROEMER. 1995. Mitochondrial perturbations define lymphocytes undergoing apoptotic depletion in vivo. Eur. J. Immunol. **25:** 3277–3284.

40. MACHO, A., M. CASTEDO, P. MARCHETTI, J.J. AGUILAR, D. DECAUDIN, N. ZAMZAMI, P.M. GIRARD, J. URIEL & G. KROEMER. 1995. Mitochondrial dysfunctions in circulating T lymphocytes from human immunodeficiency virus-1 carriers. Blood **86:** 2481–2487.

41. COSSARIZZA, A., C. MUSSINI, N. MONGIARDO, V. BORGHI, A. SABBATINI, B. DERIENZO & C. FRANCESCHI. 1997. Mitochondria alterations and dramatic tendency to undergo apoptosis in peripheral lymphocytes during acute HIV syndrome. AIDS **11:** 19–26.

42. PIEDIMONTE, G., D. GUETARD, M. MAGNANI, D. CORSI, I. PICERNO, P. SPATARO, L. KRAMER, M. MONTRONI, G. SILVESTRI, J.F.T. ROCA & L. MONTAGNIER. 1997. Oxidative protein damage and degradation in lymphocytes from patients infected with human immunodeficiency virus. J. Infect. Dis. **176:** 655–664.

43. Carbonari, M., M. Pesce, M. Cibati, A. Modica, L. Dellanna, G. Doffizi, A. Angelici, S. Uccini, A. Modesti & M. Fiorilli. 1997. Death of bystander cells by a novel pathway involving early mitochondrial damage in human immunodeficiency virus-related lymphadenopathy. Blood **90:** 209–216.
44. Lewis, W., I.L. Grupp, G. Grupp, B. Hoit, R. Morris, A.M. Samarel, L. Bruggeman & P. Klotman. 2000. Cardiac dysfunction occurs in the HIV-1 transgenic mouse treated with zidovudine. Lab. Invest. **80:** 187–197.
45. Brinkman, K., H.J.M. ter Hofstede, D.M. Burger, J.A.M. Smeitink & P.P. Koopmans. 1998. Adverse effects of reverse transcriptase inhibitors: mitochondrial toxicity as common pathway. AIDS **12:** 1735–1744.
46. Praksch, O., S. Teng, M. Ali, X.Z. Zhu, R. Coleman, R.A. Dabdoub, R. Chambers, T.Y. Aw, S.C. Flores & B.H. Joshi. 1997. The human immunodeficiency virus type 1 Tat protein potentiates zidovudine-induced cellular toxicity in transgenic mice. Arch. Biochem. Biophys. **343:** 173–180.
47. Park, C.J., A. Cooper & J. Sodroski. 1996. Molecular determinants of acute single-cell lysis by human immunodeficiency virus type 1. J. Virol. **70:** 1340–1354.
48. Plymale, D.R., D.S.N. Tang, A.M. Comardelle, C.D. Fermin, D.E. Lewis & R.G. Garry. 1999. Both necrosis and apoptosis contribute to HIV-1-induced kiling of CD4 cells. AIDS **13:** 1827–1839.
49. Scheuring, U.J., J. Corbeil, D.E. Mosier & A.N. Theofilopoulos. 1998. Early modifications of host cell gene expression induced by HIV-1. AIDS **12:** 563–570.
50. Ryo, A., Y. Suzuki, K. Ichiyama, T. Wakatsuki, N. Kondoh, A. Hada, M. Yamamoto & N. Yamamoto. 1999. Serial analysis of gene expression in HIV-1-infected T cell lines. FEBS Lett. **462:** 182–186.
51. Osborne, B.A., S.W. Smith, Z.-G. Liu, K.A. McLaughlin, L. Grimm & L.M. Schwartz. 1994. Identification of genes induced during apoptosis in T cells. Immunol. Rev. **142:** 301–320.
52. Banki, K., E. Hutter, N.J. Gonchoroff & A. Perl. 1998. Molecular ordering in HIV-induced apoptosis—oxidative stress, activation of caspases, and cell survival are regulated by transaldolase. J. Biol. Chem. **273:** 11944–11953.
53. Forte, S.E., J.L. Sullivan & M. Somasundaran. 1996. In vitro characterization of HIV type 1 biological clones from asymptomatic and symptomatic pediatric patients. AIDS Res. Hum. Retroviruses **12:** 1585–1593.
54. Ferri, K.F., E. Jacotot, J. Blanco, J.A. Esté, S.A. Susin, J.C. Reed, J.M. Penninger & G. Kroemer. 2000. Apoptosis control in syncytia induced by the HIV-1-envelope glycoprotein complex. Role of mitochondria and caspases. J. Exp. Med. **192:** 1081–1092.
55. Aillet, F., H. Masutani, C. Elbim, H. Raoul, L. Chene, M.T. Nugeyre, C. Paya, F. Barre-Sinoussi, M.A. Gougerot-Pocidalo & N. Israel. 1998. Human immunodeficiency virus induces a dual regulation of Bcl-2, resulting in persistent infection of $CD4^+$ T- or monocytic cell lines. J. Virol. **72:** 9698–9705.
56. Marshall, W.L., R. Datta, K. Hanify, E. Teng & R.W. Finberg. 1999. U937 cells overexpressing bcl-Xl are resistant to human immunodeficiency virus-1-induced apoptosis and human immunodeficiency virus-1 replication. Virology **256:** 1–7.
57. Sandstrom, P.A., D. Pardi, C.S. Goldsmith, C.Y. Duan, A.M. Diamond & T.M. Folks. 1996. Bcl-2 expression facilitates human immunodeficiency virus type 1-mediated cytopathic effects during acute spreading infections. J. Virol. **70:** 4617–4622.
58. Fermin, C.D. & R.F. Garry. 1992. Membrane alterations linked to early interaction of HIV with cell surface. Virology **191:** 941–946.
59. Hashimoto, F., N. Oyaizu,V.S. Kalyanaraman & S. Pahwa. 1997. Modulation of Bcl-2 protein by CD4 cross-linking: a possible mechanism for lymphocyte apoptosis in human immunodeficiency virus infection and for rescue of apoptosis by interleukin-2. Blood **1997:** 745–753.
60. Yusim, A., L. Franklin, S. Brooke, O. Ajiloe & R. Sapolsky. 2000. Glucocorticoids exacerbate the deleterious effects of gp120 in hippocampal and cortical explants. J. Neurochem. **74:** 1000–1007.

61. LIFSON, J.D., G.R. REYES, M.S. MCGRATH, B.S. STEIN & E.G. ENGLEMAN. 1986. AIDS retrovirus-induced cytopathology: giant cell formation and involvement of CD4 antigen. Science **232:** 1123–1127.

62. STEWART, S.A., B. POON, J.B.M. JOWETT & I.S.Y. CHEN. 1997. Human immunodeficiency virus type 1 vpr induces apoptosis following cell cycle arrest. J. Virol. **71:** 5579–5592.

63. YAO, X.J., A.J. MOULAND, R.A. SUBBRAMANIAN, J. FORGET, N. ROUGEAU, D. BERGERON & E.A. COHEN. 1998. Vpr stimulates viral expression and induces cell killing in human immunodeficiency virus type 1-infected dividing Jurkat T cells. J. Virol. **72:** 4686–4693.

64. STEWART, S.A., B. POON, J.Y. SONG & I.S.Y. CHEN. 2000. Human immunodeficiency virus type 1 Vpr induces apoptosis through caspase activation. J. Virol. **74:** 3105–3111.

65. PILLER, S.C., P. JANS, P.W. GAGE & D.A. JANS. 1998. Extracellular HIV-1 virus protein R causes a large inward current and cell death in cultured hippocampal neurons: Implications for AIDS pathology. Proc. Natl. Acad. Sci. USA **95:** 4595–4600.

66. AYYAVOO, V., A. MAHBOUBI, S. MAHALINGAM, R. ROMALINGAM, S. KUDCHODKAR, W.V. WILLIAMS, D.R. GREEN & D.B. WEINER. 1997. HIV-1 Vpr suppresses immune activation and apoptosis through regulation of nuclear factor kappa B. Nat. Med. **10:** 1117–1123.

67. HRIMECH, M., X.-J. YAS, F. BACHAND, N. ROUGEAU & E.A. COHEN. 1999. Human immunodeficiency virus type 1 (HIV-1) Vpr functions as an immediate-early protein during HIV-1 infection. J. Virol. **73:** 4101–4109.

68. JACOTOT, E., L. RAVAGNAN, M. LOEFFLER, K.F. FERRI, H.L.A. VIEIRA, N. ZAMZAMI, P. COSTANTINI, S. DRUILLENNEC, J. HOEBEKE, J.P. BRIAN, T. IRINOPOULOS, E. DAUGAS, S.A. SUSIN, D. COINTE, Z.H. XIE, J.C. REED, B.P. ROQUES & G. KROEMER. 2000. The HIV-1 viral protein R induces apoptosis via a direct effect on the mitochondrial permeability transition pore. J. Exp. Med. **191:** 33–45.

69. SASTRY, K.J., M.C. MARIN, P.N. NEHETE, K. MCCONNELL, A.K. ELNAGGAR & T.J. MCDONNELL. 1996. Expression of human immunodeficiency virus type 1 tat results in downregulation of bcl-2 and induction of apoptosis in hematopoietic cells. Oncogene **13:** 487–493.

70. MACHO, A., M.A. CALZADO, L. JIMENEZREINA, E. CEBALLOS, J. LEON & E. MUNOZ. 1999. Susceptibility of HIV-1-TAT transfected cells to undergo apoptosis. Biochemical mechanisms. Oncogene **18:** 7543–7551.

71. LI, C.J., D.J. FRIEDMAN, C. WANG, V. METELEV & A.B. PARDEE. 1995. Induction of apoptosis in uninfected lymphocytes by HIV-1 Tat protein. Science **268:** 429–431.

72. WESTENDORP, M.O., V.A. SHATROV, K. SCHULZE-OSTHOFF, R. FRANK, M. KRAFT, M. LOS, P.H. KRAMMER, W. DRÖGE & V. LEHRMANN. 1995. HIV-1 Tat potentiates TNF-induced NF-κB activation and cytotoxicity by altering the cellular redox state. EMBO J. **14:** 546–554.

73. CREAVEN, M., F. HANS, V. MUTSKOV, E. COL, C. CARON, S. DIMITROV & S. KHOCHBIN. 1999. Control of the histone-acetyltransferase activity of Tip60 by the HIV-1 transactivator protein, Tat. Biochemistry **38:** 8826–8830.

74. WESTENDORP, M.O., R. FRANK, C. OCHSENBAUMER, K. STRICKER, J. DHEIN, H. WALCZAK, K.-M. DBATIN & P.H. KRAMMER. 1995. Sensitization of T cells to CD95-mediated apoptosis by HIV-1 Tat and gp120. Nature **375:** 497–500.

75. LI-WEBER, M., O. LAUR, K. DERN & P.H. KRAMMER. 2000. T cell activation-induced HIV Tat-enhanced CD95 (APO-1/Fas) ligand transcription involves NF-kappa B. Eur. J. Immunol. **30:** 661–670.

76. KRUMAN, I.I., A. NATH & M.P. MATTSON. 1998. HIV-1 protein Tat induces apoptosis of hippocampal neurons by a mechanism involving caspase activation, calcium overload and oxidative stress. Exp. Neurol. **154:** 276–288.

77. KRUMAN, I.I., A. NATH, W.F. MARAGOS, S.L. CHAN, M. JONES, V.M. RANGNEKAR, R.J. JAKEL & M.P. MATTSON. 1999. Evidence that Par-4 participates in the pathogenesis of HIV encephalitis. Am. J. Pathol. **39–46:**

78. KONVALINSKA, J., M.A. LITTERST, R. WELKER, G. KOTTLER, F. RIPPMANN, A.M. HEUSER, & H.G. KRAUSSLICH. 1995. An active-site mutation in the human immunodeficiency virus type 1 proteinase (PR) causes reduced PR activity and loss ofPR-mediated cytotoxicity without apparent effect on virus maturation and infectivity. J. Virol. **1995:** 7180–7186.
79. STRACK, P.R., M.W. FREY, C.J. RIZZO, B. CORDOVA, H.J. GEORGE, R. MEADE, S.W. JO, J. CORMAN, R. TRITCH & B.D. KORANT. 1996. Apoptosis mediated by HIV protease is preceded by cleavage of Bcl-2. Proc. Natl. Acad. Sci. USA **93:** 9571–9576.
80. CHENG, E.H.Y., D.G. KIRSCH, R.J. CLEM, R. RAVI, M.B. KASTAN, A. BEDI, K. UENO & J.M. HARDWICK. 1997. Conversion of Bcl-2 to a Bax-like death effector by caspases. Science **278:** 1966–1968.
81. KIRSCH, D.G., A. DOSEFF, B.N. CHAU, D.S. LIM, N.C. DE SOUZA PINTO, R. HANSFORD, M.B. KASTAN, Y.A. LAZEBNIK & J.M. HARDWICK. 1999. Caspase-3-dependent cleavage of Bcl-2 promotes release of cytochrome c. J. Biol. Chem. **274:** 21155–21161.
82. BRENNER, C., H. CARDIOU, H.L.A. VIEIRA, N. ZAMZAMI, I. MARZO, Z. XIE, B. LEBER, D. ANDREWS, H. DUCLOHIER, J.C. REED & G. KROEMER. 2000. Bcl-2 and Bax regulate the channel activity of the mitochondrial adenine nucleotide translocator. Oncogene **19:** 329–336.
83. GOLDMACHER, V.S., L.M. BARTLE, S. SKLETSKAYA, C.A. DIONNE, N.L. KEDERSHA, C.A. VATER, J.W. HAN, R.J. LUTZ, S. WATANABE, E.D.C. MCFARLAND, E.D. KIEFF, E.S. MOCARSKI & T. CHITTENDEN. 1999. A cytomegalovirus-encoded mitochondria-localized inhibitor of apoptosis structurally unrelated to Bcl-2. Proc. Natl. Acad. Sci. USA **96:** 12536–12541.
84. TSAI, S.C., K.B.S. PASUMARTHI, L. PAJAK, M. FRANKLIN, B. PATTON, H. WANG, W.J. HENZEL, J.T. STULTS & L.J. FIELD. 2000. Simian virus 40 large T antigen binds a novel Bcl-2 homology domain 3-containing proapoptosis protein in the cytoplasm. J. Biol. Chem. **275:** 3239–3246.
85. WOODFIN, B.M. & A.L. KAZIM. 1993. Interaction of the amino-terminus of an influenza-virus protein with mitochondria. Arch. Biochem. Biophys. **306:** 427–430.
86. CIMINALE, V., L. ZOTTI, D.M. D'AGOSTINI, T. FERRO, L. CASARETO, G. FRANCHINI, P. BERNARDI & L. CHIECO BIANCHI. 1999. Mitochondrial targeting of the p13(II) protein coded by the x-II ORF of human T-cell leukemia/lymphotropic virus type I (HTLV-1). Oncogene **18:** 4505–4514.
87. SCHUSTER, R., W.H. GERLICH & S. SCHAEFER. 2000. Induction of apoptosis by the transactivating domains of the hepatitis B virus X gene leads to suppression of oncogenic transformation of primary rat embryo fibroblasts. Oncogene **19:** 1173–1180.
88. RAHMANI, Z., K.W. HUH, R. LASHER & A. SIDDIQUI. 2000. Hepatitis B virus X protein colocalizes to mitochondria with a human voltage-dependent anion channel, HVDAC3, and alters its transmembrane potential. J. Virol. **74:** 2840–2846.
89. SHIMIZU, S., M. NARITA & Y. TSUJIMOTO. 1999. Bcl-2 family proteins regulate the release of apoptogenic cytochrome c by the mitochondrial channel VDAC. Nature **399:** 483–487.
90. CROMPTON, M. 1999. The mitochondrial permeability transition pore and its role in cell death. Biochem. J. **341:** 233–249.

Morphologic, Biochemical and Molecular Mitochondrial Changes during Reperfusion Phase following Brief Renal Ischemia

LUIS BENÍTEZ-BRIBIESCA,[a,b] MADAÍ GÓMEZ-CAMARILLO,[c] EMILIO CASTELLANOS-JUÁREZ,[b] ELIZABETH MRAVKO,[b] AND PATRICIA SÁNCHEZ-SUÁREZ[b]

[b]*Unidad de Investigación Médica en Enfermedades Oncológicas, Hospital de Oncología, Centro Médico Nacional S-XXI, Instituto Mexicano del Seguro Social, México, D.F., México*

[c]*Departamento de Odontología, Universidad Autónoma de Tlaxcala, Tlaxcala, México*

ABSTRACT: Ischemia/reperfusion of organs and cells induces apoptosis through a complicated series of changes in mitochondria, mainly the generation of oxygen free radicals, permeability transitions, calcium translocations, and release of apoptogenic factors such as cytochrome *c* and Bcl-2 family members. The liberation of these factors occurs very early after reoxygenation and it has been assumed that it takes place without any structural alteration of the mitochondrial membranes. The aim of this study was to detect ultrastructural changes of mitochondria in the initial stages of reperfusion at the time when Bcl-2 and succinic dehydrogenase, located in the outer and inner membranes, respectively, were released. Ischemia/reperfusion was produced in adult rats by clamping one renal artery for 60 min and reoxygenating for 60, 120, 180, and 240 min. A model of chemical hypoxia with intra-arterial 50 mM sodium azide served as comparison, allowing free blood flow for 30, 60, 120 and 180 min. Light and electron microscopy, immunostaining for Bcl-2, and enzyme histochemistry for succinic dehydrogenase were performed. Our results showed mitochondrial swelling, rupture of inner and outer membranes, and leakage of mitochondrial matrix into the cytoplasm in ischemia after 120 min of reperfusion. Bcl-2 immunoreactivity and focal lowering of SDH reactivity were also noted and became more pronounced at the same time that the mitochondrial ultrastructure demonstrated more evident changes including rupture of the inner and outer membranes. Our studies seem to indicate that in early ischemia-reperfusion and in chemical hypoxia-induced apoptosis, the earliest ultrastructural changes take place in mitochondria and that swelling and rupture of mitochondrial membranes occur in parallel with the loss of Bcl-2 and SDH activity.

KEYWORDS: Mitochondria; Apoptosis; Bcl-2; Succinic dehydrogenase; Ischemia/reperfusion

[a]Address for correspondence: Dr. L. Benítez-Bribiesca, Oncological Research Unit, Oncology Hospital, National Medical Center S-XXI, IMSS, Av. Cuauhtemoc 330, 06720 Mexico, D.F., México.
luisbenbri@mexis.com

INTRODUCTION

It is well known that prolonged ischemia in highly aerobic tissues, such as brain, heart, and kidney produces necrotic cell death.[1] It is now clear however, that short periods of arterial blood flow suppression, not long enough to induce necrosis, followed by reoxygenation, can induce apoptotic cell death, but the mechanisms that induce apoptosis in reoxygenated tissues are not clear.[2] Considerable evidence points to the generation of oxygen free radicals during reperfusion of ischemic organs both from the parenchymal cells themselves or from the inflammatory cells that usually invade these tissues.[3–5] Other studies have reported the expression of Fas ligand during hypoxia/reoxygenation, which could trigger Fas-dependent death pathways.[6,7] Recently, the role of mitochondria in apoptotic cell death has been extensively studied, and the findings have placed this organelle as the central integrator in apoptotic signaling with the capacity of directly activating the death pathway. Programmed cell death can be conveniently divided into three different stages: signaling, control and execution, and structural alterations. Multiple signal pathways lead from death-triggering extracellular agents to the central control and execution stage in which mitochondria undergo a series of changes that trigger downstream a cascade of caspase activation, which in turn are responsible, directly or indirectly, for the characteristic morphological changes described in apoptosis.[8] A number of biochemical and molecular studies have demonstrated various alterations in the mitochondria of hypoxia/reperfused cells, such as electron conduction defects, generation of oxygen free radicals, mitochondrial permeability transitions (MPT), calcium ion translocations, and release of apoptogenic factors such as cytochrome-*c* and members of the Bcl-2 family.[1,9–12] Most of these studies have been conducted in cultured cells or in isolated mitochondria only a few in intact, reperfused parenchymal organs, and little attention has been paid to the initial, subtle morphological changes that occur in the mitochondria of cells from these organs. In fact, Yang *et al.*[13] found that cytochrome *c* release from the mitochondria, one of the earliest events, occurs in the absence of any structural change of this organelle. It has been shown, however, that MPT can become rapidly irreversible, initiating mitochondrial swelling that could rupture the external mitochondrial membrane, facilitating the release of intermembrane proteins to the cytoplasm.[14] In ischemia-reperfusion injury, swelling of the mitochondria is a very early event so the rupture of any of their membranes should be detected by ultrastructural techniques before the classical morphology of apoptosis develops, probably when the execution cascade of molecular events begins.

Among the various organs amenable to study reperfusion injury in laboratory animals, the kidney probably is the best suited and the most investigated. The ultrastructure of renal tubules and their mitochondria after various types of ischemia and reperfusion has been the subject of a number of investigations, mostly performed before programmed cell death was recognized as a unique form of death; hence, no mention of the relationship of these changes with apoptosis was made.[15] A recent study by Schumer *et al.* was aimed at demonstrating the occurrence of apoptosis during reperfusion after brief periods of ischemia.[2] They found the presence of apoptotic bodies as early as 12 h after reperfusion with increasing numbers at 24 hours. They did not, however, report on the ultrastructural changes of mitochondria that take place long before the classical apoptotic bodies appear. In view of the recent wealth of biochemical and molecular information on the central role of mitochondria in apoptosis, it could be useful to refocus the study of mitochondrial morphology on the mitochondria-rich cells of the cortical renal tubules of rat kidneys subjected to ischemia/reperfusion, particularly on the integrity of mitochondrial membranes and proteins bound to them, such as Bcl-2 and SDH.

The purpose of the present investigation was to study the mitochondria of renal tubules from rat kidneys subjected to 60-min occlusion of the renal artery during the first 4 h of reperfusion. Succinic dehydrogenase was assessed as a marker for mitochondrial function and part of the inner membrane and Bcl-2 as a mitochondrial anti-apoptotic protein bound to the external membrane. A model of chemical hypoxia was used for comparison.

MATERIALS AND METHODS

Animals and Tissues

Adult male Sprague-Dawley rats (300–350 g) were maintained with food and water *ad libitum*. The animals were fasted the night prior to the surgical procedure. Under general anesthesia with ketamine-xylacine a midline laparotomy was performed. The right renal artery was exposed and clamped for 60 min. Reperfusion was allowed for 60, 120, 180, and 240 min, leaving the abdominal incision partially closed and covered with gauze humidified with saline. Both kidneys were removed after the different reperfusion times and the animals were sacrificed with an extra dose of anesthesia. Five animals were used for each time of reperfusion and another set of five for time zero after 60 min arterial occlusion. A model of chemical hypoxia was used for comparison. Animals were anesthetized and operated on the same manner as the previous model. The right renal artery was dissected and 0.5 ml of 50 mM sodium azide in isotonic solution was injected. Both kidneys were removed after 30, 60, 120, and 180 min of normal arterial blood flow. No washing solutions were used to avoid osmolar or direct pressure damage to renal tubules. Free blood flow was allowed for washing out azide. Five animals were used in each experiment. Sham-operated animals received similar anesthesia and operative procedures.

Histology and Ultrastructure

Full kidney sections were obtained and fixed in 10% buffered formalin overnight, dehydrated and embedded in paraffin. Sections 5 μm thick were obtained on poly-l-lysine coated slides. H+E and PAS-diastase stains were used for light microscopy. Fresh kidney sections were snap-frozen in isopentane cooled in liquid nitrogen at –160°C and cut in a Leica cryostat at –30°C in 6 μm thick sections. Small blocks of about 1 mm^3 were sliced from the renal cortex, fixed in 3% glutaraldehyde for 2 h, rinsed and followed by 1% osmium tetroxide in phosphate buffer at pH 7.4 for 2 h. They were then dehydrated and embedded in Epon 812. Sections were obtained in a Porter-Blum MT-1 ultramicrotome, placed in copper grids, and stained with the Reynolds procedure[16] (all reagents were obtained from Sigma Chemical Co., St. Louis, MO, USA). Observation was carried out on a Zeiss EM-910 electron microscope.

Succinic Dehydrogenase

Cryostat sections of frozen tissues were incubated with the method of Nachlas *et al.*[17] using the following mixture: succinic acid (5.4 g/dl), nitro blue tetrazolium (1 mg/ml) in 0.2M phosphate buffer at pH 7.6. Incubation time was 15 min and the reaction was stopped by washing and immersion in 10% buffered formalin and counterstained with 0.5% safranin O. Slides were dehydrated in cold acetone and mounted on balsam.

Bcl-2 Detection

For immunocytochemistry a Bcl-2 C 21 rabbit polyclonal antibody (Santa Cruz Biotechnology, Inc., Santa Cruz, California, USA) was used as a primary antibody. A secondary biotinylated antibody and streptavidin-peroxidase complex were added and developed with H_2O_2 AEC according to the supplier's instructions (Histostain TM plus kit, Zymed's LAB-SA, Inc., San Francisco, California, USA)

RESULTS

Our model of 60 min of ischemia was chosen following previous studies by Reimer et al that demonstrated that this period of suspension of arterial blood flow was sufficient to induce irreversible injury to the cortical tubules.[15] Schumer *et al.* observed that apoptosis could be morphologically detected as soon as 12 h after a 45-min clamping of the renal artery.[2] But Kerr and Wyllie estimated that apoptosis is a fast phenomenon taking about 3 h to develop after the triggering stimulus.[18] Hence, reperfusion times were set at 1, 2, 3 and 4 h in an attempt to grasp sequentially the initial changes of apoptosis, in particular those that could take place in mitochondria. With regard to chemical hypoxia with azide, a 3 h observation period after free arterial blood flow was chosen, following the reversible initiation phase of Chen *et al.*[19]

Light Microscopy

After 1 h reperfusion, slight histological changes were seen. The cortical tubules, mainly the proximal convoluted and to a lesser degree the distal convoluted tubules, showed moderate cellular swelling, a few vacuoles, and some luminal blebs. There were a few tubules with cellular debris and amorphous eosinophilic material in their lumen. With PAS stain, some diastase-resistant intracytoplasmic granules were observed in the tubular epithelial cells. The brush border stained with PAS uniformly and only slight changes were seen where intraluminal blebs formed. These changes were more evident, between 2 and 3 hours. Some cells remained swollen while others were compact. Vesicles and blebs were more abundant and the intraluminal cellular debris and eosinophilic material appeared in more convoluted tubules. The intracytoplasmic PAS positive granules increased and the brush border appeared thinner or frayed. At 4 h after reperfusion, all the changes described were more apparent and apoptotic cells and apoptotic bodies were found in tubular lumina. In all groups, the changes were spotty, occurring in areas that alternated with histologically unaffected portions of the renal cortex. No necrotic cells could be identified and no inflammatory infiltrate or interstitial hemorrhage occurred. Glomeruli were unaffected.

Ultrastructure Findings

The ultrastructure of kidney tubules has been thoroughly described. It is well known that epithelial cells in the proximal convoluted tubules are rich in mitochondria, which are located in the basal portion of the cell and have a typical rod-shape appearance with abundant cristae.[20] This characteristic makes this tissue particularly suited to the study of mitochondrial damage. Our attention was focused mainly on the changes occurring in these cells and their mitochondria although other tubular structures also showed changes with increasing times of reperfusion or azide treatment. Glomeruli were unaffected. The renal medulla and papillae were not considered in this study.

Mitochondrial Changes

The earliest changes were noted after 60 min reperfusion following 60 min ischemia. These consisted of a separation of the mitochondria from their basal location and from the basal cell membrane infoldings, and they appeared shortened. Some had adopted a round shape. They were slightly swollen and distributed throughout the cytoplasm reaching the apical portion of the cell (FIG. 1 A). Between 120 and 180 min reperfusion times, the mitochondria exhibited marked swelling, diminished matrix density, slightly disorganized cristae, and some electron-dense granules. Focal fading of the outer and inner membranes was seen, particularly after 180 min, when some mitochondria showed a complete disruption of both membranes (FIG. 1 B). After 240 min reperfusion time, the mitochondria appeared more swollen and exhibited bizarre shapes, frequently lobulated or with bleb formation. At this time, the rupture of both membranes was more generalized and leakage of the mitochondrial matrix into the cytoplasm was seen (FIG. 1 C).

In contrast, with azide treatment the mitochondria showed a marked increased matrix density of the mitochondrial matrix and a striking separation of the intracrista space (FIG. 1 D and E). Only after 3 h was on occasional rupture of the outer or inner mitochondrial membranes noted.

Apoptotic Nuclei

The presence of apoptotic nuclei with heavily condensed chromatin and picnotic appearance was estimated in H+E, PAS, and in semi-thin sections. In control kidneys, after 60 min of ischemia and following 60 min reperfusion time their number ranged from 1 to 3 per 500 cells. After 120 min, we found 5 to 10 per 500 cells, and after 180 to 240 min, 10 to 12 apoptotic nuclei per 500 cells were counted. Some apoptotic cells and apoptotic bodies were found in the tubular lumina during the later reperfusion times.

Other Histological Features

Other ultrastructural changes due to ischemia/reperfusion that have been described in many previous publications were observed.[15] These included swelling

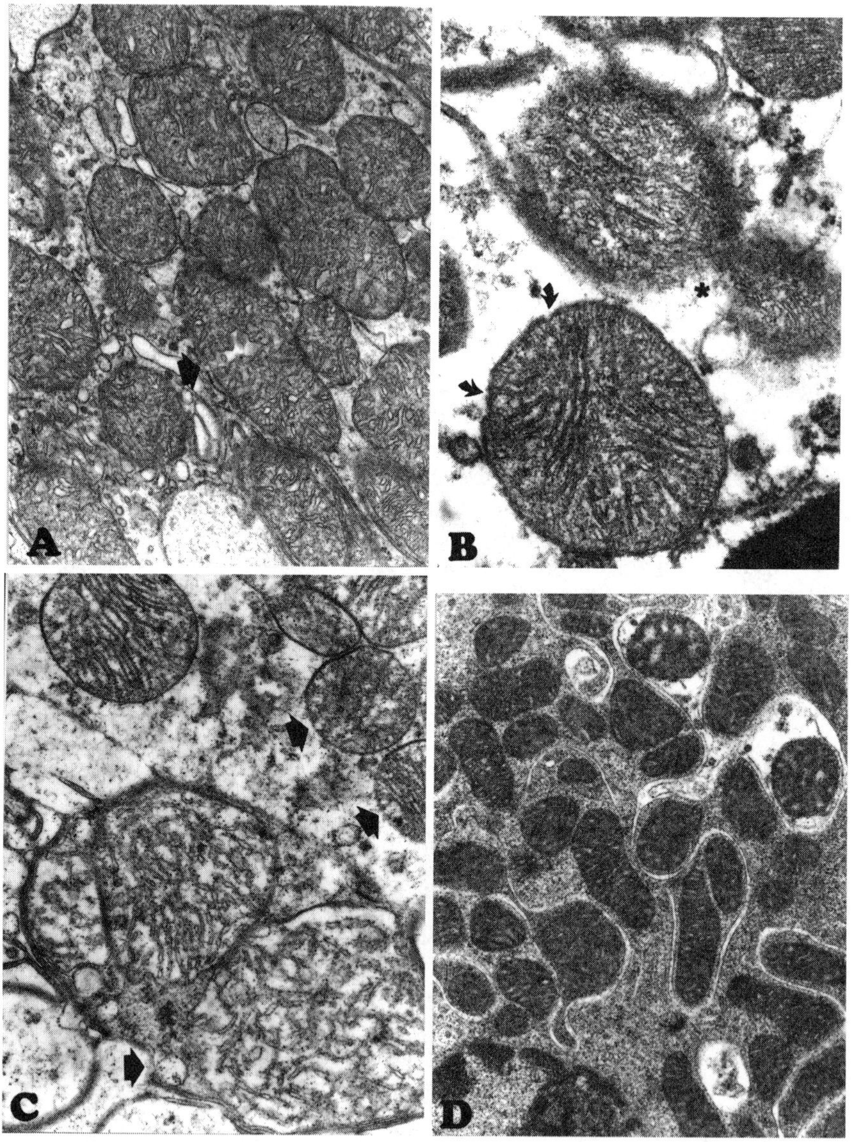
A
B
C
D

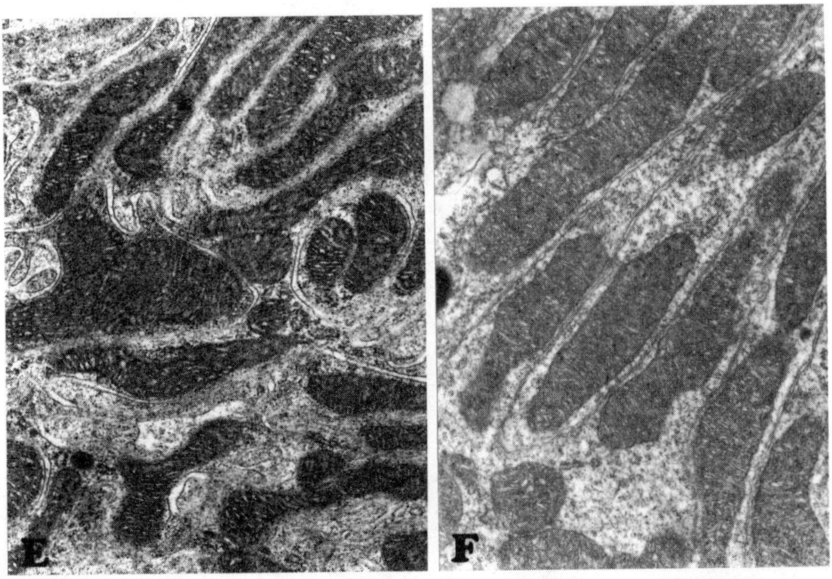

FIGURE 1. Ultrastructural morphology of mitochondria from proximal convoluted tubules in kidneys subjected to different experimental conditions (see MATERIAL AND METHODS). **A:** 120 min after ischemia/reperfusion. Mitochondria are swollen, the matrix is less dense and the shape is rounded compared to control (**F**). There is a gap and infolding of both mitochondrial membranes in one of the mitochondria (*arrow*); **B:** 180 min after ischemia/reperfusion. Mitochondria are round and swollen. There is complete fading of the outer and inner mitochondrial membranes in two, and small gaps in another (* and *arrow*); **C:** 240 min after ischemia-reperfusion. Pleomorphism and disorganization of the cristae are apparent; focal fading of mitochondrial membranes persist (*arrows*); **D:** 60 min after sodium azide. The mitochondria show pleomorphism, dense matrix, and separation of some cristae; **E:** 180 min after sodium azide. Pleomorphism is more apparent and separation of cristae and density of mitochondrial matrix are more marked; **F:** normal mitochondria of proximal convoluted tubule (enlarged from 12, 930× original negatives).

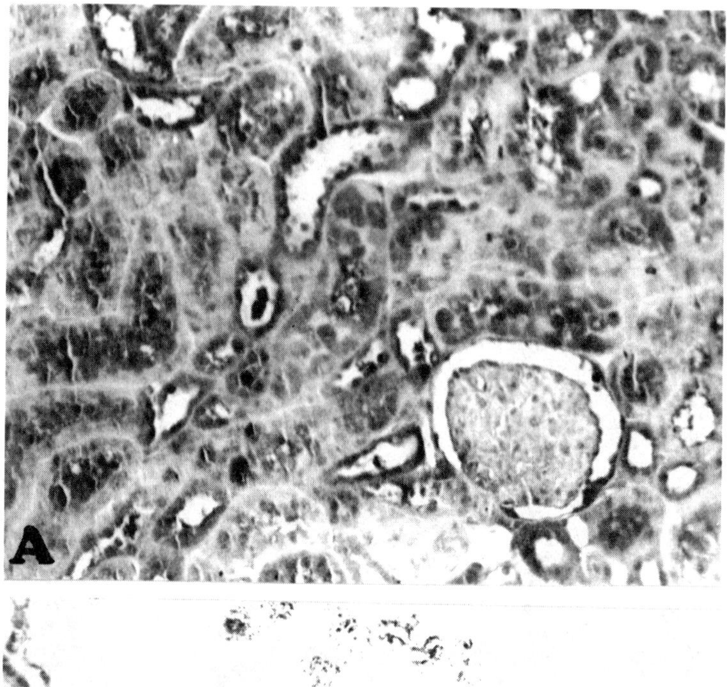
A
B

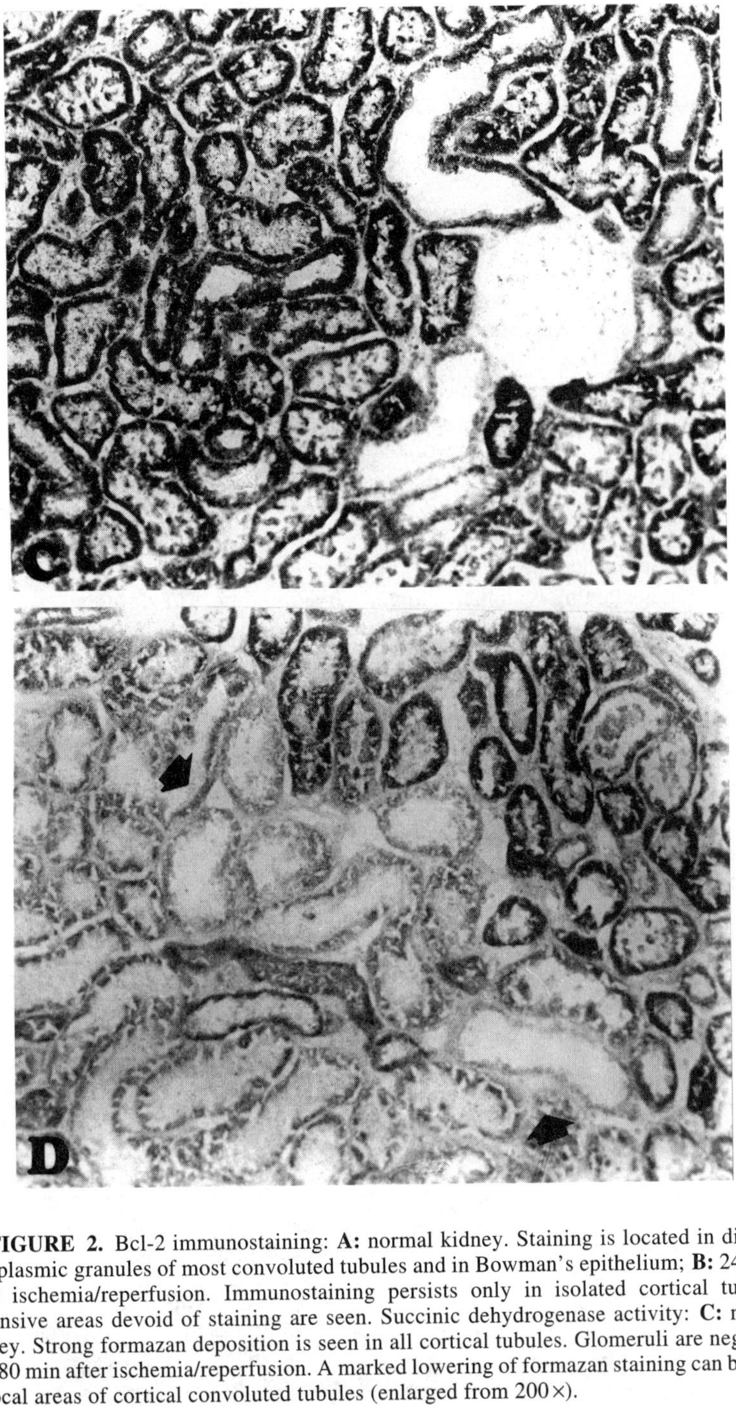

FIGURE 2. Bcl-2 immunostaining: **A:** normal kidney. Staining is located in discrete cytoplasmic granules of most convoluted tubules and in Bowman's epithelium; **B:** 240 min after ischemia/reperfusion. Immunostaining persists only in isolated cortical tubules. Extensive areas devoid of staining are seen. Succinic dehydrogenase activity: **C:** normal kidney. Strong formazan deposition is seen in all cortical tubules. Glomeruli are negative; **D:** 180 min after ischemia/reperfusion. A marked lowering of formazan staining can be seen in focal areas of cortical convoluted tubules (enlarged from 200×).

of epithelial cells, apical bleb formation, progressive disappearance of the brush border, abundant lysosomes, and focal rupture of the basement membrane. In the lumina of proximal and distal tubules, cellular debris was found, composed mainly of cell membranes and partially ruptured mitochondria.

Bcl-2 Immunolocalization

In control kidneys, the immunoreactivity for Bcl-2 was seen in all cortical tubules in discrete intracytoplasmic granules located toward the basal portion of the epithelium. These Bcl-2 positive granules had the same appearance and location as those revealed by SDH staining, suggesting that they corresponded to mitochondria. The glomeruli showed a slight staining of the Bowman's capsule and the renal medulla had a strong immunoreactivity (FIG. 2A).

With increasing times of reperfusion, the intensity of the staining in cortical tubules diminished progressively in focal areas. After 180 and 240 min reperfusion time, about 80% of proximal convoluted tubules had lost Bcl-2 immunoreactivity. The remaining 20% had markedly diminished staining and the location of the Bcl-2 positive granules was mainly in the apical portion of the cell. Tubular lumina also contained positive granules (FIG. 2B).

In the sodium azide model a similar phenomenon was noted, with a progressive loss of immunostaining with increasing observation time. The difference was more quantitative, because after 120 min, about 40% of cortical tubules retained a strong immunostaining.

Succinic Dehydrogenase

This enzyme is part of the electron transport chain and is located in the inner mitochondrial membrane. In control renal cryostat sections stained for SDH, positive formazan deposition was abundant and well localized in basal intracytoplasmic granules of most tubular epithelial cells, clearly corresponding to mitochondria. Staining was stronger in proximal convoluted tubules and almost absent in glomeruli and in non-epithelial structures such as blood vessels (FIG. 2C).

With increasing times of reperfusion, the intensity of SDH staining diminished in patchy areas that corresponded to the morphologically altered epithelium of convoluted tubules. In addition, the ordered basal localization of formazan-positive granules changed becoming irregular and diffuse. In the later stages, these granules were found in the tubular lumina. These changes were more prominent after 120 min reperfusion. In the sodium azide model, the changes observed were similar with increasing times of observation, but there was a greater distortion of the cellular architecture than in the reperfusion model (FIG. 2D).

DISCUSSION

The histological and ultrastructural changes found in our model of ischemia/reperfusion coincide in general terms with previous reports. In fact, the 60 min time of arterial occlusion was chosen because Reimer *et al.* had found this to be the ischemia capable of inducing irreversible damage.[15] The ultrastructural changes they describe occur in three progressive stages, with increasing alterations in the brush border, cell density, mitochondrial swelling, mitochondrial dense granules, lysosomes, and nuclear condensation. Although these researchers do not mention the presence of apoptosis, probably because at that time this phenomenon was not known, there are some cells in their illustrations with ultrastructural changes that are clearly those described later by Kerr and Wyllie in apoptotic cells.[21] Furthermore, the mitochondrial alterations they describe are those of progressive swelling but no reference to any specific changes in the mitochondrial membranes is made. In a more recent study of ischemia/reperfusion in rat kidney, Schumer *et al.* describe the presence of apoptotic cells and apoptotic bodies with light microscopy after 12–24 h reperfusion following a 30–45 min arterial clamping, but their investigation was not aimed at the observation of ultrastructural changes in the mitochondria.[2] In another study of ischemia/reperfusion in rat liver, Sasaki *et al.* found the largest amount of apoptotic bodies, TUNEL-stained nuclei, and DNA electrophoretic ladders 3–6 h after a 30 min hepatic artery occlusion. The number diminished progressively after 12 h.[22] There are a number of studies demonstrating that reperfusion of organs such as brain, myocardium, and liver after relatively brief periods of ischemia induces apoptotic cell death of various degrees.[1] Most of these studies rely on histology, TUNEL assay and DNA ladder formation for the identification of apoptosis. These methods, however, identify only the latest stages of programmed cell death. It is well known that the apoptotic process takes about 3 h to be completed, and it is then that the classical morphology of programmed cell death can be detected.[18] But biochemical and molecular changes triggered by the initial apoptotic signal take place much earlier. One of the earliest changes is the release of cytochrome *c* from the intermembrane space of mitochondria occurring as soon as 1 h after the induction of cell death, before mitochondrial depolarization and the activation of caspases. According to Yang *et al.*, the release of cytochrome *c* appears to be independent of any noticeable structural changes in the mitochondria.[13] It was not surprising, therefore, that in our reperfusion group of 60 min the histological and ultrastructural changes found were minimal, and there were no typical apoptotic cells or apoptotic bodies.

The mitochondrion is now considered as the organelle in which apoptotic transduction signals converge and from which the secondary signals that trigger the proteolytic cascade emerge, leading to the final path of programmed cell death.[8,14] Our objective was to investigate whether these striking biomolecular changes occurring before the DNA fragmentation and the production of apoptotic bodies could alter mitochondrial morphology in the early stages of reoxygenation after supression of arterial blood flow. Our results show that after 60 min reperfusion the mitochondria of most epithelial cells from the cortical convoluted tubules show clear-cut alterations in form, density, and intracytoplasmic localization. Already at 120 min, but mostly after 180 min reperfusion, these organelles show rupture of the external and sometimes of the internal membranes, as well as disorganization of the cristae. These changes occur in cells that show minimal chromatin condensation or no nuclear change at all, thus pointing out that they precede the late stages of apoptosis. Even more, the immunohistochemical localization of Bcl-2 in discrete granules that clearly correspond to mitochondria faded progressively with increasing reprefusion times, indicating that the loss of mitochondrial Bcl-2 parallels the morphological changes found in these organelles, in particular the rupture of mitochondrial membranes.

The pivotal role of mitochondria in cell death has been studied extensively.[23] Three general mechanisms are known to be severely altered as a consequence of the initial death signals: a) disruption of electron transport; b) release of mitochondrial membrane proteins capable of activating the caspase proteases, and c) alterations in cellular redox potential. These changes are interrelated and all can in principle disrupt the integrity of mitochondrial membranes.[24] The mitochondrial inner transmembrane potential collapses, opening a permeability transition pore. The structure of this pore includes constituents from both the inner and the outer membrane. The opening of this pore results in volume dysregulation, which causes the mitochondrial matrix to expand.[14] This phenomenon could explain the early swelling and change of shape observed in the 60 min reperfusion model and the subsequent alterations in the mitochondrial cristae. But, because the inner membrane possesses a larger surface area than the outer membrane, the matrix expansion eventually could cause the rupture of the outer membrane, which in turn releases caspase-activating proteins into the cytosol and dislocates the antiapoptotic Bcl-2 family proteins.[25] The rupture of mitochondrial membranes was observed in our reperfusion model as early as 120 min after restoration of blood flow and became more widespread by 180 min. It remains to be established whether this rupture coincides with the release of apoptogenic factors such as cytochrome *c*, or whether this molecule is indeed released before any structural alteration of the mitochondrial membranes takes place.

The Bcl-2 family proteins are bound to mitochondrial membranes. Bcl-2 in particular is located in the outer mitochondrial membrane but projects into the mitochondrial intermembrane space where cytochrome *c* is located and probably reaches the inner membrane.[24,26] It is not yet clear how this protein protects mitochondria and exerts its antiapoptotic effect, but most investigations suggest that Bcl-2 and Bcl-XL act as suppressors of caspase-activating proteins such as Bax.[27,28] Whatever the mechanism, the overexpression of Bcl-2 confers resistance to apoptosis in many cells and it is thought to promote growth of malignant tumors in both experimental animals and humans.[29] Downregulation of Bcl-2 makes cells highly susceptible to apoptotic cell death. Our observations show that immunolocalization of Bcl-2 protein in the tubular epithelium is abundant in the discrete basal granules that correspond to mitochondria. With increasing reperfusion times after renal artery ligation and to a lesser extent with chemical hypoxia the immunostaining diminishes and completely disappears in many tubules. This indicates that Bcl-2 is dislodged from mitochondrial membrane and probably is bound to other molecules or degraded, thus becoming undetectable by immunostaining.

The antiapoptotic proteins Bcl-2 and Bcl XL prevent cytochrome *c* release and thus prevent activation of procaspase-3.[8,30] If Bcl-2 is dislodged from its mitochondrial outer membrane location and then inactivated, this would allow a free release of cytochrome *c* to the cytoplasm and, in conjunction with other cofactors, the conversion of procaspase-3 to its active form. Recently, Mancini *et al.* found that caspase-3 precursor is located in the intermembrane space where it could be bound to Bcl-2, which is known to reach this space.[31] This precursor might have a unique upstream function, perhaps becoming activated in mitochondria before the mitochondrial physiology is altered. Our observations of the early loss of Bcl-2 immunostaining in mitochondria suggest a possible mechanism by which mitochondrial procaspase-3 might become activated. The stability of procaspase-3 in the intermembrane space of mitochondria from uninjured cells provided by Bcl-2 family proteins could be lost if these proteins are released from their mitochondrial location.

There is some indirect evidence as to how the disruption of the normal molecular distribution of these molecules might occur. Marzo *et al.* have proposed that a subtle derangement in volume homeostasis induced by Bax could provoke the local mechanical disruption of the outer mitochondrial membrane.[27] This may explain why opening the permeability transition pore can trigger the release of intermembrane proteins through the outer membrane. The transition pore may contain several channel-forming proteins such as Bcl-2, which under conformational changes or translocation might open these channels.[1,14,23] In azide chemical hypoxia, the loss of Bcl-2 immunoreactivity was also present though less pronounced and rupture of the mitochondrial membranes was also observed. This suggests a functional relationship of this antiapoptotic protein with the electron transport chain. Azide inhibits the terminal portion of the electron transport chain, and this system is needed for the transfer of electrons to the tetrazolium salts in the Nachlas succinic dehydrogenase histochemical assay. The diminished activity of SDH in cryostat sections could be explained in these terms. It is known, however, that Bcl-2 is found in fractions with mitochondrial succinate dehydrogenase activity[26] and that oxidative phosphorylation and electron and metabolic transport are involved in the cell survival mechanism.[32] It has been shown that chemical hypoxia induced with azide triggers apoptosis in cardiac neonatal myocytes[19] with a defined series of kinetic events including mitochondrial dysfunction. Cyanide, another terminal electron transport blocker, also induces and accelerates cell death, probably inducing a large increase in mitochondrial superoxide formation, which is one of the most favored mechanisms to explain cell damage after hypoxia reperfusion.[33] The simultaneous loss of Bcl-2 and diminished SDH activity observed in both models of ischemia offer other, albeit indirect evidence of a strong association of this anti-apoptotic protein with oxidative phosphorylation. Recently, Cheng *et al.*[34] found that Bcl-2 is cleaved by caspase-3 inside cells and that the Bcl-2 cleavage product loses its antiapoptotic activity unleashing a proapoptotic activity that amplifies the caspase cascade. This suggests the existence of a feedback loop between Bcl-2 and caspases. In view of these findings it is likely that in our model of ischemia-reperfusion and chemical hypoxia the progressive loss of Bcl-2 immunoreactivity could have been due in part to Bcl-2 cleavage by caspases, in addition to its dislocation from the membrane. Our studies seem to indicate that in early ischemia-reperfusion and in chemical hypoxia-induced apoptosis, the earliest ultrastructural changes take place in mitochondria and that swelling and rupture of mitochondrial membranes occur in parallel with loss of Bcl-2 and SDH activity.

REFERENCES

1. SAIKUMAR, P., A. DONG, J.M. WEINBERG & M.A. VENKATACHALAM.1998. Mechanisms of cell death in hypoxia/reperfusion injury. Oncogene **17:** 7–15.
2. SCHUMER, M., M.COLOMBEL, I.S. SAWCZUK , G. GOBÉ , J. CONNOR, K.M. O'TOOLE, C.A. OLSSON, G.J. WISE & R. BUTTYAN. 1992. Morphologic, biochemical and molecular evidence of apoptosis during the reperfusion phase after brief periods of renal ischemia. Am. J. Pathol. **140:** 831–838.

3. KUZUYA, T., H. FUJI, S. HOSHIDA, M. NISHIDA, K. GOSHIMA, M. HORI, T. KAMADA & M. TADA. 1994. Polymorphonuclear leukocytes-induced injury in hypoxic cardiac myocytes. Free Rad. Biol. Med. **17:** 501–510.
4. DE GROOT, H. 1992. Isolated cells in the study of molecular mechanisms of reperfusion injury. Toxicol. Lett. **63:** 111–112.
5. HESS, M.L & N.H. MANSON. 1984. Molecular oxygen: friend and foe. The role of oxygen-free radical system in the calcium paradox, the oxygen paradox and ischemia/reperfusion injury. J. Mol. Cell. Cardiol. **16:** 969–985.
6. NOGAE, S., M. MIYAZAKI, N. KOBAYASHI, T. SAITO, K. ABE, H. SAITO, P.K. NAKANE, Y. NAKANISHI & T. KOJI.1998. Induction of apoptosis in ischemia reperfusion model of mouse kidney: possible involvement of Fas. J. Am. Soc. Nephrol. **9:** 620–631.
7. VOGT, M., M.K. BAUER, D. FERRARI & K. SCHULZE-OSTHOFF. 1998. Oxidative stress and hypoxia/reoxygenation triggers CD95 (APO-1/Fas) ligand expression in microglial cells. FEBS Lett. **429:** 67–72.
8. GOLDSTEIN, P. 1997. Controlling cell death. Science **275:** 1081–1082.
9. SCHILD, L., T. REINHECKEL, I. WISWEDELI & W. AUGUSTIN. 1997. Short term impairment of energy production in isolated rat liver mitochondria by hypoxia/reoxygenation: involvement of oxidative protein modification. Biochem. J. **328:** 205–210.
10. LEMASTERS, J., A.L. NIEMINEN, T. QIAN , L.C. TROST & B. HERMAN. 1997. The mitochondrial permeability transition in toxic, hypoxic and reperfusion injury. Mol. Cell. Biochem. **174:** 159–165.
11. RICHTER, C., V. GOGVADZE, R. LAFFRANCHI, M. SCHLAPBACH, M. SCHWEIZER, M. SUTER, P. WALTER & M. YAFFEE. 1995. Oxidants in mitochondria: from physiology to diseases. Biochim. Biophys. Acta **1271:** 67–74.
12. MURPHY, A.N. 1999. Ca^{2+} Mediated mitochondrial disfuntion and the protective effects of Bcl-2. Ann. N.Y. Acad. Sci. **893:** 19–32.
13. YANG, J., X. LIU, K. BHALLA, C.N. KIM, A.M. IBRADO, J. CAI, T.I. PENG, P. JONES, X. WANG. 1997. Prevention of apoptosis by Bcl-2: release of cytochrome c from mitochondria blocked. Science **275:** 1129–1132.
14. GREEN, D.R. & J.C. REED. 1998. Mitochondria and apoptosis. Science **281:** 1309–1312.
15. REIMER, K.A., C.E. GANOTE & R. JENNINGS. 1972. Alterations in renal cortex following ischemic injury. Lab. Invest. **26:** 347–363.
16. BOZZOLA, J.J. & L.D. RUSSELL. Specimen preparation for transmission electron microscopy. *In* Electron Microscopy. Bartlett and Jones publishers. London, UK, pp. 16–38.
17. NACHLAS, M.M., K.C. TSOU, E. DE SOUZA, C.S. CAHNG & A.M. SELIGMAN. 1957. Cytochemical demonstration of succinic acid dehydrogenase by use of a new p-nitrophenyl substituted tetrazole. J. Histochem. Cytochem. **5:** 420–436.
18. WYLLIE, A.H., J.F.R. KERR & A.R. CURRIE. 1980. Cell death. The significance of apoptosis. Int. Rev. Cytol. **68:** 251–306.
19. CHEN, S.J., M.E. BRADLEY & T. LEE. 1998. Chemical hypoxia triggers apoptosis of cultured neonatal rat cardiac myocytes: modulation by calcium-regulated proteases and protein kinases. Mol. Cell. Biochem. **178:** 141–149.
20. HEPINSTALL, R.H. 1982. Anatomy of the kidney. *In* Pathology of the Kidney, 2nd edit. Little Brown and Company. Boston, MA, pp. 1–50.
21. KERR, J.F.R., A.H. WYLLIE & A. CURRIE. 1972. Apoptosis: a basic biological phenomenon with wide ranging implications in tissue kinetics. Br. J. Cancer. **26:** 239–257.
22. SASAKI, H., T. MATSUNO, K. NAKAGAWA & N. TANAKA. 1997 Induction of apoptosis during the early phase of reperfusion after rat liver ischemia. Acta Med. Okayama **51:** 305–312.
23. KROEMER, G. & J.C. REED. 2000. Mitochondrial control of cell death. Nature Med. **6:** 513–519.
24. ADACHI S., A.R. CRPSS, B.M. BABIOR & R.A. GOTTLIEB. 1997. Bcl-2 and the outer mitochondrial membrane in the inactivation of cytochrome c during fas-mediated apoptosis. J Biol. Chem. **272:** 21878–21882.

25. KLUCK, R.M., E. BOSSY-WETZEL, D.R. GREEN & D.D. NEWMEYER. 1997. The release of cytochrome c from mitochondria: a primary site for Bcl-2 regulation of apoptosis. Science **275:** 1132–1136.
26. HOCKENBERY, D., G. NUÑEZ, C. MILLIMAN, R.D. SCHREIBER & S.J. KORSMEYER. 1990. Bcl-2 is an inner mitochondrial membrane protein that blocks programmed cell death. Nature **348:** 334–336.
27. MARZO I., C. BRENNER, N. ZAMZAMI, J.M. JURGENSMEIER, S.A. SUSIN, H.L.A. VIEIRA, M.C. PRÉVOST, Z. XIE, S. MATSUYAMA, J.C. REED & G. KROEMER. 1998. Bax and adenine nucleotid translocator cooperate in the mitochondrial control of apoptosis. Science **281:** 2027–2031.
28. KHARBANDA, S., P. PANDEY, L. SCHOFIELD, S. ISRAELS, R. RONCINSKE, K. YOSHIDA, A. BHARTI, Z.M. YUAN, S. SAXENA, R. WEICHSELBAUM, C. NALIN & D. KUFE. 1997. Role for Bcl-xl as an inhibitor of cytosolic cytochrome c accumulation in DNA damage-induced apoptosis. Proc. Natl. Acad. Sci. USA **94:** 6939–6942.
29. BENITEZ-BRIBIESCA, L. 1998. Assessment of apoptosis in tumor growth: importance in clinical oncology and cancer therapy. *In* When Cells Die. R.A. Lockshin, Z. Zakeri & J.L. Tilly, Eds. Wiley-Liss, Inc. New York.
30. MANON, S., B. CHAUDHURI & M. GUÉRIN. 1997. Release of cytochrome c and decrease of cytochrome c oxidase in Bax-expressing yeast cells, and prevention of these effects by coexpression of Bcl-xl. FEBS Lett. **415:** 29–32.
31. MANCINI, M., D.W. NICHOLSON, S. ROY, N.A. THORNBERRY, E.P. PETERSON, L.A. CASCIOLA-ROSEN & A. ROSEN. 1998. The caspase-3 precursor has a cytosolic and mitochondrial distribution: implications for apoptotic signaling. J. Cell. Biol. **140:** 1485–1495.
32. LA PIANA, G., E. FRANSVEA, D. MARZULLI & N.E. LOFRUMENTO. 1998. Mitochondrial membrane potential supported by exogenous cytochrome c oxidation mimics the early stages of apoptosis. Biochem. Biophys. Res. Commun. **246:** 556–561.
33. VANDEN HOEK, T.L., Z. SHAO, C.H. LI, R. ZAK, P.T. SCHUMACKER & L.B. BECKER. 1996. Reperfusion injury in cardiac myocytes after simulated ischemia. Am. Physiol. Soc. **270:** 1334–1341.
34. CHENG, E.H., D.G. KIRSCH, R.J. CLEM, R. RAVI, M.B. KASTAM, A. BEDI, K. UENO & J.M. HARDWICK. 1997. Conversion of Bcl-2 to a Bax-like death effector by caspases. Science, **278:** 1966–1988.

Role of ECM Remodeling in Thyroid Hormone-Dependent Apoptosis during Anuran Metamorphosis

SASHKO DAMJANOVSKI,[a] TOSIKAZU AMANO,[a] QING LI,[a] SHUICHI UEDA,[b] YUN-BO SHI,[a,c] AND ATSUKO ISHIZUYA-OKA[b,c]

[a]*Laboratory of Molecular Embryology, National Institute of Child Health and Human Development, National Institutes of Health, Bethesda, Maryland 20892-5431, USA*

[b]*Department of Histology and Neurobiology, Dokkyo University School of Medicine, Mibu, Tochigi 321-02, Japan*

ABSTRACT: Programmed cell death or apoptosis is an important aspect in organogenesis and tissue remodeling. It is precisely controlled both temporally and spatially during development. Amphibian metamorphosis is an excellent model to study developmental control of apoptosis in vertebrates. This process involves the transformation of essentially every organ/tissue as tadpoles change to frogs, yet is controlled by a single hormone, thyroid hormone (TH). Although different organs and tissues undergo vastly different developmental changes, including *de novo* development and total resorption, most require apoptotic elimination of at least some cell types. Such properties and the dependence on TH make frog metamorphosis a unique model to isolate and functionally characterize genes participating in the regulation of tissue specific cell death during organ development in vertebrates. Indeed, molecular studies of the TH-dependent gene regulation cascade have led to the discovery of a group of genes encoding matrix metalloproteinases (MMPs) participating in metamorphosis. *In vivo* and *in vitro* studies have provided strong evidence to support a role of MMP-mediated remodeling of the extracellular matrix in regulating apoptotic tissue remodeling during metamorphosis.

KEYWORDS: Thyroid hormone; Amphibian metamorphosis; Matrix metalloproteinases; Apoptosis

INTRODUCTION

Programmed cell death or apoptosis is critical for proper balance among different cell types in different organs and during development. Misregulation of apoptosis can be of detrimental consequences to organ function, leading to diseases and tumor development. Recent advances in cell death detection methodologies have allowed

[c]Address for correspondence: Dr. Yun-Bo Shi, Building 18T, Rm. 106, NICHD, NIH, Bethesda, MD, 20892. Voice: 301-402-1004; fax: 301-402-1323.
Shi@helix.nih.gov; or
Atsuko Ishizuya-Oka, Department of Histology and Neurobiology, Dokkyo University School of Medicine, Mibu, Tochigi 321-02, Japan. Voice: 81-282-87-2124; fax: +81-282-86-1463.
i-oka@dokkyomed.ac.jp

the demonstration that cell death occurs naturally in a wide variety of developmental and pathological processes.[1–3]

In vitro studies have shown that many intra- and extra-cellular factors influence cell fate. Of particular interest here is the fact that most cells in an organism are in constant contact with the extracellular matrix (ECM). Although alterations of the ECM undoubtedly affect cellular events, how and what roles are played by ECM remodeling in various physiological and developmental settings remain largely unknown. Here we will use the thyroid hormone (TH)-induced apoptosis during metamorphosis of South African toad *Xenopus laevis* as an example to demonstrate a role for ECM remodeling in apoptosis during organ development.

THYROID HORMONE AND ANURAN METAMORPHOSIS

Nearly nine decades ago, J.F. Gundernatsch[4] found that a substance(s) in the thyroid gland could induce amphibian metamorphosis. Shortly thereafter, E.C. Kendall[5] showed that the active ingredient is thyroid hormone (TH). These studies led to the isolation and structural characterization of two natural thyroid hormones, 3, 5, 3′, 5′-tetraiodothyronine (T_4), commonly known as thyroxine, and 3, 5, 3′-triiodothyronie (T_3), with T_3 being the more active TH.

Three independent lines of evidence firmly establish the causative role of TH in anuran metamorphosis.[6] First, elevations in the circulating plasma concentrations of the thyroid hormone T_3 and T_4 correlate with metamorphosis. In *Xenopus laevis*, there is little TH before stage 54 when tadpoles grow rapidly but exhibit little morphological changes.[7,8] During prometamorphosis (stage 54–58), endogenous T_3 and T_4 levels rise in the plasma. Accompanying this, tadpoles undergo both growth and morphological transformation, most noticeable of which is the development of the hind limbs. Then, at the climax of metamorphosis (stages 58–66), TH is at peak levels and the tadpoles stop feeding and undergo a rapid metamorphic transition. Upon the completion of metamorphosis at stage 66, the plasma TH levels also decrease. The second line of evidence supporting the role of thyroid hormone in metamorphosis comes from the ability of TH to induce precocious metamorphosis. Thus, precocious metamorphosis can be induced by simply adding physiological concentrations of TH to the rearing water of premetamorphic tadpoles or even the culture medium of tadpole organs such as the limb, tail and intestine. Finally, amphibian metamorphosis can be prevented by blocking the synthesis of endogenous TH. This can be accomplished by thyroidectomy or using chemical inhibitors (goitrogens). Such actions can be reversed by treating the resulting tadpoles with exogenous TH.

MORPHOLOGICAL CHANGES DURING METAMORPHOSIS

There are three major types of changes taking place during metamorphosis.[6,9] The first one is the complete resorption of tadpole-specific organs such as the tail and gill. At the other extreme, frog-specific organs such as the limb develop *de novo* from undifferentiated blastemal cells.

Most of the organs undergo the third type of change, partial but drastic remodeling of the existing larval organs into their adult forms. One of the best studied is the intestine (FIG. 1).[9,10] During metamorphosis, the larval epithelium undergoes complete degeneration. Concurrently, adult epithelial cells, whose origin is yet unknown, rapidly proliferate. Toward the end of metamorphosis, the adult epithelial cells differentiate to form a multiply folded epithelium, which is underlined by a well-developed connective tissue and outer muscle layers. Such an organization resembles that in adult mammalian and avian intestine.[10] The less differentiated or stem cells of the epithelium are localized and proliferate at the troughs of the folds. As the daughter cells of these stem cells migrate toward the crest, they gradually differentiate into the absorptive epithelium possessing the brush border. After a finite period of time, the fully differentiated epithelial cells at the crest are sloughed off into the lumen and are replaced by the newly arrived differentiated epithelial cells. Such a self-renewal process is similar to that in higher vertebrates.[10]

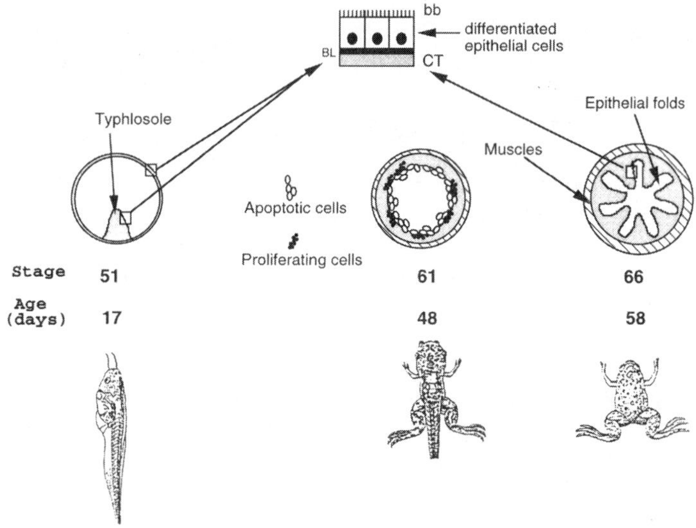

FIGURE 1. Schematic diagram outlining intestinal transformation during *Xenopus* metamorphosis. Tadpole intestine has a single fold, the typhlosole, where connective tissue (CT) is abundant. The frog intestine has a multiply folded epithelium. The outer muscle layers are thin in tadpoles but much thicker in frogs. *Open circles*: apoptotic epithelial cells; *filled circles*: proliferating adult epithelial cells; bb: brush border of the epithelium; BL: basal lamina (the ECM separating the epithelium and CT). Stages 51, 61, 66, correspond to premetamorphic tadpole stage, climax of metamorphosis, and end of metamorphosis, respectively.

TH-INDUCED APOPTOSIS DURING TISSUE REMODELING

Selective removal of specific cells plays important roles in many developmental and pathological processes. However, such cells are quickly removed upon death and it had been difficult to detect them. On the other hand, the resorption of the tadpole tail involves complete removal of all cell types within a short developmental period. This offers an excellent model system to study cell death. Kerr *et al.*[11] were the first to report that cell death during tail resorption occurs with the distinct morphological changes of apoptosis, including membrane blebbing, chromatin condensation close to the nuclear envelope, and cytoplasmic and nuclear fragmentation to form membrane-enclosed residues (apoptotic bodies) containing nuclear and cytoplasmic fragments, typical of apoptosis in mammals.

Apoptosis is not unique to resorbing organs like the tail.[6] Studies on remodeling intestine during *Xenopus laevis* metamorphosis have revealed that the degeneration of larval intestinal epithelial cells also occurs through apoptosis.[12,13] The resulting apoptotic bodies are removed at least in part through phagocytosis by macrophages migrated from the underlying connective tissue.[14] While the mechanism that triggers these macrophages to migrate across the basal lamina into the degenerating larval epithelium is unclear, these macrophages engulfing the apoptotic bodies appear to be excreted into the intestinal lumen.

The apoptosis observed during metamorphosis is thyroid hormone-dependent. In fact, it can be induced even when individual organs such as the tail and intestine isolated from premetamorphic tadpoles are cultured *in vitro* in the presence of physiological levels of TH.[15,16] This *in vitro* TH-induced apoptosis accompanies specific tissue transformations, i.e., the resorption of the tail and remodeling of the intestine, that closely reflect the *in vivo* events.

Our studies on intestinal cells support a cell autonomous response to TH. We have isolated partially purified intestinal epithelial cells and fibroblasts from tadpoles of *Xenopus laevis* and cultured them *in vitro*.[17,18] Both cell types can proliferate *in vitro* but with limited life spans, especially the epithelial cells. When the cells are treated with TH, both the epithelial and fibroblastic cells increase their DNA synthesis, suggesting that TH can stimulate the proliferation of both cell types. However, the epithelial cells are induced to undergo rapid apoptosis by TH while the fibroblasts are refractory to this TH-induced cell death. Thus, TH-treatment of primary cells in culture appears to accurately reproduce the *in vivo* behaviors of these cells during metamorphosis.

MOLECULAR MECHANISM OF TH ACTION

Thyroid Hormone Receptors

TH is known to regulate gene transcription through thyroid hormone receptors (TRs). TRs belong to the superfamily of nuclear receptors that also include steroid receptors, retinoic acid receptors, and 9-*cis* retinoic acid receptors (RXRs).[19] Like most receptors of this family, TR has a DNA binding domain in its N-terminal half that recognizes thyroid hormone response elements (TREs), and a TH-binding domain in its C-terminal half.[19–22] Although TRs can bind to TREs as monomers or homodimers, they most likely function as heterodimers formed with RXRs *in vivo*. Upon binding to TREs, TR/RXR heterodimers can regulate the transcription of the target genes in a TH-dependent manner.

Thyroid Hormone Receptor Expression during Intestinal Metamorphosis

As in higher vertebrates, both TRα and TRβ are present in *Xenopus laevis* as well as *Rana catesbeiana*.[23–25] We and others have analyzed the expression of TR genes in different organs at different developmental stages in *Rana catesbeiana* and *Xenopus laevis*.[24–26] In general, TRα and TRβ genes are upregulated in an organ when metamorphosis occurs. For example, in the hindlimb of *Xenopus laevis*, the receptor mRNAs are high when limb morphogenesis, i.e., digit formation, takes place around stages 52–56, while they are upregulated only after stage 62 in the tail because tail resorption occurs mainly after stage 62. Moreover, we have found that RXR genes (RXRα and γ) are coordinately expressed with the TR genes in various organs,[27] supporting a role of RXR in the formation of TR/RXR heterodimers to mediate the effects of TH during metamorphosis.

In the *Xenopus* intestine, the expression of TRα and RXRα genes do not change substantially. However, TRβ and RXRγ genes are upregulated during intestinal remodeling,[27] paralleling the rise in plasma TH levels.[8] By using *in situ* hybridization, we have carried out a detailed analysis of TRβ gene expression during intestinal metamorphosis.[28] Little or no TRβ mRNA is present in any tissue within the premetamorphic intestine. Around stage 57, TRβ mRNA becomes detectable in some larval epithelial cells and by stage 59, just prior to the onset of larval epithelial apoptosis, all larval epithelial cells express high levels of TRβ mRNA. Subsequently, as apoptosis takes place in the larval epithelium, the TRβ mRNA levels are downregulated in the larval epithelium (Stage 61). In contrast, TRβ expression is high in the proliferating adult epithelial cells as soon as they can be identified as small islets between the degenerating larval epithelium and the connective tissue. The TRβ expression in the adult epithelial cells remains high until stage 62 when adult epithelial cells begin to differentiate. Likewise, in the connective tissue and muscles, TRβ mRNA levels are high when these cells proliferate but are downregulated as these cells differentiate to form the adult connective tissue and muscles, respectively. Thus, TRβ appears to be involved in regulating genes that promote both cell proliferation and apoptosis, depending upon the cell types in which it is expressed. In particular, high levels of TRβ mRNA appears to be incompatible with high degrees of differentiation. Thus, in the differentiated larval epithelial cells, high levels of TRβ expression are associated with apoptosis. On the other hand, as the cells of the adult epithelium, connective tissue, and muscles begin to differentiate, they need to shut down their TRβ expression to prevent likely deleterious (apoptotic) consequences associated with high levels of TRβ mRNA.

Thyroid Hormone Response Genes in the Intestine

The key to understand the TH-dependent metamorphosis is to identify and functionally characterize the TH-regulated genes within metamorphosing tissues. Of particular interest are the direct TH-response genes as they are likely the ones that initiate, together with other existing factors, the tissue specific responses to TH. A subtractive screen has isolated over 20 intestinal genes that are regulated within 24 h of TH treatment of premetamorphic tadpoles.[6] Most of these genes appear to be direct TH-response genes, i.e., being regulated directly by TRs. Cloning and sequence analysis of the full-length cDNAs have revealed that these TH-response genes belong to several different groups.[6]

The first group includes genes encoding transcription factors, such as TRβ and NF-I, etc. These genes are expected to directly regulate the expression of downstream genes, thus influencing cell behavior during metamorphosis. The second group contains genes that can affect cell-cell and cell-ECM (extracellular matrix) interactions, e.g., genes encoding matrix metalloproteinases (MMPs), which degrade various components of the ECM. The rest of the genes regulated by TH within the first 24 h of treatment of premetamorphic tadpoles include genes encoding various other enzymes and cell specific proteins. Their functions during metamorphosis are likely to vary depending on the cell types in which they are expressed.

The existence of multiple classes of TH-response genes suggest that TH simultaneously induces many intra- and extra-cellular events during metamorphosis. It is unclear how these events cooperate to affect the expression of downstream indirect TH-response genes, many of which have been identified through various means.[6]

Expression studies have provided evidence to support the involvement of the TH-response genes in intestinal remodeling. Furthermore, the developmental expression profiles of both the direct and indirect response genes can be reproduced by treating premetamorphic tadpoles with physiological concentrations of TH, supporting the participation of these genes in intestinal remodeling.

ROLES OF MMPS AND ECM IN LARVAL CELL DEATH AND ADULT TISSUE MORPHOGENESIS DURING INTESTINAL REMODELING

MMPs are extracellular enzymes that are capable of degrading various components of the ECM.[29,30] This growing family of enzymes includes collagenases, gelatinases, and stromelysins, etc., which have different but often overlapping substrate specificity. They are secreted into the ECM as proenzymes with the exception of stromelysin-3 (ST3) and membrane-type MMPs, which appear to be secreted in the active form and membrane-bound, respectively.[31,31a,32] The proenzymes can be activated in the ECM or on the cell surface[32a] through the proteolytic removal of their propeptide. Thus differential expression and activation of various MMPs can result in specific remodeling of the ECM.

The identification of MMPs as TH-response genes suggests that they participate in ECM remodeling to influence cell transformation. The ECM provides the essential physical support for most types of cells within an organism. It consists of multiple proteins and non-protein components that are connected through complicated interactions.[33,34] Among the common components are proteins such as collagen, fibronectin, and laminin. These proteins interact with each other and/or other components such as proteoglycans to form an extracellular network. Cells can directly interact with this extracellular network through various cell surface receptors for ECM components.[35,36] In addition, the ECM also serves as a reservoir of extracellular factors such as growth factors and morphogens.[37,38] These molecules are trapped at various sites within ECM. Some can interact with cells through their receptors while others are kept inactive by ECM. Thus a changes in the ECM can result in two types of signals for nearby cells, direct signaling through ECM receptors and indirect effects through the alteration in the availability of extracellular factors that interact with cells.

ECM Remodeling during Intestinal Metamorphosis

The intestinal epithelium is separated from the connective tissue by a distinct ECM, the basal lamina, which is composed of laminin, entactin, collagens, and proteoglycans, etc.[33,34] As the epithelium undergoes metamorphic transformations, the basal lamina remodels concurrently. In premetamorphic *Xenopus laevis* tadpoles, the intestinal basal lamina lining the larval epithelium remains thick until the larval epithelium finally disappears through massive apoptosis.[10] Interestingly, the basal lamina appears to be much more permeable at the climax of metamorphosis (stage 60–63) in spite of the increased thickness. This permeability is reflected by the frequently observed migration of macrophages across the ECM into the degenerating epithelium, where they participate in the removal of apoptotic epithelial cells. In addition, extensive contacts are present between the proliferating adult epithelial cells and the fibroblasts on the other side of the basal lamina.[10] Larval epithelial cells are completely replaced by the adult epithelial cells around stages 62–63. After stage 63, with the progress of intestinal morphogenesis, i.e. intestinal fold formation, the adult epithelial cells differentiate into absorptive epithelium. Concurrently, the basal lamina underlining the differentiated adult epithelium becomes thin and flat again. Such changes support a role of ECM in intestinal transformation.

Direct evidence for a role of ECM in influencing cell fate has come from *in vitro* studies. As described above, tadpole intestinal epithelial cells can be cultured *in vitro* and induced to undergo apoptosis by TH just as *in vivo*.[17,18] When the plastic culture dishes are coated with various ECM proteins, the cells become resistant to TH-induced cell death. Consistent with these apoptosis-inhibiting effects of the ECM coatings, when proliferating/differentiating adult epithelial cells of the intestine at stage 64 are cultured *in vitro* on plastic dishes, they also undergo TH-induced apoptosis. *In vivo*, these adult cells proliferate and differentiate instead of undergoing apoptosis in the presence of high levels of circulating plasma TH. Thus, dissociating the adult cells from the ECM alters their response to TH.

Involvement of MMPs in Metamorphosis

The participation of MMPs in amphibian metamorphosis was first implicated over 30 years ago by the drastic increases in collagen degradation activity in the resorting tadpole tail.[39] However, the cloning of frog MMP genes came only after two *Xenopus* MMP genes were identified as thyroid hormone-inducible genes from subtractive differential screens.[6] The full-length proteins encoded by these two genes are highly homologous to their mammalian counterparts, collagenase-3 (Col3) and stromelysin-3 (ST3), respectively,[40,41] suggesting a functional conservation as well. Another frog MMP, the *Rana catesbeiana* collagenase-1, was cloned by screening an expression cDNA library with an antiserum against purified *Rana* tail collagenase.[42] All three frog MMPs as well as the subsequently cloned *Xenopus* collagenase-4 (Col4) are upregulated by TH in the tail while only ST3 is significantly activated during intestinal remodeling.[40–45]

In situ hybridization analyses have revealed that ST3 is highly expressed in the fibroblastic cells underlying the degenerating larval epithelium and proliferating adult epithelial cells.[44,46] In contrast, Col3 and Col4 have little or sporadic expression in these fibroblasts.[44] Furthermore, ST3 expression is temporally correlated with epithelial apoptosis and ECM remodeling.[46] Moreover, the ST3 gene is activated prior to cell death.[12,41] In contrast, Col3 and Col4 expression do not change significantly.[43] Using a human cDNA as a probe, we also found that *Xenopus* gelatinase A (Gel A) is not upregulated until stage 62,[41] when larval cell death is essentially complete. Such correlation argues that ST3 directly or indirectly causes specific degradation/cleavage of certain ECM components that facilitates ECM remodeling to alter cell-cell and/or cell-ECM interactions. Such changes in turn facilitate larval epithelial cell death. The other MMPs, especially gelatinase A, may be involved in the removal of the ECM associated with the degenerated larval epithelium.

The expression of these MMPs at stage 62 or later in the intestine suggests a role for them in adult tissue morphogenesis. In particular, both ST3 and GelA are highly expressed at stage 62 but repressed after stage 64. During this period, adult epithelial cells and cells of the connective tissue and muscles differentiate. The morphogenesis of intestinal folds also progress through migration of the epithelial and/or fibroblastic cells, thereby establishing a crest-trough axis of the fold that is similar to the villus-crypt axis in higher vertebrates.[10] Thus, it is highly likely that the MMPs participates in the establishment of new adult basal lamina to facilitate differentiation and/or cell migration.

Direct support for a role of ST3 in intestinal metamorphosis has come from *in vitro* organ culture studies (unpublished observations). By adding an antibody against the catalytic domain of ST3 in culture medium of tadpole intestinal fragments in the presence or absence of TH, we have shown that the antibody inhibits both larval epithelial cell death and the invagination of the adult epithelial cells into the connective tissue, a process critical for the adult epithelial morphogenesis. The mechanism underlying this role of ST3 is yet unknown. Interestingly, mammalian ST3 is secreted in its enzymatically active form when over-expressed in tissue culture cells.[31] This has allowed the identification of one potential physiological substrate, the α1-proteinase inhibitor, a non-ECM derived serine proteinase inhibitor.[47] Although no natural ECM substrates for ST3 have been identified, such substrates may still exist. On the other hand, the ability of ST3 to cleave a non-ECM substrate raises the possibility that ST3 may affect cell behavior through both ECM and non-ECM mediated pathways.

The other frog MMPs, such as GelA, Col3, and Col4, are expected to digest specific ECM substrates just like their mammalian counterparts, thus leading to the remodeling and degradation of the ECM during metamorphosis. For example, the up-regulation of GelA at stage 62 in the intestine when cell death is mostly completed implies that GelA may participate in ECM removal after cell death. It should be pointed out that direct demonstration of the ECM degradation activity for these frog MMPs is lacking with the exception of Col4, which has been shown to be capable of cleaving native type I collagen similar to human collagenase-1.[43] However, the

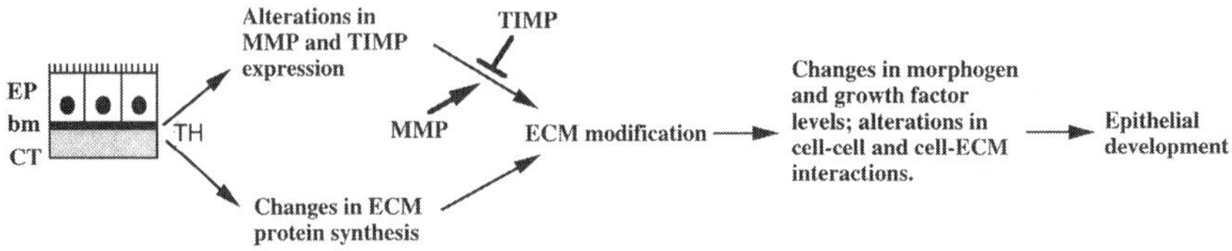

FIGURE 2. A model for ECM remodeling and its effect on intestinal epithelial development. During metamorphosis, thyroid hormone regulates directly or indirectly the expression of genes encoding MMPs. In addition, some proteins of the ECM are indirectly upregulated by TH in the intestine.[50] Furthermore, the expression of tissue inhibitors of metalloproteinases (TIMPs) may also change in response to TH. These will result in ECM remodeling, which in turn may lead to the alterations in local concentrations of growth factors or morphogens that are trapped/stored in ECM, or direct alterations of cell-cell and cell-ECM interactions, thus influencing cell fate and behavior to facilitate tissue transformation.

sequence conservation of the frog MMPs with the mammalian homologues argues for conservation at the function level.

CONCLUSION

Programmed cell death or apoptosis is critical for proper development of animals ranging from *C. elegans*, *Drosophila*, to human,[1,3,48,48a,49] downstream execution pathway seems to employ several families of evolutionarily conserved proteins. The upstream signals to induce cell death are much more complex and diverse. They vary extensively in different cell death events. ECM remodeling seems to be such an upstream signal that can facilitate or inhibit apoptosis depending upon the developmental system. The remodeling of the ECM is manifested in part through the action of MMPs and in part through new synthesis of ECM components (Fig. 2). Selective regulation of different combinations of MMPs and their inhibitors (TIMPs), which remain to be studied in amphibians, can lead to distinct modification of the ECM and thus potentially different effects on cell fate (Fig. 2). In the case of the intestinal transformation during frog metamorphosis, strong evidence supports a role of MMPs, especially stromelysin-3, in ECM remodeling, which in turn facilitates larval epithelial cell death and adult epithelial development. How such ECM remodeling alters the downstream cell death execution steps, especially the regulation of cell death genes, remains to be deciphered.

REFERENCES.

1. Wyllie, A.H., J.F.R. Kerr & A.R. Curie. 1980. Cell death: the significance of apoptosis. Int. Rev. Cytol. **68:** 251–306.
2. Schwartman, R.A. & J.A. Cidlowski. 1993. Apoptosis: the biochemistry and molecular biology of programmed cell death. Endocr. Rev. **14:** 133–151.
3. Jacobson, M.D., M. Weil & M.C. Raff. 1997. Programmed cell death in animal development. Cell **88:** 347–354.

4. GUDERNATSCH, J.F. 1912. Feeding experiments on tadpoles I: The influence of specific organs given as food on growth and differentiation: a contribution to the knowledge of organs with internal secretion. Arch. Entwicklungsmech. Org. **35:** 457–483.
5. KENDALL, K.C. 1919. Physiological action of the thyroid hormone. Am. J. Physiol. **49:** 136–137.
6. SHI, Y.-B. 1999. Amphibian Metamorphosis: From Morphology to Molecular Biology. John Wiley & Sons, Inc. New York.
7. NIEUWKOOP, P.D. & J. FABER. 1956. Normal Table of *Xenopus laevis.* North Holland Publishing. Amsterdam.
8. LELOUP, J. & M. BUSCAGLIA. 1977. La triiodothyronine: hormone de la métamorphose des amphibiens. C.R. Acad. Sci. **284:** 2261–2263.
9. GILBERT, L.I. & E. FRIEDEN. 1981. Metamorphosis: A Problem in Developmental Biology, 2nd edit. Plenum Press. New York.
10. SHI, Y.-B. & A. ISHIZUYA-OKA. 1996. Biphasic intestinal development in amphibians: Embryogenesis and remodeling during metamorphosis. Curr. Topics Devel. Biol. **32:** 205–235.
11. KERR, J.F.R., B. HARMON & J. SEARLE. 1974. An electron-microscope study of cell deletion in the anuran tadpole tail during spontaneous metamorphosis with special reference to apoptosis of striated muscle fibers. J. Cell Sci. **14:** 571–585.
12. ISHIZUYA-OKA, A. & S. UEDA. 1996. Apoptosis and cell proliferation in the *Xenopus* small intestine during metamorphosis. Cell Tissue Res. **286:** 467–476.
13. ISHIZUYA-OKA, A., S. UEDA, S. DAMJANOVSKI, Q. LI, V.C.-T. LIANG, & Y.-B. SHI. 1997. Anteroposterior gradient of epithelial transformation during amphibian intestinal remodeling: immunohistochemical detection of intestinal fatty acid-binding protein. Devel. Biol. **192:** 149–161.
14. ISHIZUYA-OKA, A. & A. SHIMOZAWA. 1992. Connective tissue is involved in adult epithelial development of the small intestine during anuran metamorphosis in vitro. Roux's Arch. Devel. Biol. **201:** 322–329.
15. ISHIZUYA-OKA, A. & A. SHIMOZAWA. 1992. Programmed cell death and heterolysis of larval epithelial cells by macrophage-like cells in the anuran small intestine in vivo and in vitro. J. Morphol. **213:** 185–195.
16. TATA, J.R., A. KAWAHARA & B.S. BAKER. 1991. Prolactin inhibits both thyroid hormone-induced morphogenesis and cell death in cultured amphibian larval tissues. Devel. Biol. **146:** 72–80.
17. SU, Y., Y. SHI & Y.-B. SHI. 1997. Cyclosporin A but not FK506 inhibits thyroid hormone-induced apoptosis in *Xenopus* tadpole intestinal epithelium. The FASEB J. **11:** 559–565.
18. SU, Y., Y. SHI, M. STOLOW & Y.-B. SHI. 1997. Thyroid hormone induces apoptosis in primary cell cultures of tadpole intestine: cell type specificity and effects of extracellular matrix. J. Cell Biol. **139:** 1533–1543.
19. MANGELSDORF, D.J., C. THUMMEL, M. BEATO, P. HERRLICH, G. SCHUTZ, K. UMESONO, B. BLUMBERG, P. KASTNER, M. MARK, P. CHAMBON & R.M. EVANS. 1995. The nuclear receptor superfamily: the second decade. Cell **83:** 835–839.
20. LAZAR, M.A. 1993. Thyroid hormone receptors: multiple forms, multiple possibilities. Endocr. Rev. **14:** 184–193.
21. TSAI, M.-J. & B.W. O'MALLEY. 1994. Molecular mechanisms of action of steroid/thyroid receptor superfamily members. Ann. Rev. Biochem. **63:** 451–486.
22. YEN, P.M. & W.W. CHIN. 1994. New advances in understanding the molecular mechanisms of thyroid hormone action. Trends Endocrinol. Metab. **5:** 65–72.
23. YAOITA, Y., Y.B. SHI & D.D. BROWN. 1990. *Xenopus laevis* α and β thyroid hormone receptors Proc. Natl. Acad. Sci. USA **87:** 7090–7094.
24. SCHNEIDER, M.J. & V.A. GALTON. 1991. Regulation of c-erbA-α messenger RNA species in tadpole erythrocytes by thyroid hormone. Mol. Endocrinol. **5:** 201–208.
25. HELBING, C.C., G. GERGELY & B.G. ATKINSON. 1992. Sequential up-regulation of thyroid hormone b receptor, ornithine transcarbamylase, and carbamyl phosphate synthetase mRNAs in the liver of *Rana catesbeiana* tadpoles during spontaneous and thyroid hormone-induced metamorphosis. Devel. Genet. **13:** 289–301.

26. SHI, Y.-B., Y. SU, Q. LI & S. DAMJANOVSKI. 1998. Auto-regulation of thyroid hormone receptor genes during metamorphosis: roles in apoptosis and cell proliferation. Int. J. Devel. Biol. **42:** 107–116.
27. WONG J. & Y.-B. SHI. 1995. Coordinated regulation of and transcriptional activation by *Xenopus* thyroid hormone and retinoid X receptors. J. Biol. Chem. **270:** 18479–18483.
28. SHI, Y.-B. & A. ISHIZUYA-OKA. 1997. Autoactivation of *Xenopus* thyroid hormone receptor β gene correlates with larval epithelial apoptosis and adult cell proliferation. J. Biomed. Sci. **4:** 9–18.
29. ALEXANDER, C.M. & Z. WERB. 1991. Extracellular matrix degradation. *In* Cell Biology of Extracellular Matrix, 2nd edit. E.D. Hay, Ed.: 255–302. Plenum Press. New York.
30. BIRKEDAL-HANSEN, H., W.G.I. MOORE, M.K. BODDEN, L.J. WINDSOR, B. BIRKEDAL-HANSEN, A. DECARLO & J.A. ENGLER. 1993. Matrix metalloproteinases: a review. Crit. Rev. Oral Biol. Med. **4:** 197–250.
31. PEI, D. & S.J. WEISS. 1995. Furin-dependent intracellular activation of the human stromelysin-3 zymogen. Nature **375:** 244–247.
31a. PEI, D. 1999. Leukolysin/MMP25/MT6-MMP: a novel matrix metalloproteinase specifically expressed in the leukocyte lineage. Cell Res. **9:** 291–303.
32. URIA, J.A. & Z. WERB. 1998. Matrix metalloproteinases and their expression in mammary gland. Cell Res. **8:** 187–194.
32a. NAGASE, H. 1998. Cell surface activation of progelatinase A (proMMP-2) and cell migration. Cell Res. **8:** 179–186.
33. HAY, E.D. 1991. Cell Biology of Extracellular Matrix, 2nd edit. Plenum Press. New York.
34. TIMPL, R. & J.C. BROWN. 1996. Supramolecular assembly of basement membranes BioEssays **18:** 123–132.
35. BROWN, K.E. & K.M. YAMADA. 1995. The role of integrins during vertebrate development. Devel. Biol. **6:** 69–77.
36. SCHMIDT, J.W., P.A. PIEPENHAGEN & W.J. NELSON. 1993. Modulation of epithelial morphogenesis and cell fate by cell-to-cell signals and regulated cell adhesion. Semin. Cell Biol. **4:** 161–173.
37. VUKICEVIC, S., H.K. KLEINMAN, F.P. LUYTEN, A.B. ROBERTS, N.S. ROCHE & A.H. REDDI. 1992. Identification of multiple active growth factors in basement membrane matrigel suggests caution in interpretation of cellular activity related to extracellular matrix components. Exp. Cell Res. **202:** 1–8.
38. WERB, Z., C.J. SYMPSON, C.M. ALEXANDER, N. THOMASSET, L.R. LUND, A. MACAULEY, J. ASHKENAS & M.J. BISSELL. 1996. Extracellular matrix remodeling and the regulation of epithelial stromal interactions during differentiation and involution. Kidney International Suppl. **54:** S68–S74.
39. GROSS, J. 1996. How tadpoles lose their tails. J. Invest. Dermatol. **47:** 274–277.
40. WANG, Z. & D. D. BROWN. 1993. Thyroid hormone-induced gene expression program for amphibian tail resorption. J. Biol. Chem. **268:** 16270–16278.
41. PATTERTON D., W.P. HAYES & Y.-B. SHI. 1995. Transcriptional activation of the matrix metalloproteinase gene stromelysin-3 coincides with thyroid hormone-induced cell death during frog metamorphosis. Devel. Biol. **167:** 252–262.
42. OOFUSA, K., S. YOMORI & K. YOSHIZATO. 1994. Regionally and hormonally regulated expression of genes of collagen and collagenase in the anuran larval skin. Int. J. Devel. Biol. **38:** 345–350.
43. STOLOW, M.A., D.D. BAUZON, J. LI, T. SEGWICK, V.C.-T. LIANG, Q.A. SANG & Y.-B. SHI. 1996. Identification and characterization of a novel collagenase in *Xenopus laevis*: possible roles during frog development. Mol. Biol. Cell **7:** 1471–1483.
44. DAMJANOVSKI, S., A. ISHIZUYA-OKA & Y.-B. SHI. 1999. Spatial and temporal regulation of collagenases-3, -4, and stromelysin-3 implicate distinct functions in apoptosis and tissue remodeling during frog metamorphosis. Cell Res. **9:** 91–105.

45. DAMJANOVSKI, S., M. PUZIANOWSKA-KUZNICKA, A. ISHIZUYA-OKA & Y.-B. SHI. 2000. Differential regulation of three thyroid hormone-responsive matrix metalloproteinase genes implicates distinct functions during frog embryogenesis. FASEB J. **14:** 503–510.
46. ISHIZUYA-OKA, A., S. UEDA & Y.-B. SHI. 1996. Transient expression of stromelysin-3 mRNA in the amphibian small intestine during metamorphosis. Cell Tissue Res. **283:** 325–329.
47. PEI, D., G. MAJMUDAR & S.J. WEISS. 1994. Hydrolytic inactivation of a breast carcinoma cell-derived serpin by human stromelysin-3. J. Biol. Chem. **269:** 25849–25855.
48. JIANG, C., E.H. BAEHRECKE & C.S. THUMMEL. 1997. Steroid regulated programmed cell death during *Drosophila* metamorphosis. Develop **124:** 4673–4683.
48a. LEE, C.-Y. & E.H. BAEHRECKE. 2000. Genetic regulation of programmed cell death in *Drosophila*. Cell Res. **10:** 193–204.
49. HENGARTNER, M.O. & H.R. HORVITZ. 1994. The ins and outs of programmed cell death during *C. elegans* development. Phil. Trans. R. Soc. **345:** 243–246.
50. AMANO, T. & K. YOSHIZATO. 1998. Isolation of genes involved in intestinal remodeling during anuran metamorphosis. Wound Rep. Reg. **6:** 302–313.

A Caspase-Independent Cell Clearance Program

The LEI/L-DNase II Pathway

ALICIA TORRIGLIA,[a] PAOLO PERANI, JEAN YVES BROSSAS, SÉVERINE ALTAIRAC, SAMIA ZEGGAI, ELISABETH MARTIN, JACQUES TRÉTON, YVES COURTOIS, AND MARIE-FRANCE COUNIS

Unité 450 INSERM, Association Claude Bernard, 75016 Paris, France

ABSTRACT: The discovery of caspase-mitochondrial pathway counts as one of the most important discovery in apoptosis biochemistry. Today, however, we begin to recognize its limits. Inhibition of caspase does not prevent cell death in many mammalian models. Targeted disruption of caspases does not impair every type of apoptosis. Other pathways, caspase independent, are now described. Here we present one of these pathways. It is a serine-protease dependent pathway and its key event is the transformation of LEI (a serine protease inhibitor) into L-DNase II (an endonuclease). When using this apoptotic pathway the cell activates, at the same time, its endonuclease activity (L-DNase II appears) and its protease activity (there is a release of inhibition of proteases).

KEYWORDS: Caspases; Apoptosis; Proteases; LEI; L-DNase II

CASPASE ACTIVATION DURING APOPTOSIS: IS THIS "THE" APOPTOTIC PATHWAY?

The genetic basis of programmed cell death was first elucidated in the worm *Caenorhabditis elegans*. Since the discovery that the cysteine protease CED-3 was essential for developmental cell death in this nematode,[1] homologous proteases were searched for in mammalian apoptosis. Fourteen of these proteases, now termed caspases[2] have been found. This is the first indication of a greater complexity of the apoptotic systems in mammals. Caspases may be divided into two groups, the caspase-1–like proteases, which are involved in cytokine processing, and all the others, which are involved in cell death.[3] The latter group can be divided according to the function of the enzyme into activators or executioners of cell death.[4] At least two caspases, caspase 3 and 7, act as executioners of cell death. Most apoptotic inducers activate one or the other enzyme and their substrate-analog inhibitor, aldehyde or fluromethylketone derivatives based on the aminoacid sequence DEVD (asp-glu-val-asp), blocks most forms of apoptosis.[5] Caspase 3 may be activated from at least two pathways (FIG. 1), the proximity pathway and the mitochondria initiated

[a]Address for correspondence : Dr. Alicia Torriglia, U 450 INSERM, 29, rue Wilhem, 75016 Paris, France. Voice: 33 (1) 45 25 21 93; fax: 33 (1)40 50 01 95.
torrigli@infobiogen.fr

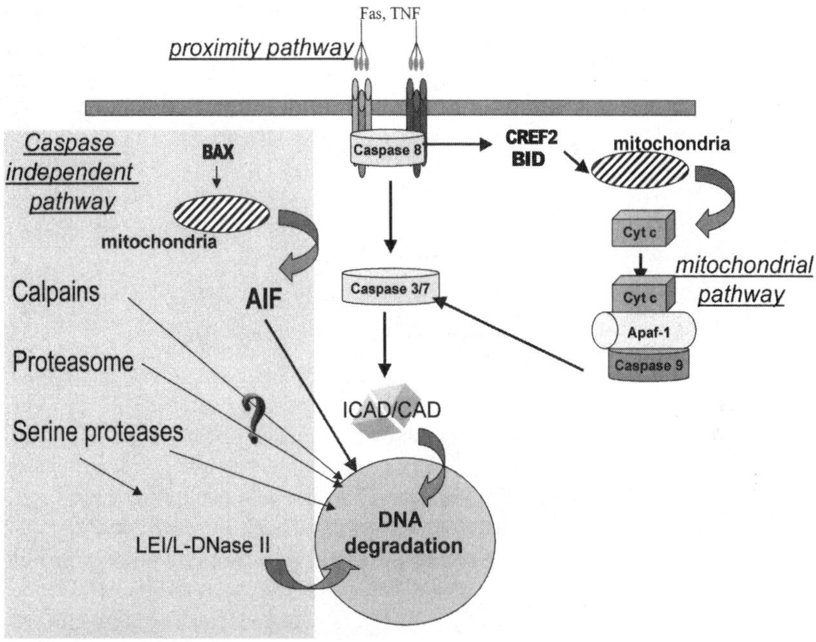

FIGURE 1. Two main apoptotic pathways are described: the caspase-dependent pathway and the caspase-independent pathway. The caspase independent pathway is not yet very clearly defined and several molecules have been described as being activated in apoptosis.

pathway. In the proximity pathway, activation of caspase-8, following its recruitment to the death-inducing signaling complex (DISC), is the critical event that transmits the death signal.[6] In the mitochondria-initiated pathway, caspase activation is triggered by the formation of a multimeric Apaf-1/cytochrome *c* complex that recruits and activates procaspase-9.[7] The existence of at least two pathways for the activation of executioner caspases reveals the lack of a common single pathway of apoptosis. Activation of caspase 3 results in degradation of many important cellular proteins. Proteolytic degradation of the inhibitor of caspase-activated DNase (ICAD) also activates caspase-activated DNase (CAD). Crosstalk exists between mitochondrial and proximity pathways. For instance, caspase 8 may cleave Bid, inducing cytochrome *c* release from the mitochondria.[8]

The discovery of this caspase-mitochondrial pathway counts as one of the most important discovery in apoptosis biochemistry. We begin, however, to recognize its limits. Inhibition of caspases does not prevent cell death in many mammalian models.[9–14] Targeted disruption of caspase 1, 2 or 11 has little effect on apoptosis.[15–18] Although mice deficient in caspase 8 die *in utero*, owing to an abnormal heart development,[19] and most mutant mice deficient in caspase 3 or caspase 9 exhibit brain malformations due to decreased apoptosis in neuroepithelium,[20–23] these caspases are dispensable in apoptosis arising during T cell selection in the thymus.[20,21,23] Mice deficient in Apaf-1 showed a phenotype more severe than the one observed in caspase 9 knockout mice.[24] For instance, Apaf-1 deficient mice show a persistence of interdigital webs not observed in caspase 9–deficient mice. Therefore, Apaf-1 might regulate other caspases, probably indirectly through adapter molecules, or have caspase-independent functions. Interestingly, Apaf-1 knockout mice have no abnormalities in thymus and most organs develop nearly normally. All these results indicate that caspases are indeed important in cell death, but they also indicate a very important redundancy of the apoptotic pathways and allow us to presume the existence of other pathways activating apoptosis.

CASPASE-INDEPENDENT PATHWAYS

Several caspase-independent pathways have been described:

1. The bax pathway: bax and bax-like are proteins of the bcl-2 family displaying a pro-apoptotic activity. They translocate to mitochondrial membranes in response to apoptotic inducers,[25,26] and liberate cytochrome *c*, AIF (Apoptosis Inducting Factor) and other pro-apoptotic molecules.[27]
2. AIF, is a mitochondrial protein located in the interstitium between mitochondrial membranes.[28] AIF is subject to mitochondrial-nuclear translocation upon induction of apoptosis.[29] In isolated HeLa cell nuclei, AIF induces apoptotic morphology and activates a nuclear DNase.[30] These phenomena seem to be all caspase independent.[31]
3. Other proteases have been implicated in apoptosis.

   a) Calpains: based on inhibitor studies, caspase-independent pathways may be mediated by calpains.[32,33]

   b) The proteasome: in *Manduca sexta*, there is a dramatic increase of polyubiquitin expression during muscle atrophy, which is correlated with an increase of protein-ubiquitin conjugates and apoptosis.[34] Proteasome inhibitors promote apoptosis in most cell lines,[35–37] but protection has been found in primary cell cultures of neurons[38,39] and thymocytes.[40] Proteasome inhibitors prevent, in some cases, disruption of mitochondrial transmembrane potential, degradation of nuclear DNA and exposure of phosphatidylserine. These data were interpreted as showing that the proteasome acted at a pre-mitochondrial and pre-caspase stage of apoptosis.[41]

   c) Serine proteases: a newly emerging group of proteases participating in apoptosis are the serine proteases. They seem to play a crucial role in the early cleavage of chromatin[42] and are activated in a very classic model of apoptosis: the induction of thymocyte cell death by glucocorticoids.[43] Granzyme A may also mediate caspase-independent CTL/target cell apoptosis, since protease inhibitors, but not caspase inhibitors blocked this cell death.[44,45] This is also the case for cathepsin D, a lysosomal serine protease: inhibitors such as pepstatin blocked apoptosis induced by TNF, Fas and interferon γ.[46] Another crucial serine protease is AP24. This protease is activated in U937 cells induced to die by TNF-α or UV radiation.[47] The specific inhibition of AP24 by DK120 (carbobenzoxy-Ala-Ala-borophe) confers better protection than inhibition by caspase 3 inhibitor.[48]

DIFFERENT PATHWAYS CAN BE ACTIVATED IN THE CELL

The data discussed above indicate that proteases other than caspases might be implicated in apoptosis. Why have these proteases attracted less attention than caspases? Perhaps several types of proteases are activated during apoptosis and the biochemical changes are more difficult to detect.

Our group, looking at activation of DNases, found that when HeLa cells were induced to undergo apoptosis by long-term culture, both Ca-Mg–dependent DNases and acid DNases were activated. These enzymes were activated through different pathways, since treatment with the tetrapeptide inhibitor YVAD-CHO completely inhibits Ca-Mg-dependent endonucleases, but had no effect on acid endonucleases[49] and no effect on apoptosis. However, we saw an increase of caspase 3/7 activity that was visualized by the cleavage of PARP. This cleavage was inhibited by the tetrapeptide inhibitor DEVD-CHO, but this inhibitor did not protect cells from apoptosis (unpublished results). These data indicate that: 1) induction of apoptosis in HeLa cells by a metabolic stress is triggered by a caspase-independent pathway; 2) activation of acid DNases during apoptosis is caspase independent, as already suggested by other authors;[50] and 3) In these cells different stimuli activate different apoptotic pathways since the induction of apoptosis by etoposide is caspase dependent.[49]

It seems then that different apoptotic pathways may exist. They can be activated in the same cell by different inducers, as seen before, but cell specificity seems also be involved, as seen in caspase 3 knockout mice. Moreover, other parameters may affect the pathway activated, like the cell cycle step or the differentiated or undifferentiated state of the cell.

THE LEI/L-DNASE II PATHWAY

Our group studies the LEI/L-DNase II pathway, which results in the activation of L-DNase II. The activation of this DNase II was first discovered in the lens during lens cell differentiation,[51] which is an apoptosis-related cellular differentiation.[52] The activation of this enzyme has also been seen in other physiological models such as neural apoptosis during retina development (unpublished results) or in cell culture.[53,54] This pathway is independent of caspases, as stated above, and appears to be dependent on serine proteases.[53]

The key molecule in this pathway is LEI (leukocyte elastase inhibitor). LEI belongs to the Serpin (serine protease inhibitors) superfamily and it can be classified among the ovalbumin serpins.[55] Like most serpins, LEI has an anti-protease activity. In its native form, it inhibits elastase, cathepsin G,[56] and probably other proteases. LEI may be post-translationally modified, either by exposure to an acidic pH[57] or by the action of proteases like elastase. Treatment of LEI by these agents is linked to a decrease of the apparent molecular weight of LEI, to a loss of its anti-protease activity, and to the appearance of an endonuclease activity.[57] We have called this endonuclease L-DNase II, since it is derived from LEI and shows almost the same pH and ion dependence as DNase II. *In vitro* experiments, using purified recombinant LEI, show that the native form has no activity on purified nuclei whereas its post-translationally activated form induces pycnosis in isolated nuclei and DNA degradation with production of oligonucleosomal ladders. Antibodies directed against L-DNase II inhibit this degradation.[57] Moreover, in different cell lines, a nuclear translocation of this enzyme is seen during apoptosis.[54] We have, therefore, a protein which has the fascinating capacity of changing its enzymatic activity and which seems implicated in some forms of apoptosis. During the last two years we have asked several questions to elucidate the role of this enzyme in apoptosis: What is the nature of the post-translational modification leading to the conversion of LEI into L-DNase II? What are the proteins involved in this conversion? How do LEI and L-DNase II work in the living and apoptotic cells?

THE POST-TRANSLATIONAL MODIFICATION LEADING TO L-DNASE II

As stated above, the postranslational modification transforming LEI into L-DNase II reduces the apparent molecular weight of LEI. Three hypotheses were evaluated to explain this observation: first LEI might be cleaved by a proteolytic enzyme and the cleaved peptide lost. We were unable to confirm this by recovery of the released peptide. Furthermore, we found that a tryptic digestion of LEI produced the same peptides as L-DNase II when studied by mass spectrometry. Finally, synthesis of a truncated protein by introducing a stop codon in the sequence of LEI yielded a protein with no DNase activity (unpublished results).

The second hypothesis was that L-DNase II was derived from LEI by a protein splicing.[58] This hypothesis was suggested by results obtained from protein sequencing. When sequencing L-DNase II, peptides from both the N-terminal and the C-terminal ends of LEI were found to be retained. Here again, we never found the spliced element.

The third hypothesis involves a conformational modification.[59] LEI is a serpin. Serpins (FIG. 2) are proteases inhibitors that act by a suicide mechanism. They interact with their target proteinases in a substrate-like manner that traps target enzymes as stable covalent acyl-enzyme complexes. This mechanism appears to require both cleavage of the serpin reactive center loop (RCL) by the proteinase and a significant conformational change in the serpin structure involving rapid insertion of the RCL into the center of the existing beta-sheet A.[60] In the native (S) state, the reactive center is exposed and is free to interact with proteinases. In the cleaved (R) state the reactive center loop forms an additional strand within the beta-sheet. We believe that this conformational modification is involved in the genesis of the endonuclease site. This determines a more compact protein conformation that may explain its higher mobility in an electrophoresis gel. This model is also in accord with the induction of DNase activity by elastase and by acidic pH which may also cut the serpin in its P1-P1′ site and induce LEI to shift to its cleaved state. The latent state is an uncleaved state in which the intact reactive center loop is integrated into the A sheet as in the cleaved form, to give an alternative R state.[61] This model is also in accord with a

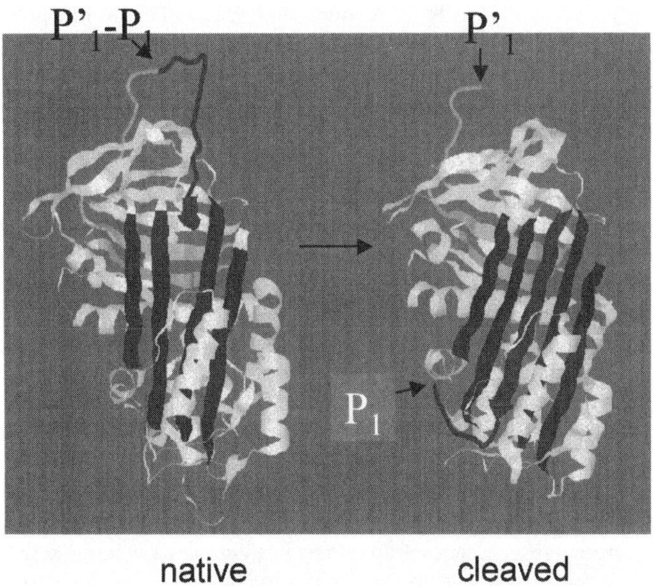

FIGURE 2. Three-dimensional structure of native and cleaved conformation of LEI. In the cleaved conformation the N-terminal portion of the reactive loop is inserted into the beta sheet A. The C-terminal peptide is retained in the molecule, inside the beta sheet B.

stabilization of the Ct peptide inside the molecule and explains why we were able to sequence this peptide from L-DNase II. Moreover, this hypothesis is supported by the behavior of mutants of LEI constructed in our laboratory (unpublished results). Using site-directed mutagenesis we mutated P10, P12, and P14 sites. This transforms LEI into a substrate of elastase and promotes the R conformation. All these mutants showed DNase activity.

PROTEINS INVOLVED IN THE CONVERSION OF LEI INTO L-DNASE II

We have shown that the conversion of LEI into L-DNase II at an acidic pH is increased by a tissue extract,[57] indicating that cellular factors might catalyze this transformation. Biochemical techniques have been used to characterize and purify these factors and two activities that have different optimum pH have been identified (unpublished results). These activities are present in some models of apoptosis activating L-DNase II, as in the embryonic retina, but absent in other models such as in HeLa cells induced to apoptosis by long-term culture. This is why we also examined some proteases that might interact with LEI. The first protease we examined was elastase. We knew that this protease is able to transform LEI but we did not know if it was implicated in apoptosis. We found that this enzyme was activated in two models—retina pigment epithelium cells induced to die by ethanol[53] and murine leukemia cells induced to die by staurosporine.[54] Other elastase-like proteases activated during apoptosis were also studied, for instance AP-24 (apoptotic protease 24 kD). This protease is activated during tumor necrosis factor or ultraviolet (UV) light–induced DNA fragmentation in U937 histiocytic lymphoma as well as UV light–induced DNA fragmentation in the BT-20 breast carcinoma, HL-60 myelocytic leukemia, and 3T3 fibroblasts.[47,48] The protease was purified by affinity chromatography with DK120, and its interaction with LEI was analyzed. We found that AP24 strongly binds LEI,[62] but we do not know if AP24 is able to activate LEI into L-DNase II. Studies are in progress to evaluate this point. This result is very interesting since it links the LEI pathway to the AP24 pathway. More work is needed to decide if both molecules, LEI and AP24 are members of the same pathway or if we have found crosstalk between two different pathways.

LEI AND L-DNASE II IN THE LIVING AND IN THE APOPTOTIC CELL

We have proposed that LEI plays a critical role in apoptosis (FIG. 3), acting as a molecular switch between living cells and apoptotic cells. Its double function is to prevent the proteolytic cascade of apoptosis in living cells while, via the transition to L-DNase, it releases this proteolytic inhibition and inducing nuclear degradation in apoptotic cells. In this model, the link between proteases and endonucleases may not be linear; that is, activation of proteases may not lead to downstream activation

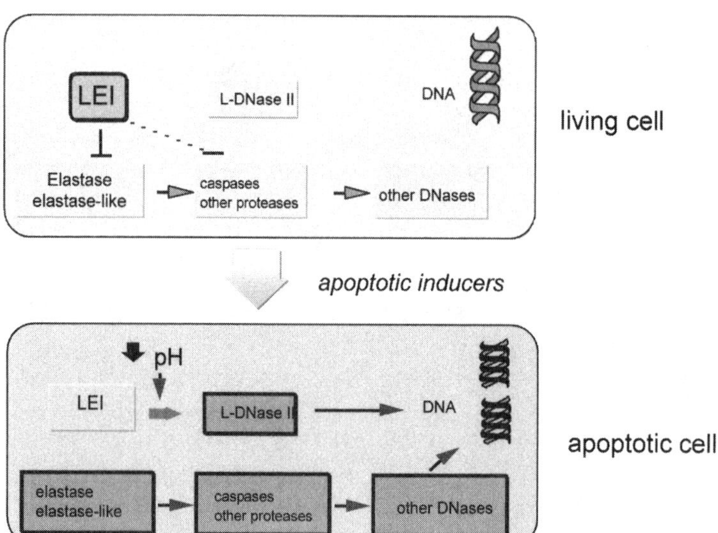

FIGURE 3. In a living cell LEI acts as an inhibitor of proteases. In an apoptotic cell, which presents a decrease of the intracellular pH, LEI is transformed into L-DNase II. An endonuclease activity appears in the cell, while the anti-protease activity decreases.

of endonucleases, but both pathways might be activated at the same time. We hypothesized that the major control element operating this switch is intracellular pH. When intracellular pH decreases in apoptotic cells the transition from LEI to L-DNase II is possible. We did several experiments to evaluate this model. Using cells stably transfected with LEI (unpublished results) we saw an increase of sensitivity of these cells to hexa-methylene-amiloride, an agent that induces apoptosis by decreasing intracellular pH. This phenomenon is related to an increase of L-DNase II in transfected cells that has been confirmed by western blot. These results are in agreement with the model proposed; we need now to know if LEI is anti-apoptotic in the living cell. Finally, we have to investigate whether this pathway is ubiquitous and may be used by any cell when the conditions required for its activation are fulfilled. Preliminary studies show that LEI is present in all tissues and the cells studied (unpublished results) but the level of expression differs from tissue to tissue.

CONCLUSION

We have described a new apoptotic pathway, the LEI–L-DNase II pathway. This is a caspase-independent pathway and may be included among the serine protease pathways, probably related to or communicating with the AP24 pathway. This pathway coexists with caspase pathways and seems to be activated by only some inductors of apoptosis. Which factors drive a cell to choose between different pathways to kill itself is one of the most exciting questions that we have to answer in the years to come.

ACKNOWLEDGMENTS

The work presented here has been supported by Retina France, Association Française pour la Recherche contre le Cancer (ARC), INSERM, contrat APEX 4X012D and Université Paris V. The authors also acknowledge their collaborators: Dr. Ivana Scovassi, Dr. Evelyne Segal-Bendirdjian, M. Chafke Belmokhtar, Dr. Hubert Hondermarck, Dr. Claudia Negri, Dr. Elisabeth Chaudun, and Dr. Susan C. Wright.

REFERENCES

1. Ellis, H.M. & H.R. Horvitz. 1986. Genetic control of programmed cell death in the nematode *C. elegans*. Cell **44:** 817–829.
2. Alnemri, E.S., D.J. Livingston, D.W. Nicholson, G. Salvesen, N.A. Thornberry, W.W. Wong & J. Yuan. 1996. Human ICE/CED-3 protease nomenclature. Cell **87:** 171.
3. Nicholson, D.W. & N.A. Thornberry. 1997. Caspases: killer proteases. Trends Biochem. Sci. **22:** 299–306.
4. Thornberry, N.A. & Y. Lazebnik. 1998. Caspases: enemies within. Science **281:** 1312–1316.
5. Budihardjo, I., H. Oliver, M. Lutter, X. Luo & X. Wang. 1999. Biochemical pathways of caspase activation during apoptosis. Annu. Rev. Cell. Dev. Biol. **15:** 269–290.

6. WAJANT, H., F.J. JOHANNES, E. HAAS, K. SIEMIENSKI, R. SCHWENZER, G. SCHUBERT, T. WEISS, M. GRELL & P. SCHEURICH. 1998. Dominant-negative FADD inhibits TNFR60-, Fas/Apo1- and TRAIL-R/Apo2-mediated cell death but not gene induction. Curr. Biol. **8:** 113–116.
7. ZOU, H., Y.LI, X. LIU & X. WANG. 1999. An APAF-1.cytochrome c multimeric complex is a functional apoptosome that activates procaspase-9. J. Biol. Chem. **274:** 11549–11556.
8. TANG, D., J.M. LAHTI & V.J. KIDD. 2000. Caspase-8 activation and bid cleavage contribute to MCF7 cellular execution in a caspase-3-dependent manner during staurosporine-mediated apoptosis. J. Biol. Chem. **275:** 9303–9307.
9. CARMODY, R.J. & T.G. COTTER. 2000. Oxidative stress induces caspase-independent retinal apoptosis in vitro. Cell Death Differ. **7:** 282–291.
10. ASSEFA, Z., A. VANTIEGHEM, M. GARMYN, W. DECLERCQ, P. VANDENABEELE, J.R. VANDENHEEDE, R. BOUILLON, W. MERLEVEDE & P. AGOSTINIS. 2000. p38 Mitogen-activated protein kinase regulates a novel, caspase-independent pathway for the mitochondrial cytochrome c release in ultraviolet b radiation-induced apoptosis. J Biol. Chem. **275:** 21416–21421.
11. MATHIASEN, I.S., U. LADEMANN & M. JAATTELA. 1999. Apoptosis induced by vitamin D compounds in breast cancer cells is inhibited by Bcl-2 but does not involve known caspases or p53. Cancer Res. **59:** 4848–4856.
12. DESHMUKH, M. & E.M. JOHNSON, JR. 2000. Staurosporine-induced neuronal death: multiple mechanisms and methodological implications. Cell Death Differ. **7:** 250–261.
13. MATEO, V., L. LAGNEAUX, D. BRON, G. BIRON, M. ARMANT, G. DELESPESSE & M. SARFATI. 1999. CD47 ligation induces caspase-independent cell death in chronic lymphocytic leukemia. Nat. Med. **5:** 1277–1284.
14. MATHIASEN, I.S., U. LADEMANN, & M. JAATTELA. 1999. Apoptosis induced by vitamin D compounds in breast cancer cells is inhibited by Bcl-2 but does not involve known caspases or p53. Cancer Res. **59:** 4848–4856.
15. KUIDA, K., J.A. LIPPKE, G. KU, M.W. HARDING, D.J. LIVINGSTON, M.S. SU & R.A. FLAVELL. 1995. Altered cytokine export and apoptosis in mice deficient in interleukin-1 beta converting enzyme. Science **267:** 2000–2003.
16. LI, P., H. ALLEN, S. BANERJEE, S. FRANKLIN, L. HERZOG, C. JOHNSTON, J. MCDOWELL, M. PASKIND, L.J. RODMAN, SALFELD, *et al.* 1995. Mice deficient in IL-1 beta-converting enzyme are defective in production of mature IL-1 beta and resistant to endotoxic shock. Cell **80:** 401–411.
17. BERGERON, L., G.I. PEREZ, G. MACDONALD, L. SHI, Y. SUN, A. JURISICOVA, S. VARMUZA, K.E. LATHAM, J.A. FLAWS, J.C. SALTER, H. HARA, M.A. MOSKOWITZ, E. LI, A. GREENBERG, J.L. TILLY & J. YUAN. 1998. Defects in regulation of apoptosis in caspase-2-deficient mice. Genes Devel. **12:** 1304–1314.
18. WANG, S., M. MIURA, Y.K. JUNG, H. ZHU, E. LI & J. YUAN. 1998. Murine caspase-11, an ICE-interacting protease, is essential for the activation of ICE. Cell **92:** 501–509.
19. VARFOLOMEEV, E.E., M. SCHUCHMANN, V. LURIA, N. CHIANNILKULCHAI, J.S. BECKMANN, I.L. METT, D. REBRIKOV, V.M. BRODIANSKI, O.C. KEMPER, O. KOLLET, T. LAPIDOT, D. SOFFER, T. SOBE, K.B. AVRAHAM, T. GONCHAROV, H. HOLTMANN, P. LONAI & D. WALLACH. 1998. Targeted disruption of the mouse Caspase 8 gene ablates cell death induction by the TNF receptors, Fas/Apo1, and DR3 and is lethal prenatally. Immunity **9:** 267–276.
20. KUIDA, K., T.S. ZHENG, S. NA, C KUAN, D. YANG, H. KARASUYAMA, P. RAKIC & R.A. FLAVELL. 1996. Decreased apoptosis in the brain and premature lethality in CPP32-deficient mice. Nature **384:** 368–372.
21. KUIDA, K., T.F. HAYDAR, C.Y. KUAN, Y. GU, C. TAYA, H. KARASUYAMA, M.S. SU, P. RAKIC & R.A. FLAVELL. 1998. Reduced apoptosis and cytochrome c-mediated caspase activation in mice lacking caspase 9. Cell **94:** 325–337.
22. WOO, M., R. HAKEM, M.S. SOENGAS, G.S. DUNCAN, A. SHAHINIAN, D. KAGI, A. HAKEM, M. MCCURRACH, W. KHOO, S.A. KAUFMAN, G. SENALDI, T. HOWARD, S.W. LOWE & T.W. MAK. 1998. Essential contribution of caspase 3/CPP32 to apoptosis and its associated nuclear changes. Genes Devel. **12:** 806–819.

23. Hakem, R., A. Hakem, G.S. Duncan, J.T. Henderson, M. Woo, M.S. Soengas, A. Elia, J.L. de la Pompa, D. Kagi, W. Khoo, J. Potter, R. Yoshida, S.A. Kaufman, S.W. Lowe, J.M. Penninger & T.W. Mak. 1998. Differential requirement for caspase 9 in apoptotic pathways in vivo. Cell **94:** 339–352.

24. Yoshida, H., Y.Y. Kong, R. Yoshida, A.J. Elia, A. Hakem, R. Hakem, J.M. Penninger & T.W. Mak. 1998. Apaf1 is required for mitochondrial pathways of apoptosis and brain development. Cell **94:**739–750.

25. Hsu, Y.T., K.G. Wolter & R.J. Youle. 1997. Cytosol-to-membrane redistribution of Bax and Bcl-X(L) during apoptosis. Proc. Natl. Acad. Sci. USA **94:** 3668–3672.

26. Gross, A. J. Jockel, M.C. Wei & S.J. Korsmeyer. 1998. Enforced dimerization of BAX results in its translocation, mitochondrial dysfunction and apoptosis. EMBO J. **17:** 3878–3885.

27. Jurgensmeier, J.M., Z. Xie, Q. Deveraux, L. Ellerby, D. Bredesen & J.C. Reed. 1998. Bax directly induces release of cytochrome c from isolated mitochondria. Proc. Natl. Acad. Sci. USA **95:** 4997–5002.

28. Susin, S.A., H.K. Lorenzo, N. Zamzami, I. Marzo, B.E. Snow, G.M. Brothers, J. Mangion, E. Jacotot, P. Costantini, M. Loeffler, N. Larochette, D.R. Goodlett, R. Aebersold, D.P. Siderovski, J.M. Penninger & G. Kroemer. 1999. Molecular characterization of mitochondrial apoptosis-inducing factor. Nature **397:** 441–446.

29. Daugas, E., S.A. Susin, N. Zamzami, K.F. Ferri, T. Irinopoulou, N. Larochette, M.C. Prevost, B. Leber, D. Andrews, J. Penninger & G. Kroemer. 2000. Mitochondrio-nuclear translocation of AIF in apoptosis and necrosis. FASEB J. **14:**729–739.

30. Zamzami, N., S.A. Susin, P. Marchetti, T. Hirsch, I. Gomez-Monterrey, M. Castedo & G. Kroemer. 1996. Mitochondrial control of nuclear apoptosis. J. Exp. Med. **183:**1533–1544.

31. Lorenzo, H.K., S.A. Susin, J. Penninger & G. Kroemer. 1999. Apoptosis inducing factor (AIF): a phylogenetically old, caspase-independent effector of cell death. Cell Death Differ. **6:** 516–524.

32. Bronk, S.F. & G.J. Gores. 1993. pH-dependent nonlysosomal proteolysis contributes to lethal anoxic injury of rat hepatocytes. Am. J. Physiol. **264:** G744–751.

33. Squier, M.K., A.C. Miller, A.M. Malkinson & J.J. Cohen. 1994. Calpain activation in apoptosis. J. Cell. Physiol. **159:** 229–237.

34. Schwartz, L.M., A. Myer, L. Kosz, M. Engelstein & C. Maier. 1990. Activation of polyubiquitin gene expression during developmentally programmed cell death. Neuron **5:** 411–419.

35. Grimm, L.M. & B.A. Osborne. 1999. Apoptosis and the proteasome. Results Probl. Cell. Differ. **23:** 209–228.

36. Shinohara, K., M. Tomioka, H. Nakano, S. Tone, H. Ito & S. Kawashima. 1996. Apoptosis induction resulting from proteasome inhibition. Biochem. J. **317:** 385–388.

37. Delic, J., P. Masdehors, S. Omura, J.M. Cosset, J. Dumont, J.L. Binet & H. Magdelenat. 1998. The proteasome inhibitor lactacystin induces apoptosis and sensitizes chemo- and radioresistant human chronic lymphocytic leukaemia lymphocytes to TNF-alpha-initiated apoptosis. Br. J. Cancer **77:** 1103–1107.

38. Sadoul, R., P.A. Fernandez, A.L. Quiquerez, I. Martinou, M. Maki, M. Schroter, J.D. Becherer, M. Irmler, J. Tschopp & J.C. Martinou. 1996. Involvement of the proteasome in the programmed cell death of NGF-deprived sympathetic neurons. EMBO J. **15:** 3845–3852.

39. Martinou I., M. Missotten, P.A. Fernandez, R. Sadoul & J.C. Martinou. 1998. Bax and Bak proteins require caspase activity to trigger apoptosis in sympathetic neurons. Neuroreport **9:** 15–19.

40. Stefanelli, C., F. Bonavita, I. Stanic, C. Pignatti, G. Farruggia, L. Masotti, C. Guarnieri & C.M. Caldarera. 1998. Inhibition of etoposide-induced apoptosis with peptide aldehyde inhibitors of proteasome. Biochem. J. **332:** 661–665.

41. HIRSCH, T., B. DALLAPORTA, N. ZAMZAMI, S.A. SUSIN, L. RAVAGNAN, I. MARZO, C. BRENNER & G. KROEMER. 1998. Proteasome activation occurs at an early, premitochondrial step of thymocyte apoptosis. J. Immunol. **161:** 35–40.
42. HUGHES, F.M. JR, R.B. EVANS-STORMS & J.A. CIDLOWSKI. 1998 Evidence that non-caspase proteases are required for chromatin degradation during apoptosis. Cell Death Differ. **5:** 1017–1027.
43. MANN, C.L., F.M. HUGHES, JR. & J.A. CIDLOWSKI. 2000. Delineation of the signaling pathways involved in glucocorticoid-induced and spontaneous apoptosis of rat thymocytes. Endocrinology **141:** 528–538.
44. SARIN, A., M.S. WILLIAMS, M.A. ALEXANDER-MILLER, J.A. BERZOFSKY, C.M. ZACHARCHUK & P.A. HENKART. 1997. Target cell lysis by CTL granule exocytosis is independent of ICE/Ced-3 family proteases. Immunity **6:** 209–215.
45. CHANG, T.W. & H.N. EISEN. 1980. Effects of N alpha-tosyl-L-lysyl-chloromethylketone on the activity of cytotoxic T lymphocytes. J. Immunol. **124:** 1028–1033.
46. DEISS, L.P., H. GALINKA, H. BERISSI, O. COHEN & A. KIMCHI. 1996. Cathepsin D protease mediates programmed cell death induced by interferon-gamma, Fas/APO-1 and TNF-alpha. EMBO J. **15:** 3861–3870.
47. WRIGHT, S.C., Q.S. WEI, J. ZHONG, H. ZHENG, D.H. KINDER & J.W. LARRICK. 1994. Purification of a 24-kD protease from apoptotic tumor cells that activates DNA fragmentation. J. Exp. Med. **180:** 2113–2123.
48. WRIGHT, S.C., U. SCHELLENBERGER, H. WANG, D.H. KINDER, J.W. TALHOUK & J.W. LARRICK. 1997 Activation of CPP32-like proteases is not sufficient to trigger apoptosis: inhibition of apoptosis by agents that suppress activation of AP24, but not CPP32-like activity. J. Exp. Med. **186:** 1107–1117
49. TORRIGLIA, A., C. NEGRI, E. CHAUDUN, E. PROSPERI, Y. COURTOIS, M.F. COUNIS & A.I. SCOVASSI. 1999. Differential involvement of DNases in HeLa cell apoptosis induced by etoposide and long term-culture. Cell Death Differ. **6:** 234–244.
50. FAMULSKI, K.S., D. MACDONALD, M.C. PATERSON & E. SIKORA. 1999. Activation of a low pH-dependent nuclease by apoptotic agents. Cell Death Differ. **6:** 281–289.
51. TORRIGLIA, A., E. CHAUDUN, F. CHANY, J. C. JEANNY, Y. COURTOIS & M.F. COUNIS. 1995 Involvement of DNase II in nuclear degeneration during lens cell differentiation J. Biol. Chem. **27:** 28579–28585.
52. COUNIS, M.F., E. CHAUDUN, C. ARRUTI, L. OLIVER, M. SANWAL, Y. COURTOIS & A. TORRIGLIA. 1998 Advances in nuclear degradation during lens cell differentiation. Cell Death Differ. **5:** 251–261.
53. BROSSAS, J.Y., R. TANGUY, P. PERANI, A. TORRIGLIA, E . CHAUDUN, M.F. COUNIS, Y. COURTOIS & J. TRÉTON. 1999. Une nouvelle voie de la mort cellulaire dans les cellules de l'épithélium pigmentaire de la rétine (ERP) humaine en culture. Les séminaires ophtalmologiques IPSEN vision, génome et cerveau. **10:** 119–130.
54. BELMOKHTAR C.A., A. TORRIGLIA, M.F. COUNIS, Y. COURTOIS, A. JACQUEMIN-SABLON & E. SÉGAL-BENDIRDJIAN. 2000. Nuclear translocation of a leukocyte elastase inhibitor/elastase complex during staurosporine-induced apoptosis: role in the generation of nuclear L-DNase II activity. Exp. Cell Res. **254:** 99–109.
55. KORPULA-MASTALERZ R. & A. DUBIN. 1996. The intracellular serpin family. Acta Biochim. Polonica **43:** 419–430.
56. POTEMPA, J., A. DUBIN, W. WATOREK & J. TRAVIS. 1988. An elastase inhibitor from equine leukocyte cytosol belongs to the serpin superfamily. Further characterization and aminoacid sequence of the reactive center. J. Biol. Chem. **263:** 7364–7369.
57. TORRIGLIA, A., P. PERANI, J.Y. BROSSAS, E. CHAUDUN, J. TRÉTON, Y. COURTOIS & M.F. COUNIS.1998. L-DNase II, a molecule that links proteases and endonucleases in apoptosis, derives from the ubiquitous serpin Leukocyte Elastase Inhibitor. Mol. Cell. Biol. **18:** 3612–3619.
58. DERBYSHIRE, V. & M. BELFORT. 1998. Lightning strikes twice: intron-intein coincidence. Proc. Natl. Acad. Sci. USA **95:**1356–1357.
59. PERANI, P., J. Y. BROSSAS, E. CHAUDUN, J. TRETON, M.-F. COUNIS, Y. COURTOIS & A. TORRIGLIA. 1999. L-DNase II, a DNase involved in apoptosis, derives from the serpin, leukocyte elastase inhibitor.[abstract] International Meeting on Serpins, Cambridge, UK.

60. SCHULZE, A.J, R. HUBER, W. BODE & R.A. ENGH. 1994. Structural aspects of serpin inhibition. FEBS Lett. **344:**117–124.
61. WHISSTOCK, J.C., R. SKINNER, R.W. CARRELL & A.M. LESK. 2000. Conformational changes in serpins: I. The native and cleaved conformations of alpha(1)-antitrypsin J. Mol. Biol. **296:** 685–699.
62. ALTAIRAC, S., S.C. WRIGHT., Y. COURTOIS & A. TORRIGLIA. 2000. Régulation de l'activité de la L-DNaseII dans l'apoptose [abstract]. French Society for Biochemistry and Molecular Biology.

Role of Apoptosis in the Response of Lung Carcinomas to Anti-Cancer Treatment

BERTRAND JOSEPH,[a] ROLF LEWENSOHN,[b] AND BORIS ZHIVOTOVSKY[a,c]

[a]*Division of Toxicology, Institute of Environmental Medicine, Karolinska Institutet, SE-171 77 Stockholm, Sweden*

[b]*Unit of Medical Radiobiology, Cancer Center Karolinska, SE-171 76 Stockholm, Sweden*

ABSTRACT: Resistance of tumor cells to treatment often accounts for the failure of traditional forms of anti-cancer therapy. It is well known that tumors from the same histological group and stage of development are highly heterogeneous in their sensitivity to therapy. Among the factors that can influence tumor sensitivity are DNA repair capacity, distribution of cells throughout the cell cycle, proliferation potential, etc. In many cases, anti-cancer therapy eliminates tumor cells via apoptosis, an active form of cell death characterized by cell shrinkage and the removal of cells in a neat, orderly fashion. However, this process is not always efficient. In the present review, the precise role that apoptosis plays in the response of lung carcinomas to chemotherapy and radiation treatment is discussed.

KEYWORDS: Lung carcinoma; Apoptosis; Cancer therapy; Bcl-2 family; p53; Ras family; c-Myc; Rb; p16^{INK4}; Cyclin D1; Fas-FasL; Caspases; IAP family

INTRODUCTION

Lung cancer, the major cause of cancer deaths in the western world, is a growing problem in developing countries.[1] Lung cancer can be divided into two broad groups: small cell lung cancer (SCLC), which accounts for about 25% of bronchogenic carcinomas, and non-small cell lung cancer (NSCLC).[2] SCLC cells normally possess neuroendocrine-like functions, whereas most NSCLC cells lack these functions and can be divided into three major subtypes: adenocarcinoma; squamous cell carcinoma; and large cell carcinoma. SCLC is one of the most virulent forms of human cancer and is characterized by earlier dissemination and a more aggressive clinical evolution as compared to NSCLC.[3] The initial responsiveness of SCLC to anti-cancer drug and radiation treatment contrasts with the relative resistance of NSCLC to the same types of intervention.[4] The radiobiological properties of SCLC and NSCLC have been characterized largely using *in vitro* cell culture models, since clonogenic assays on tumor cells obtained by biopsy have been difficult to perform owing to the low growth yields and low cloning efficiencies.[5,6] As compared to

[c]Address for correspondence: Dr. Boris Zhivotovsky, Division of Toxicology, Institute of Environmental Medicine, Karolinska Institutet, Box 210, SE-171 77 Stockholm, Sweden. Voice: +46 8 728 75 88; fax: +46 8 32 90 41.
Boris.Zhivotovsky@imm.ki.se

SCLC cell lines, NSCLC cell lines are generally less sensitive to radiation, as evidenced by higher radiation survival curve extrapolation numbers, surviving fraction values following a 2 Gy dose (SF2) and differences in mean inactivation dose (D) values.[6] The spectrum of *in vitro* radiation responses observed among these different cell lines was similar to that reported in clinical practice. Therefore, cell lines derived from tumors taken from lung cancer patients provide a powerful tool to investigate the mechanism of their intrinsic sensitivity/resistance to anti-cancer treatments. The reasons for the differences between SCLC and NSCLC cell lines are not fully understood. One of the main cellular mechanisms of anti-cancer therapy is the killing of tumor cells via apoptosis.[7] Thus, one possible explanation to account for this difference may involve disparities in the susceptibility of SCLC and NSCLC to undergo apoptosis. This review discusses the role of spontaneous and induced apoptosis in the sensitivity of SCLC and NSCLC to treatment.

Does Spontaneous Apoptosis Reflect Radiation Sensitivity?

Spontaneous apoptosis has been described in almost all types of tumors. However, there is considerable variability in the amount of apoptosis observed among different types of tumors, as well as within a tumor type.[8] Moreover, it is not entirely clear whether there is a correlation between levels of spontaneous apoptosis, DNA damage-induced apoptosis and drug- and radiosensitivity in lung carcinomas. In experimental tumors, the level of spontaneous apoptosis correlates well with the degree of sensitivity to radiation-induced apoptosis.[9] These observations are consistent with the idea that spontaneous apoptosis accounts for cell loss in untreated tumors, a process that is enhanced by radiation treatment. Since radiosensitive tumors show greater spontaneous apoptosis than radioresistant tumors, this raises the possibility that basal or untreated level of apoptosis may be an indicator of tumor sensitivity to radiotherapy. Indeed, several studies, using paraffin-embedded histological material from primary lung tumors, have examined a potential link between the apoptotic index and clinical outcome of patients to various forms of anti-cancer therapy. In pathological stage N1 NSCLC tumors, spontaneous apoptosis was found to range from 0.2% to 2.8%.[10] Specifically, spontaneous apoptotic index was significantly higher in squamous cell carcinoma than in adenocarcinoma, and this difference correlated with a better survival rate of patients with the squamous cell carcinoma.[10] To our knowledge, no investigations have been performed comparing the level of apoptosis in SCLC tumors and the clinical outcome of patients with this form of cancer. Furthermore, in both types of primary lung tumors a potential link between the incidence of spontaneous apoptosis and the response of the tumors to radiotherapy is not clearly understood and requires further study.

Recently, using a large number of lung carcinoma cell lines (10 SCLC and 6 NSCLC), with previously characterized intrinsic radiosensitivities, we investigated the occurrence of spontaneous apoptosis.[11–13] We observed a significantly higher apoptotic index in SCLC (2–20%) compared with NSCLC lines (0.2–1%) in untreated cells. This is consistent with the high apoptotic index seen in samples from 40 resected SCLC tumors.[14] It is important to add that the level of spontaneous apoptosis in lung cancer cell lines positively correlated with the intrinsic radiosensitivity of these cells. Indeed, spontaneous apoptosis was detected at a high level in the radiosensitive SCLC U1285, at an intermediate level in the less radiosensitive SCLC

U1906 and not at all in the radioresistant NSCLC U1810 cell line.[12] Therefore, it seems that the radiosensitive SCLC cells are more prone to undergo spontaneous apoptosis as compared to the radioresistant NSCLC cells.

TREATMENT-INDUCED APOPTOSIS

As mentioned above, the higher level of spontaneous apoptosis in SCLC as compared to NSCLC correlates with their increased radiosensitivity. Anti-cancer treatment with radiation, etoposide (topoisomerase II inhibitor) or melphalan (DNA intrastrand crosslinker) induces DNA damage resulting in apoptosis.[7] However, it is unclear whether the two types of lung carcinomas, which respond differently to these treatments, display different propensities to undergo apoptosis upon DNA damage.

Using the same panel of cell lines (see above), we were able to show that all SCLC cell lines undergo apoptosis upon radiation or etoposide treatment. In contrast, the NSCLC cell lines were resistant to apoptosis induced by both treatments.[12,15] The increased sensitivity of the SCLC cell lines to DNA damage-induced apoptosis has been reported by other research groups.[16,17] These results taken together with the data presented above suggest a positive correlation among the following factors: the extent of spontaneous apoptosis in untreated cells, the propensity to undergo apoptosis upon DNA damage, and the intrinsic radio- and drug-resistance in lung cancer cells.

APOPTOTIC MACHINERY IN LUNG CANCER CELLS

Apoptosis, is an evolutionarily conserved multistage process, which maintains homeostasis in adult tissues.[18] Most, if not all, mammalian cells, including those which eventually give rise to cancer, contain the intrinsic machinery necessary to carry out cell suicide.[19] Many cancer treatments eliminate cells by the activation of the apoptotic machinery. However this killing is not always efficient and in cases where tumor cells are resistant to treatment, the activation of the apoptotic machinery is defective. Consequently, the absence of spontaneous and treatment-induced apoptosis in NSCLC suggests that deficiencies in the apoptotic process may be responsible for their resistance to anti-cancer therapy. To this end, different steps and effectors of the apoptotic process have been characterized in lung carcinomas.

p53 and Related Proteins

The p53 tumor suppressor gene plays a critical role at the G1/S transition, where it can either block entry into S phase or activate apoptosis in response to DNA damage. Experimental studies have implicated normal or "wild type" p53 protein (i.e., wt-p53) in the cellular response to ionizing radiation and other DNA damaging agents.[20] Although wt-p53 usually sensitizes cells to radiotherapy,[20] the effect that p53 mutations have on radiation treatment are more varied. For example, p53 mutations in bone and breast tumors have been associated with resistance to radiation. In contrast, mutated p53 was reported to increase radiosensitivity of colorectal and bladder tumors.[21]

It is believed that loss of p53 function causes increased resistance to radiation and that p53 mutational status is an important determinant of cancer prognosis. However, one copy of the chromosomal region 17p13, which contains p53, is frequently deleted in both SCLC and NSCLC, and mutational inactivation of the remaining allele occurs in 75~100% of the SCLC and ~50% of NSCLC.[22] p53 mutations do not predict survival or response to anti-cancer therapy of patients with NSCLC.[23] Finally, there seems to be no relationship between the mutational status of p53 and the susceptibility of either SCLC or NSCLC cells to undergo apoptosis.[12]

Bcl-2 Family

Much of our current understanding of the regulation of apoptosis came from studies of the Bcl-2 gene family. Bcl-2 family proteins are important regulators of apoptotic signaling, acting to either inhibit (Bcl-2, Bcl-X_L) or promote (Bax, Bcl-Xs) cell death.[24] The intracellular ratio between these proteins seems to determine how a cell will respond to an apoptotic stimulus.[24] Several studies have shown that Bcl-2 inhibits apoptotic cell death, and overexpression of this protein is associated with enhanced radio-[25] and chemoresistance[26] in some cell systems. The expression of the anti-apoptotic protein Bcl-2 in lung carcinomas has been investigated in detail. A Bcl-2-transfected SCLC cell line showed increased resistance to apoptosis induced by some anti-cancer agents,[27] whereas SCLC or NSCLC cells transfected with an oligonucleotide Bcl-2 antisense construct showed reduced overall viability and an increased susceptibility to apoptosis.[28] These reports suggest that the extent of apoptosis in lung cancer cells could be linked to the expression of this anti-apoptotic protein. However, several studies have shown that the magnitude of Bcl-2 protein expression is much higher in SCLC than in NSCLC. Immunohistochemical analysis on primary tumors revealed that Bcl-2 protein is expressed in most SCLC (65–90%, average 75%) but only in 15%–46% (average 27%) of NSCLC. This leads to an apparent paradox since SCLC is usually much more sensitive to chemotherapy than NSCLC.

Among the different types of NSCLC, Bcl-2 expression was detected more frequently in squamous cell carcinoma (25–35%) than in adenocarcinoma (~10%). Although several studies have reported improved survival in Bcl-2-immunopositive NSCLC,[29,30] a different report contradicts this finding.[31,32] A detailed analysis of 427 patients for whom information on long-term follow-up was available revealed that Bcl-2 immunopositivity does not appear to act as an independent prognostic indicator in NSCLC.[33] In the same manner, studies on the correlation between Bcl-2 expression and the extent of apoptosis in SCLC have produced conflicting results,[14,32] suggesting that the expression of Bcl-2 on its own does not appear to be a prognostic factor for the outcome of lung carcinomas.

Since Bax, a pro-apoptotic protein, antagonizes Bcl-2 function it is likely that the Bcl-2/Bax ratio determines the apoptotic susceptibility of the cells. In biopsies of patients affected by NSCLC, a high expression of Bax was detected in 47–73% of the cases,[34,35] and the Bcl-2/Bax ratio was found to be a significant positive factor for prognosis of patients with NSCLC.[35] In SCLC tumors, a low expression of Bax, resulting in a high Bcl-2/Bax ratio has been described. However, this ratio did not correlate with a decreased apoptotic index in these samples.[36]

Using large panels of lung cancer cell lines, the absence of any significant differences in the expression of Bax and Bcl-X_L between SCLC and NSCLC has been reported.[11,13,37] Both proteins were detected in all cell lines and in similar quantities. In contrast, Bcl-2 was detected at the highest level in SCLC.[11,13,37] It appears that, as in tumor biopsies, the ratio Bcl-2/Bax is higher in the SCLC than in the NSCLC cell lines. Taken together, it seems that the expression of Bcl-2 and Bax proteins and their ratio cannot be used as a prognostic factor for lung cancer therapy.

Ras Family

Various proto-oncogene products, such as Ras, Myc and Rb, have been shown to be involved in proliferation as well as in the apoptotic process in different tumors. Ras proteins have been implicated in both promotion and protection from apoptosis based on the ability to regulate multiple signaling pathways. Ras mutations resulted in a continued and inappropriate growth signal.[38] Ras protein was found overexpressed at a high frequency in NSCLC compared to the SCLC.[39] Moreover, this protein was overexpressed more often in adenocarcinoma (~60%) than in squamous cell carcinomas (~35%).[40] Alterations in the ras gene family was more specific for NSCLC as compared to SCLC, although conflicting results have been presented.[41,42] The possible importance of Ras alterations for apoptosis and resistance of NSCLC has not yet been explored. Moreover, recent investigations indicate the expression of Ras is not associated with the clinical outcomes.[39,40]

c-Myc and Other Nuclear Proto-Oncogenes

The Ras signal transduction cascade ultimately results in the activation of nuclear proto-oncogenes, such as the transcription factor c-Myc. This protein emerges as an oncogenic transcription factor that integrates proliferation, differentiation and apoptosis.[43] Of the myc gene family, c-myc is the most frequently activated in SCLC and NSCLC, whereas abnormalities of N-myc and L-myc usually only occur in SCLC. In 18% of the SCLC tumors and 31% of derived cell lines, as well as in 8% of the NSCLC tumors and 20% of derived cell lines, Myc amplification was observed.[44] The median survival time for patients with c-Myc-negative SCLC or NSCLC was significantly longer as compared to the patients with c-Myc-positive tumors.[45,46] However, c-Myc protein status does not correlate with the apoptotic index in SCLC and NSCLC cell lines.[11] Furthermore, a possible relationship between *in vivo* radiosensitivity and activity of myc and ras proto-oncogenes has been investigated in a SCLC xenografts model, but it appears that SCLC xenografts with different radiosensitivities express similar amounts of these proto-oncogenes. These results suggest that some other factors are responsible for the tumor radiosensitivity.[47] There are also several reports concerning the expression of other nuclear proto-oncogene, such as the immediate early transcription factors c-Jun and c-Fos. However their role in chemo- and radiosensitivity of SCLC and NSCLC remains obscure.

Rb, p16^{INK4}, and Cyclin D1

The retinoblastoma protein (Rb), p16^{INK4}, and cyclin D1 are major components of the Rb pathway that control the G1 checkpoint of the cell cycle. Rb protein, as well as cyclin D1, have been reported to be frequently mutated or deleted in primary

SCLC tumors (~90%), whereas both mutations are less common in NSCLC tumors (~15%).[48,49] On the other hand, NSCLC shows a high frequency of p16^{INK4} deletions (~50%).[48,49] Although, the Rb pathway is disrupted in the vast majority of lung cancers, prognostic values of Rb, p16^{INK4} and Cyclin D1 proteins are uncertain.[11,48,50]

Fas-FasL

Even if anticancer treatments that induce DNA damage are not supposed to directly act through the plasma membrane receptors, several receptors (e.g.,TNF/Fas-receptor family) are important for apoptosis. Their implications in DNA damage-induced apoptosis via an autocrine/paracrine loop that has been described in brain cancer tumors and lymphomas can not be excluded.[51]

Fas (Apo-1/CD95) is a transmembrane protein, which through its binding to Fas ligand (FasL), induces a cell death signal. Expression of the Fas ligand has been detected in cell lines of both types of lung carcinomas.[52] Immunohistochemical analysis of Fas and FasL in NSCLC primary tumors revealed Fas expression in 36–54% of the cases and FasL expression in 67% of the cases.[53–55] Thus, the FasL pathway may be operative in some or most of NSCLC tumors. In accordance with this idea, analysis of the entire coding region and all splice sites of the Fas gene showed Fas mutations in only 7.7% of NSCLC tumors.[56] Fas expression and its correlation with prognosis of NSCLC patients has also been examined. Several reports suggest that positive staining for Fas in the tumor specimens could be considered a good prognostic indicator for the patient.[53,55] However, in cases where Fas was detected in ~50% of NSCLC tumors, an immunohistochemical study disclosed that this protein is expressed in the cytoplasm and not on the cell surface, suggesting that the detection of Fas expression may not explain its functionality in these cells.[54] Another mechanism in lung cancer cells that describes a non-functional Fas/FasL pathway is based on the discovery of a soluble decoy receptor, termed decoy receptor 3 (DcR3). This receptor binds to FasL and inhibits FasL-induced apoptosis. The DcR3 gene was amplified in about half of primary lung tumors studied. Thus, certain tumors may escape FasL-dependent immune-cytotoxic attack by expressing a decoy receptor that blocks FasL.[57]

Expression of Fas in SCLC is less clear. Although a recent study demonstrated that drug-induced apoptosis in lung cancer cells is not mediated by the Fas/FasL signaling pathway,[58] information is lacking in order to implicate the Fas/FasL pathway in the sensitivity of both types of lung carcinomas to various kinds of treatment. Furthermore, it is unknown whether there is any correlation between Fas/FasL and the clinical outcome of patients undergoing radiation therapy.

Caspases: Their Presence and Activation

The role of caspases in the activation and execution of apoptosis is well-described.[59] These enzymes are constitutively expressed in the cell as pro-caspases and converted into mature enzymes by a cleavage cascade. During receptor-mediated apoptosis the activation of pro-caspase-8 plays a key role. In non–receptor-mediated apoptosis, the release of cytochrome *c* from mitochondria and the activation of pro-caspase-9 within the apoptosome complex commit a cell to apoptotic cell death. In both cases, the activation of the executioner caspases (caspase-3, -6 and -7) leads to the appearance of morphological and biochemical hallmarks of apoptosis.

A recent study of primary NSCLC tumors showed that the median survival time for patients with caspase-3-positive carcinomas was almost two times longer as compared to the patients with caspase-3-negative carcinomas.[46] Using a panel of 16 lung cancer cell lines, no differences in the expression of pro-caspase-3 were detected between the SCLC and NSCLC cells.[11] However, using the same panel of cell lines, differences in the expression of some other pro-caspases in SCLC and NSCLC was observed.[13] While pro-caspases-2, -3, -6, -7, and -9 were found in SCLC and NSCLC cells, pro-caspases-1, -4, and -10 were expressed only in NSCLC cell lines. The presence of pro-caspase-8 also seems to be limited to the NSCLC type, with the exception of one SCLC cell line, U1285, suggesting a deficiency of several pro-caspases in SCLC.[13] These results also suggest that pro-caspases-8 and -10, which are involved in the receptor-mediated apoptotic pathway, are not important for the occurrence of spontaneous apoptosis in SCLC cells. It has been shown that paclitaxel-induced apoptosis in NSCLC[60] or vinblastine-induced apoptosis in SCLC[61] are associated with caspase-3 activation and that DEVD-CHO, a caspase-3 inhibitor, can inhibit the occurrence of vinblastine-induced apoptosis in the SCLC cell line. Activation of caspase-8 in NSCLC cell lines, as well as caspase-9 and -3 (all cell lines), was observed upon treatment with etoposide or radiation.[15] In addition, DNA-damaging agents induce a release of cytochrome *c* into the cytoplasm in both types of lung cancer cells. It is worth noting that whereas the activation of these caspases was observed in both types of lung cancer, apoptosis occurred only in the SCLC cells. Thus, the activation of expressed pro-caspases is efficient in lung cancer, suggesting that the inhibition of the apoptotic pathway in NSCLC occurs downstream of mitochondrial and caspase events.

Survivin (IAP)

Tumor cells express several proteins that render them resistant to apoptosis. Increased expression of protective proteins, like anti-apoptotic members of the Bcl-2 protein family, as well as IAPs, such as survivin, may result in therapy-resistant tumors.[62] The strongest evidence for IAP involvement in cancer progression comes from studies on survivin, which is not expressed in differentiated tissues but is present in many types of cancers, including high levels in lung cancers.[63] Previous studies demonstrated that the IAP protein family can directly bind and inhibit active caspase-3 and -7. Survivin, like the other IAP proteins, probably inhibits the terminal effector caspase-3 and -7.[63] The expression of survivin was found to correlate with the resistance of tumors to treatment and with shorter survival of patients with colorectal and gastric carcinomas and neuroblastoma.[64]

In tumors of patients with NSCLC, the survivin gene transcript was identified in a vast majority (85.5%) of the tumor samples but in only 12% of the paired, histopathologically normal lung samples. Patients with suppressed survivin expression had significantly better overall survival than patients with high survivin expression, suggesting that survivin could be considered as a prognostic factor in NSCLC.[65] However, since SCLC and NSCLC cell lines express a high and rather equal level of survivin protein,[15] the possibility that survivin expression can explain the different sensitivity of the two types of lung carcinomas to undergo apoptosis upon anti-cancer treatments is doubtful. Since survivin belongs to a rather large family of proteins, including cIAP1, cIAP2 or XIAP, it is necessary to examine the anti-apoptotic effects of these other members to more fully grasp this family's potential effect in NSCLC.

CONCLUDING REMARKS

Radioresistance is an intrinsic or acquired phenomenon, which apparently cannot be explained by a deficiency in any single factor (see TABLE 1). SCLC cells are

TABLE 1. Comparative characteristics of SCLC and NSCLC cells

	SCLC		NSCLC	
Incidence	~25%		~75%	
	Primary tumors	Cell lines	Primary tumors	Cell lines
Radiation sensitivity	High	High	Low	Low
Anti-tumor drug sensitivity	High	High	Low	Low
Spontaneous apoptosis		2–20%		0.1–1%
Bcl-2 expression	75~90%	High	10~30%	Low
p53 mutation	75~100%	100%	~50%	100%
Ras over-expression	~30%	0	~60%	~20%
Myc amplification	~20%	~30%	~10%	~20%
Rb mutation	~90%	nd	~15%	nd
Cyclin D1 amplification	~90%	nd	~15%	nd
$p16^{INK4}$ mutation	rare	nd	~50%	nd
Fas Ligand expression	nd	100%	~50%	100%
Fas receptor expression	nd	nd	~70%	nd
Procaspase-8, -10 expression	nd	~20%	nd	100%
Procaspase-3, -6, -7	nd	100%	nd	100%
Upon radiation or VP16 treatment:				
Caspase-3 like activity	nd	yes	nd	yes
Caspase-9 activity	nd	yes	nd	yes
Caspase-8 activity	nd	yes	nd	yes
Cytochrome *c* release	nd	yes	nd	yes
PARP processing	nd	yes	nd	no
DFF45 processing	nd	yes	nd	no
DNA-PK activity	nd	decrease	nd	stable

initially sensitive to treatment with anti-cancer drugs and radiation, and only a small population of these cells remains resistant. This resistant cell population is characterized by fast dissemination and a more aggressive clinical evolution. Although during the past few decades several hypotheses were suggested to explain this acquired resistance, the entire mechanism is still unclear. In contrast to SCLC, NSCLC are comparably more resistant to death induced by the same types of intervention. Since resistant NSCLC cells do not undergo spontaneous or chemo- and radiation-induced apoptosis, it is very likely that the apoptotic machinery is impaired in cells with an intrinsic resistance to treatment.

Plasma membrane receptors and mitochondria serve as two of the main initiation sites during cell killing. Activation of the apoptotic process at these sites involves the sequential activation of the caspase cascade. Pro-caspases are activated in all investigated tumor cells undergoing apoptosis after treatment with radiation and chemotherapeutic agents. A more detailed understanding of the mechanisms of caspase activation in tumor cells can be expected to provide additional information concerning potential molecular targets that could be useful in improving tumor cells' susceptibility to radiation treatment. The absence of several pro-caspases in SCLC that are involved in receptor-mediated killing suggests that this pathway is not required for spontaneous and inducible apoptosis in these cells. The activation of pro-caspases is tightly regulated. For instance, the release of cytochrome *c* from mitochondria, which is important for the activation of the apoptosome complex in cytoplasm, and is an early event in radiation-induced apoptosis of tumor cells, is controlled by Bcl-2 family proteins. However, despite a higher level of Bcl-2 protein in SCLC than in NSCLC cell lines, the release of cytochrome c occurs in both types of tumors, suggesting that at least in lung cancer cells, the level of Bcl-2 is not the only or primary factor determining their resistance to treatment.

In most experimental models, the release of cytochrome *c* has been described as a point of no return in caspase activation. However, in several tumors, including NSCLC, cells may survive the release of mitochondrial cytochrome *c* and maintain their clonogenicity even in the presence of active caspases. This suggests that in these cells the apoptotic process is impaired downstream of mitochondrial events and caspase activation. Which factors might influence apoptosis at this stage? It has been shown that caspase activity downstream of mitochondria can be controlled by heat shock proteins, different intracellular IAPs, as well as by some activity subordinate to the kinase system (Akt and p21-Ras). The precise contribution of these factors to differential sensitivity of lung cancer cells to treatment remains to be elucidated. This knowledge will contribute to our understanding of the mechanism of resistance of NSCLC and will likely provide us with some tools to modulate the apoptotic machinery in lung cancer cells in order to increase their sensitivity to anticancer treatment.

ACKNOWLEDGMENTS.

We thank Dr. John D. Robertson for critical reading of the manuscript. Work from the authors' laboratories was supported by grants from the Swedish Cancer Society

(Cancerfonden) and Stockholm Cancer Society (Cancerföreningen). B.J. was also supported by INSERM-MRC and ARC Fellowships.

REFERENCES

1. SCHOTTENFELD, D. 1996. Epidemiology of lung cancer. *In* Lung Cancer—Principles and Practice. H.I. Pass, J.B. Mitchel, D.H. Johnson & A.T. Turrisi, Eds.: 305–322. Lipincott-Raven Publ.
2. TRAVIS, W.D., J. LINDER & B. MACKAY. 1996. Classification, histology, cytology and electron microscopy. *In* Lung Cancer—Principles and Practice. H.I. Pass, J.B. Mitchel, D.H. Johnson & A.T. Turrisi, Eds.: 361–395. Lipincott-Raven Publ.
3. COMIS, R.L., D.M. FRIEDLAND & B.C. GOOD. 1998. Small-cell lung cancer: a perspective on the past and a preview of the future. Oncology **12:** 44–50.
4. OWENS, A.H., JR. & M.D. ABELOFF. 1985. Neoplasms of the lung. *In* Medical Oncology. P. Calabresi, P.S. Schein, S.A. Rosenberg, Eds.: 715–757. Macmillan Press.
5. SULLIVAN, F.J., M.S. KRISHNA & J.B. MITCHELL. 1996. Radiation biology of lung cancer. *In* Lung cancer—Principles and Practice. H.I. Pass, J.B. Mitchel, D.H. Johnson, & A.T. Turrisi, Eds.: 219–249. Lipincott-Raven Publ.
6. BRODIN, O., L. LENNARTSSON & S. NILSON. 1991. Single-dose and fractionated irradiation of four human lung cancer cell lines in vitro. Acta. Oncol. **30:** 967–974.
7. ZHIVOTOVSKY, B., B. JOSEPH & S. ORRENIUS. 1999. Tumor radiosensitivity and apoptosis. Exp. Cell Res. **248:** 10–17.
8. SOINI, Y., P. PAAKKO & V.P. LEHTO. 1998. Histopathological evaluation of apoptosis in cancer. Am. J. Pathol. **153:** 1041–1053.
9. HENDRY, J.H. & C.M. WEST. 1997. Apoptosis and mitotic cell death: their relative contributions to normal-tissue and tumor radiation response. Int. J. Radiat. Biol. **71:** 709–719.
10. KOMAKI. R., L. MILAS, J.Y. RO, *et al.* 1998. Prognostic biomarker study in pathologically staged N1 non-small cell lung cancer. Int. J. Radiat. Oncol. Biol. Phys. **40:** 787–796.
11. SIRZÉN, F., B. ZHIVOTOVSKY, A. NILSSON, *et al.* 1998. Higher spontaneous apoptotic index in small cell compared with non small cell lung carcinoma cell lines; lack of correlation with Bcl-2/Bax. Lung Cancer. **22:** 1–13.
12. SIRZEN, F., B. ZHIVOTOVSKY, A. NILSSON, *et al.* 1998. Spontaneous and radiation-induced apoptosis in lung carcinoma cells with different intrinsic radiosensitivities. Anticancer Res. **18:** 695–699.
13. JOSEPH, B., J. EKEDAHL, R. LEWENSOHN, *et al.* 1999. Differences in expression of pro-caspases in small cell and non-small cell lung carcinoma. Biochem. Biophys. Res. Commun. **262:** 381–387.
14. EEROLA, A.K., U. TORMANEN, P. RAINIO, *et al.* 1997. Apoptosis in operated small cell lung carcinoma is inversely related to tumor necrosis and p53 immunoreactivity. J. Pathol. **181:** 172–177.
15. JOSEPH, B., J. EKEDAHL, R. LEWENSOHN, *et al.* 2000. Defective caspase-3 relocalization in non small cell lung carcinoma: mechanism for a nuclear apoptosis resistance. [abstract 64]. Mechanisms of cell death 2000. 3rd symposium of the International Cell Death Society. Madrid, May 6–10, 2000.
16. OKAMOTO-KUBO, S., K. NISHIO, Y. HEIKE, *et al.* 1994. Apoptosis induced by etoposide in small-cell lung cancer cell lines. Cancer Chemother. Pharmacol. **33:** 385–390.
17. DRAG-ZALESINSKA, M., T. WYSOCKA, E. GEBAROWSKA, *et al.* 1999. Evaluation of apoptosis and necrosis induction in cultured cells of lung carcinomas under effect of cytostatic drugs in vitro. Folia Histochem. Cytobiol. **37:** 133–134.
18. BOWEN, I.D. & S.M. BOWEN. 1990. Programmed cell death in tumors and tissues. Chapman and Hall. London, UK.
19. FADEEL, B., S. ORRENIUS & B. ZHIVOTOVSKY. 1999. Apoptosis in human disease: a new skin for the old ceremony? Biochem. Biophys. Res. Commun. **266:** 699–717.

20. BRISTOW, R.G., S. BENCHIMOL & R.P. HILL. 1996. The p53 gene as a modifier of intrinsic radiosensitivity: implications for radiotherapy. Radiother. Oncol. **40:** 197–223.
21. CHIARUGI, V., L. MAGNELLI & M. CINELLI. 1998. Role of p53 mutations in the radiosensitivity status of tumor cells. Tumori **84:** 517–520.
22. GREENBLATT, M.S., W.P. BENNETT, M. HOLLSTEIN, *et al.* 1994.Mutations in the p53 tumor suppressor gene: clues to cancer etiology and molecular pathogenesis. Cancer Res. **54:** 4855–4878.
23. SAFRAN, H., T. KING, H. CHOY, *et al.* 1996. p53 mutations do not predict response to paclitaxel/radiation for nonsmall cell lung carcinoma. Cancer **78:** 1203–1210.
24. REED, J.C. 1998 Bcl-2 family proteins. Oncogene **17:** 325–336.
25. FUKUNAGA-JOHNSON, N., J.J. RYAN, M. WICHA, *et al.* 1995. Bcl-2 protects murine erythroleukemia cells from p53-dependent and –independent radiation-induced cell death. Carcinogenesis **16:** 1761–1767.
26. CAMPOS, L., J.P. ROUAULT, O. SABIDO, *et al.* 1993. High expression of bcl-2 protein in acute myeloid leukemia cells is associated with poor response to chemotherapy. Blood **81:** 3091–3096.
27. OHMORI, T., E.R. PODACK, K. NISHIO, *et al.* 1993. Apoptosis of lung cancer cells caused by some anti-cancer agents (MMC, CPT-11, ADM) is inhibited by bcl-2. Biochem. Biophys. Res. Commun. **192:** 30–36.
28. KOTY, P.P., H. ZHANG & M.L. LEVITT. 1999. Antisense bcl-2 treatment increases programmed cell death in non-small cell lung cancer cell lines. Lung Cancer **23:** 115–127.
29. SILVESTRINI, R., A. COSTA, C. LEQUAGLIE, *et al.* 1998. Bcl-2 protein and prognosis in patients with potentially curable non-small-cell lung cancer. Virchows Arch. **432:** 441–444.
30. EEROLA, A.K., H. RUOKOLAINEN, Y. SOINI, *et al.* 1999. Accelerated apoptosis and low bcl-2 expression associated with neuroendocrine differentiation predict shortened survival in operated large cell carcinoma of the lung. Pathol. Oncol. Res. **5:**179–186.
31. O'NEILL, A.J., M.J. STAUNTON & E.F. GAFFNEY. 1996. Apoptosis occurs independently of bcl-2 and p53 over-expression in non-small cell lung carcinoma. Histopathology **29:** 45–50.
32. MAITRA, A., R.H. AMIRKHAN, M.H. SABOORIAN, *et al.* 1999. Survival in small cell lung carcinoma is independent of Bcl-2 expression. Hum. Pathol. **30:** 712–717.
33. ANTON, R.C., R.W. BROWN, M. YOUNES, *et al.* 1997. Absence of prognostic significance of bcl-2 immunopositivity in non-small cell lung cancer: analysis of 427 cases. Hum. Pathol. **28:** 1079–1082.
34. CAPUTI, M., A.M. GROEGER, V. ESPOSITO, *et al.* 1999. Frequent high expression of bax pro-apoptotic protein in non-small cell lung cancer. Anticancer Res. **19:** 825–827.
35. HUANG, C., N. KOHNO, H. INUFUSA, *et al.* 1999. Overexpression of bax associated with mutations in the loop-sheet-helix motif of p53. Am. J. Pathol. **155:** 955–965.
36. BRAMBILLA, E., S. GAZZERI, S. LANTUEJOUL, *et al.* 1998. p53 mutant immunophenotype and deregulation of p53 transcription pathway (Bcl2, Bax, and Waf1) in precursor bronchial lesions of lung cancer. Clin. Cancer Res. **4:** 1609–1618.
37. REEVE, J.G., J. XIONG, J. MORGAN, *et al.* 1996. Expression of apoptosis-regulatory genes in lung tumor cell lines: relationship to p53 expression and relevance to acquired drug resistance. Br. J. Cancer. **73:** 1193–1200.
38. WITTINGHOFER, A. 1998. Signal transduction via Ras. Biol. Chem. **379:** 933–937.
39. KIM, J.O., M.M. NAU, K.A. ALLIKIAN, *et al.* 1998. Co-amplification of a novel cyclophilin-like gene (PPIE) with L-myc in small cell lung cancer cell lines. Oncogene. **17:** 1019–1026.
40. MOLDVAY, J., P. SCHEID, P. WILD, *et al.* 2000. Predictive survival markers in patients with surgically resected non-small cell lung carcinoma. Clin. Cancer Res. **6:** 1125–1134.
41. WODRICH, W. & M. VOLM. 1993. Overexpression of oncoproteins in non-small cell lung carcinomas of smokers. Carcinogenesis. **14:** 1121–1124.

42. CHO, J.Y., J.H. KIM, Y.H. LEE, *et al.* 1997. Correlation between K-ras gene mutation and prognosis of patients with nonsmall cell lung carcinoma. Cancer. **79:** 462–467.
43. DANG, C.V., L.M. RESAR, E. EMISON, *et al.* 1999. Function of the c-Myc oncogenic transcription factor. Exp. Cell Res. **253:** 63–77.
44. RICHARDSON, G.E. & B.E. JOHNSON. 1993. The biology of lung cancer. Semin. Oncol. **20:** 105–127.
45. VAN WAARDENBURG, R.C., C. MEIJER, S.J. PINTO-SIETSMA, *et al.* 1998. Effects of c-myc oncogene modulation on differentiation of human small cell lung carcinoma cell lines. Anticancer Res. **18:** 91–95.
46. VOLM, M. & K. KOOMAGI. 2000. Prognostic relevance of c-Myc and caspase-3 for patients with non-small cell lung cancer. Oncol. Rep. **7:** 95–98.
47. RYGAARD, K., R.J. SLEBOS, & M. SPANG-THOMSEN. 1991. Radiosensitivity of small-cell lung cancer xenografts compared with activity of c-myc, N-myc, L-myc, c-raf-1 and K-ras proto-oncogenes. Int. J. Cancer **49:** 279–284.
48. BRAMBILLA, E., D. MORO, S. GAZZERI, *et al.* 1999. Alterations of Rb pathway (Rb-p16^{INK4}-cyclin D1) in preinvasive bronchial lesions. Clin. Cancer Res. **5:** 243–250.
49. DOSAKA-AKITA, H., P.T. CAGLE, H. HIROUMI, *et al.* 2000. Differential retinoblastoma and p16(INK4A) protein expression in neuroendocrine tumors of the lung. Cancer **88:** 550–556.
50. HOMMURA, F., H. DOSAKA-AKITA, I. KINOSHITA, *et al.* 1999. Predictive value of expression of p16^{INK4A}, retinoblastoma and p53 proteins for the prognosis of non-small-cell lung cancers. Br. J. Cancer. **81:** 696–701.
51. BELKA, C., P. MARINI, W. BUDACH, *et al.* 1998. Radiation-induced apoptosis in human lymphocytes and lymphoma cells critically relies on the up-regulation of CD95/Fas/APO-1 ligand. Radiat. Res. **149:** 588–595.
52. NIEHANS, G.A., T. BRUNNER, S.P. FRIZELLE, *et al.* 1997. Human lung carcinomas express Fas ligand. Cancer Res. **57:** 1007–1012.
53. KOOMAGI, R. & M. VOLM. 1999. Expression of Fas (CD95/APO-1) and Fas ligand in lung cancer, its prognostic and predictive relevance. Int. J. Cancer. **84:** 239–343.
54. NAMBU, Y., S.J. HUGHES, A. REHEMTULLA, *et al.* 1998. Lack of cell surface Fas/APO-1 expression in pulmonary adenocarcinomas. J. Clin. Invest. **101:** 1102–1110.
55. URAMOTO, H., T. OSAKI, M. INOUE, *et al.* 1999. Fas expression in non-small cell lung cancer: its prognostic effect in completely resected stage III patients. Eur. J. Cancer. **35:** 1462–1465.
56. LEE, S.H., M.S. SHIN, W.S. PARK, *et al.* 1999. Alterations of Fas (Apo-1/CD95) gene in non-small cell lung cancer. Oncogene **18:** 3754–3760.
57. PITTI, R.M., S.A. MARSTERS, D.A. LAWRENCE, *et al.* 1998. Genomic amplification of a decoy receptor for Fas ligand in lung and colon cancer. Nature **396:** 699–703.
58. FERREIRA, C.G., C. TOLIS, S.W. SPAN, *et al.* 2000. Drug-induced apoptosis in lung cancer cells is not mediated by the Fas/FasL (CD95/APO1) signaling pathway. Clin. Cancer Res. **6:** 203–212.
59. RICH, T., C.J. WATSON & A.H. WYLLIE. 1999. Apoptosis: the germs of death. Nature Cell Biol. **1:** E69–E71.
60. WEIGEL, T.L., M.T. LOTZE, P.K. KIM, *et al.* 2000. Paclitaxel-induced apoptosis in non-small cell lung cancer cell lines is associated with increased caspase-3 activity. J. Thorac. Cardiovasc. Surg. **119:** 795–803.
61. TASHIRO, E., M. IDEISHI, S. MIURA, *et al.* 1998. Caspase-3 activation is not responsible for vinblastine-induced Bcl-2 phosphorylation and G2/M arrest in human small cell lung carcinoma Ms-1 cells. Jpn. J. Cancer Res. **89:** 940–946.
62. JÄÄTTELÄ, M. 1999. Escaping cell death: survival proteins in cancer. Exp. Cell Res. **248:** 30–43.
63. TAMM, I., Y. WANG, E. SAUSVILLE, *et al.* 1998. IAP-family protein survivin inhibits caspase activity and apoptosis induced by Fas (CD95), Bax, caspases, and anticancer drugs. Cancer Res. **58:** 5315–5320.
64. ADIDA, C., D. BERREBI, M. PEUCHMAUR, *et al.* 1998. Anti-apoptosis gene, survivin, and prognosis of neuroblastoma. Lancet **351:** 882–883.

65. MONZO, M., R. ROSELL, E. FELIP, *et al.* 1999. A novel anti-apoptosis gene: re-expression of survivin messenger RNA as a prognosis marker in non-small-cell lung cancers. J. Clin. Oncol. **17:** 2100.

The Central Role of Phosphatidylserine in the Phagocytosis of Apoptotic Thymocytes

ROBERT A. SCHLEGEL,[a,b] MELISSA K. CALLAHAN,[a]
AND PATRICK WILLIAMSON[c]

[a]*Department of Biochemistry and Molecular Biology, Penn State University, University Park, Pennsylvania 16802, USA*

[c]*Department of Biology, Amherst College, Amherst, Massachusetts 01002, USA*

ABSTRACT: Apoptotic thymocytes inactivate the aminophospholipid translocase, which transports phosphatidylserine (PS) to the inner leaflet of the plasma membrane, and activate the scramblase, which randomizes phospholipids across the membrane and brings PS to the cell surface. Although different macrophages use at least two different systems to recognize and engulf apoptotic thymocytes, both systems recognize PS on the apoptotic target. Thymocytes treated with Ca^{2+} and ionophore to inactivate the translocase and activate the scramblase immediately expose PS on their surface and are immediately recognized and phagocytosed. These targets, on which PS has been artificially exposed, are recognized by the PS exposed on their surface. However, they apparently also engage the vitronectic receptor, a lectin-like receptor and CD14. All of these receptors are implicated in the phagocytosis of apoptotic thymocytes, suggesting that loss of asymmetry and/or exposure of PS is sufficient to generate the ligands recognized by those receptors. The role of PS is not confined to the target cell surface, however. PS is constitutively exposed on the surface of macrophages and is as necessary for apoptotic cell engulfment as is recognition of PS on the target cell surface.

KEYWORDS: Apoptosis; Phosphatidylserine; Phagocytosis; Thymocytes; Translocase; Scramblase

OVERVIEW

Virtually every cell in the body confines phosphatidylserine (PS) to the inner leaflet of the plasma membrane bilayer. There, it serves as a cofactor or provides the proper milieu for the optimal functioning of membrane enzymes. But satisfying these requirements could be accomplished just as satisfactorily with an equal amount of PS in the outer leaflet. Yet, the cell consumes a large amount of ATP fueling the enzyme that restricts PS to the inner leaflet: in the brain, 8% of the energy budget may be devoted to this process.[1] These considerations suggest that the functional importance of an asymmetric distribution of PS lies not in concentrating it in the internal leaflet, but rather in keeping it *from* the outer leaflet of the plasma membrane

[b]Address for correspondence: Dr. Robert A. Schlegel, Department of Biochemistry and Molecular Biology, Pennsylvania State University, University Park, PA 16802. Voice: 814-865-6974; fax: 814-863-7024.
UR3@psu.edu

bilayer. In fact, preventing PS exposure on the cell surface is necessary to maintain the normal associations between cells and their neighbors. When PS is allowed to reach the outer leaflet, it is detected there by surveillance systems present in professional phagocytes and other neighboring cells, initiating the cell's destruction.

PHOSPHOLIPID ASYMMETRY

An asymmetric phospholipid distribution across the plasma membrane bilayer is not specific to PS; the other major classes of phospholipids are also asymmetrically distributed with the neutral phospholipids, phosphatidylcholine and sphingomyelin, found predominantly in the outer leaflet, and the other aminophospholipid, phosphatidylethanolamine, found predominantly in the inner leaflet. PS is, however, the only phospholipid restricted exclusively to the internal leaflet.[2] This scrupulous maintenance of PS in the inner leaflet is accomplished by the aminophospholipid translocase, a P-type ATPase which transports PS, and to a lesser extent phosphatidylethanolamine, from the outer to the inner leaflet of the plasma membrane.[3, 4] This enzyme is apparently responsible for restricting PS to the inner leaflet of all normal animal cells, from *Drosophila* to man, with the exception of cells in which massive membrane remodeling is occurring, such as fusing myoblasts, megakaryocytes and macrophages.[5]

This asymmetric distribution, which is actively and constantly maintained by the return to the inner leaflet of PS molecules that spontaneously flip to the outer leaflet, can be destroyed in a comparative instant by the scramblase, a non-specific lipid flip-site which allows all classes of phospholipids to rapidly flip from one leaflet of the bilayer to the other.[6] Activation of the scramblase can abolish asymmetry, producing a random distribution of phospholipids across the bilayer in a matter of several minutes. Activation of the scramblase thereby brings PS to the cell surface. In erythrocytes, the scramblase can be activated by elevating cytosolic levels of Ca^{2+}. The resulting lipid-symmetric erythrocytes are preferentially phagocytosed *in vitro* by macrophages.[7] Artificial lipid vesicles containing PS competitively inhibit this preferential uptake,[8] indicating the erythrocytes are being recognized by the PS exposed on their surface. Importantly, artificial lipid vesicles containing as little as 3 mole percent PS are promptly cleared from the circulation of mice, demonstrating that PS itself serves as a recognition signal *in vivo*.[9]

The scramblase can also be activated in other cells, such as lymphocytes, either artificially by elevating cytosolic Ca^{2+} [10] or physiologically in lymphocytes undergoing programmed cell death; concomitantly, the aminophospholipid translocase is inactivated.[11] This combination results in exposure of PS on the surface of apoptotic cells and provokes recognition and phagocytosis by macrophages.

DIFFERENT PS RECOGNITION SYSTEMS

Macrophages employ two different PS-dependent systems to recognize apoptotic cells (TABLE 1). The phagocytosis of apoptotic thymocytes by so-called activated macrophages is inhibited by PS vesicles.[12] On the other hand, phagocytosis by so-called unactivated macrophages is inhibited by PS exposed on lipid-symmetric erythrocytes but not by PS vesicles.[13] Sensitivities to other inhibitors also distinguish the different types of macrophage recognition systems. Phagocytosis of apoptotic thymocytes by activated macrophages, but not unactivated macrophages, is inhibited by N-acetylglucosamine (GlcNAc), implying recognition of a carbohydrate ligand on the apoptotic cell surface by a lectin-like receptor on macrophages;[13,14] and the tetrapeptide RGDS inhibits phagocytosis by unactivated macrophages, but not activated macrophages, by competitively inhibiting the vitronectin receptor, an integrin expressed on macrophages.[15] Under normal circumstances, recognition of apoptotic thymocytes utilizes either one or the other of these two recognition systems. However, with at least one macrophage cell line, competitive inhibition by RGDS persists after activation of the GlcNAc-inhibitable system,[16] suggesting that inactivation of recognition by the vitronectin receptor is not required for activation of the lectin-mediated recognition system normally used by activated macrophages.

Despite these differences in receptors employed, the fact remains that the recognition systems used by both activated and unactivated macrophages depend on exposure of PS on the apoptotic cell surface for recognition: masking PS on thymocytes by pretreatment with annexin V, a protein which binds specifically to PS, inhibits phagocytosis by both activated and unactivated macrophages.[17] In addition to sharing the requirement for PS on the apoptotic cell surface, both activated and unactivated macrophages also have in common some receptors required for recognition. The 61D3 monoclonal antibody, which recognizes the LPS receptor CD14,[18] inhibits phagocytosis of apoptotic lymphocytes by unactivated[19] and activated macrophages.[13] Similarly, treating macrophages with phosphatidylinositol-specific phospholipase C to remove the GPI-linked CD14 compromises the ability of both activated and unactivated macrophages to phagocytose apoptotic thymocytes.[20] The

TABLE 1. Comparison of apoptotic lymphocytes and ionophore-treated DO11.10 T cells as targets for macrophages

		Inhibition by					
Macrophage	Target	PS	GlcNAc	RGDS	LS	61D3	AnnV
Unactivated	Apoptotic Thymocytes	No	No	Yes	Yes	Yes	Yes
	Ionophore/DO11.10	No	No	Yes	Yes	Yes	Yes
Activated	Apoptotic Thymocytes	Yes	Yes	No*	No	Yes	Yes
	Ionophore/DO11.10	Yes	Yes	No*	No	Yes	Yes

Either primary thymocytes, induced to undergo apoptosis by dexamethasone, or DO11.10 T lymphocytes, treated with ionophore and Ca^{2+} to artificially expose PS on their surface, were presented as targets to either unactivated or activated macrophages in the presence of various inhibitors. After 30 min at 37°C, the number of phagocytosed cells was enumerated and compared to controls without inhibitor to establish whether inhibition had occurred.

PS: PS vesicles; GlcNAc: N-Acetylglucosamine; RGDS: arginine-glycine-aspartate-serine; LS: lipid-symmetric erythrocytes; 61D3: monoclonal antibody 61D3 to CD14; AnnV: annexin V

*Primary macrophages; phagocytosis of apoptotic thymocytes (S. Krahling, unpublished observation) and ionophore-treated DO11.10 cells[16] by activated J774 macrophages is not inhibited by RGDS.

macrophage scavenger receptor CD36 may also be used by both activated and unactivated macrophages in the recognition of apoptotic neutrophils.[21]

THE CENTRALITY OF PS

The participation of multiple macrophage receptors in the recognition process raises the question of whether the ligand(s) presented on apoptotic cells is PS itself or PS in association with other cell surface molecules. This question can be addressed by examining the temporal development of recognition signals. When cells are induced to undergo apoptosis, the signals by which they are recognized appear only after a period of time. In the case of apoptotic thymocytes or the model T cell hybridoma line DO11.10, for example, a significant fraction of cells with PS exposed on their surface does not appear until 4–5 h after induction of apoptosis *in vitro*.[10] That time is not required for PS accumulation on the cell surface; rather it is the time required before the induction of the regulatory events by which the aminophospholipid translocase is inactivated and the scramblase activated;[11] the combination of these two events rapidly brings PS to the cell surface. However, during the extended time before PS is exposed, the cell could acquire on its surface other signals for recognition or other molecules, altered or newly presented, which combine with PS to generate a signal. The ability to rapidly and artificially bring PS to the cell surface in the absence of apoptosis allows an evaluation of the contribution of PS to the various signaling systems which have been identified.

Treating DO11.10 cells with ionophore in the presence of Ca^{2+} inactivates the translocase and activates the scramblase, bringing PS to the surface of essentially all cells in the population within 10 min.[10] These treated cells are phagocytosed by macrophages as quickly as uptake can be measured, indicating that loss of phospholipid asymmetry generates a functional phagocytosis signal. As presented in TABLE 1, that PS provided the signal was shown by the ability of PS vesicles to inhibit phagocytosis by activated peritoneal macrophages.[10] That the treatment generated a PS signal recognizable by unactivated macrophages as well was shown by the ability of PS-exposing lipid-symmetric erythrocytes to inhibit the uptake by unactivated J774 macrophages.[16] Conclusively, pretreating target cells with annexin V inhibited phagocytosis by both activated and unactivated macrophages (ref. 16; FIG. 1).

The immediacy of PS exposure induced by Ca^{2+} and ionophore limits the time available for the development of other independent recognition signals or components which do not depend upon PS exposure to mediate recognition. Accordingly, it was of interest to ask whether the recognition of these ionophore-treated cells entailed the same macrophage receptors that participate in recognition of apoptotic cells. To examine whether the integrin recognition system was invoked, the ability of RGDS to inhibit the phagocytosis of ionophore-treated T cells by unactivated J774 macrophages was measured; the tetrapeptide did inhibit.[16] To examine the engagement of the lectin-like receptor system, the ability of GlcNAc to inhibit phagocytosis by activated J774 macrophages was measured; the sugar did inhibit.[16] Finally, the question of whether CD14 on activated or unactivated macrophages was engaged during recognition of ionophore-treated target cells was examined. As shown in FIGURE 1A, the monoclonal antibody 61D3 inhibited phagocytosis by unactivated macrophages; as shown in FIGURE 1B, the same was true for activated macrophages. These data indicate that protein receptors specific to recognition by either activated or unactivated macrophages, as well as a receptor common to both types of macrophages, are all engaged during phagocytosis of ionophore-treated target cells. Collectively, these results suggest either that PS itself is the ligand for these

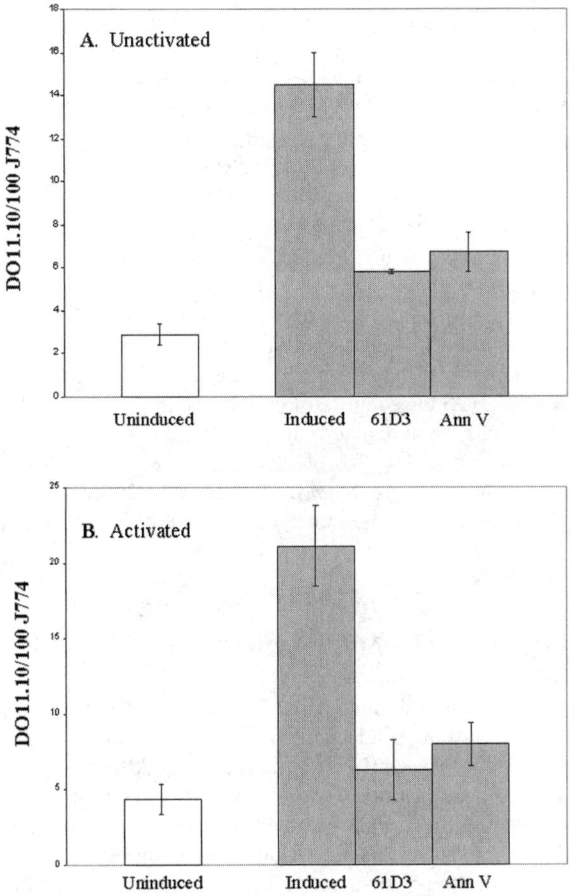

FIGURE 1. Phagocytosis of DO11.10 T cells induced to express PS by incubating with ionophore in the presence of Ca^{2+}. To induce PS expression, log phase DO11.10 cells were incubated for 10 min at 37°C with 1 mM $CaCl_2$ plus 10 mg/mL of the Ca^{2+} ionophore A23187 or with 1 mM $CaCl_2$ alone as a control.[16] Cells were then resuspended in buffer (10 mM HEPES/NaOH, pH 7.4, 140 mM NaCl, 2.5 mM $CaCl_2$) alone or buffer containing either 35 mg/ml of 61D3 monoclonal antibody or 0.1 mM annexin V, incubated for 15 min at room temperature, then presented to (**A**) unactivated J774 macrophages or (**B**) J774 macrophages activated by culturing in the presence of 100 μg/mL of β-glucan for at least 3 days.[16] Phagocytosis assays were performed as described.[13]

recognition systems, or that PS, or the loss of lipid asymmetry, reorganizes components on the surface of the target cells to generate a signal which engages these constitutively expressed receptors.

DUAL ROLE FOR PS: EXPRESSION ON MACROPHAGES

Prior to the development of annexin V as a probe for surface exposure of PS, the fluorescent lipophilic dye merocyanine 540 (MC540) was used as an indirect indicator of loss of plasma membrane asymmetry based on its ability to detect alterations in lipid packing that accompany loss of asymmetry. Curiously, both cultured primary macrophages and macrophage cell lines stained with MC540,[22] suggesting they did not maintain an asymmetric lipid distribution. Just recently, that conclusion was confirmed by the observation that these cells bind fluoresceinated annexin V.[16, 23] Given that masking PS on the target cell surface inhibits phagocytosis, the relevance of the PS exposed on the macrophages surface was similarly tested by pre-treating macrophages with annexin V. Blocking the PS on the phagocyte surface proved to be as efficient in inhibiting phagocytosis as blocking on the target cell surface.[16] This surprising result brings to mind the demonstration in *C. elegans* that expression of the *ced-7* gene is required in both target and phagocyte for clearance of apoptotic cells.[24] The mammalian homolog of the *ced-7* gene is ABC1, a member of the ATP-binding cassette (ABC) ATPase family of membrane transporters, whose expression is required in both targets and macrophages for phagocytosis.[25,26] Interestingly, inhibiting the activity of ABC1 also blocks induced exposure of PS in either cell type.[26] It is tempting to speculate that ABC1 is the scramblase or is a protein whose activity is required for scrambling and thus exposure of PS on both the target and macrophage surface.

LOSS OF ASYMMETRY VERSUS PS EXPOSURE

The inhibition by annexin V implies that PS itself is part of the signal on apoptotic cells recognized by macrophages. However, whether activation of the systems utilizing the integrin, lectin-like or CD14 receptors requires PS itself is less certain. Activation of the scramblase not only brings PS to the surface of apoptotic cells and macrophages, it also randomizes all the other phospholipids across the plasma membrane bilayer. In erythrocytes, this alteration in the composition of the outer leaflet produces alterations in the physical properties of the bilayer, including an increase in fluidity[27] and lateral mobility[28] of phospholipids, a decrease in ordering,[29] an alteration in lipid packing[30] and an increase in surface hydrophobicity;[7] any one of these might contribute to the generation of signals.

Leukocyte function-associated antigen-1 (LFA-1) is an integrin (CD11a (α1)/CD18 (β2)) expressed on the surface of lymphocytes. When lymphocytes are stimulated with phorbol ester, they aggregate in an LFA-1-dependent manner. However, the expression of neither LFA-1 nor ICAM-1, its ligand, increases, indicating that aggregation is not a simple consequence of induced expression of either of these surface molecules; rather, stimulation activates the molecules already present on the lymphocyte surface. Since stimulation potentiates adhesion of lymphocytes to immobilized ICAM-1 but not LFA-1,[31] activation appears to affect the LFA-1 side of the couple. To test whether changes in the lipid environment of LFA-1 might be responsible for its activation, alterations in lipid packing of stimulated lymphocytes were measured using MC540. Progression of cells into aggregates was paralleled by a decrease in lipid packing of the population as a whole, measured by an increase in MC540 staining. More importantly, cells from aggregates stained more intensely than nonaggregated cells, indicating that aggregates are preferentially formed from cells in the population in which the external phospholipids are more loosely packed.[32] The association of this alteration in the physical properties of the membrane with the functional activation of a cell adhesion molecule is consistent with the possibility that a change in the lipid environment might expose or create new recognition motifs in preexisting surface receptors.

Although PS is the only known novel ligand on the apoptotic lymphocyte surface, ICAM-3 (CD50), expressed on apoptotic and non-apoptotic lymphocytes alike, has been implicated in the phagocytosis of apoptotic lymphocytes by macrophages.[33] A member of the Ig-superfamily, ICAM-3 on apoptotic cells is altered such that the first Ig-like domain can participate in recognition. It is possible, as with LFA-1, that this alteration is induced by a change in the lipid milieu of the protein brought about by the disruption of phospholipid asymmetry. It would be interesting to know if ionophore-treated lymphocytes display this alteration in ICAM-3 recognition, or whether traversing the apoptotic program is required for its development.

It is worth noting that in nearly every instance inhibition by any agent or treatment is never 100% complete; nor are combinations thereof generally more effective at reducing the rate of phagocytosis of apoptotic cells to background levels. These results suggest that perhaps yet undiscovered mechanisms may also be operating. As mentioned above, loss of asymmetry by erythrocytes renders their surface more hydrophobic as determined by two-phase aqueous partitioning. Surface hydrophobicity has been suggested to play a dominant role in nonimmune phagocytosis of bacteria,[34] and this feature of the apoptotic cell surface could account for the residual phagocytosis seen in the presence of a variety of inhibitors. To our knowledge, the relative surface hydrophobicity of normal and apoptotic lymphocytes has not been examined. It is also interesting to note that in certain situations neighboring cells, rather than professional phagocytes, engulf and destroy apoptotic cells. This result implies that all cells must have a means of recognizing cells as apoptotic. If recognition is by a specific receptor, all cells must express that receptor. On the other hand, it is difficult to envision a receptor capable of sensing a change in hydrophobicity. Rather, a change in this global feature could perhaps result in an increase in hydrophobic interactions between neighbors, leading to the engulfment of the one by the other.

SUMMARY

All animal cells, regardless of origin or lineage, expose PS on their surface as part of the program of apoptotic cell death. Most evidence that apoptotic cells are recognized by macrophages by this exposed PS has come from blood cells. However, evidence that PS is a universal recognition signal is accumulating. The rapidity with which the recognition signal is generated once PS appears on the surface suggests that PS alone is the signal, or that PS combines with other molecules already present of the cell surface to generate functional recognition complexes, although alterations resulting from loss of phospholipid asymmetry rather than exposure of PS cannot be ruled out. The dual requirement for exposure of PS on target cells and on macrophages emphasizes the central role PS plays in apoptotic cell clearance.

REFERENCES

1. PURDON, A D. & S.I. RAPOPORT. 1998. Energy requirements for two aspects of phospholipid metabolism in mammalian brain. Biochem. J. **335:** 313–318.
2. WILLIAMSON, P. & R.A. SCHLEGEL. 1994. Back and forth: the regulation and function of transbilayer phospholipid movement in eukaryotic cells. Mol. Membr. Biol. **11:** 199–216.
3. SEIGNEURET, M. & P.F. DEVAUX. 1984. ATP-Dependent asymmetric distribution of spin-labeled phospholipids in the erythrocyte membrane: Relation to shape changes. Proc. Natl. Acad .Sci. USA **81:** 3751–3755.
4. TANG, X., M.S. HALLECK, R.A. SCHLEGEL & P. WILLIAMSON. 1996. A subfamily of P-type ATPases with aminophospholipid transporting activity. Science **272:** 1495–1497.
5. VAN DEN EIJNDE, S.M., L. BOSHART, E.H. BAEHRECKE, C.I. DEZEEUW, C.P.M. REUTELINGSPERGER & C. VERMEIJKEERS. 1998. Cell surface exposure of phosphatidylserine during apoptosis is phylogenetically conserved. Apoptosis **3:** 9–16.
6. WILLIAMSON, P., A. KULICK, A. ZACHOWSKI, R.A. SCHLEGEL & P.F. DEVAUX. 1992. Ca^{2+} induces transbilayer redistribution of all major phospholipids in human erythrocytes. Biochemistry **31:** 6355–6360.
7. MCEVOY, L., P. WILLIAMSON & R.A. SCHLEGEL. 1986. Membrane phospholipid asymmetry as a determinant of erythrocyte recognition by macrophages. Proc. Natl. Acad. Sci. USA **83:** 3311–3315.
8. PRADHAN, D., P. WILLIAMSON & R.A. SCHLEGEL. 1994. Phosphatidylserine vesicles inhibit phagocytosis of erythrocytes with a symmetric transbilayer distribution of phospholipids. Mol. Membr. Biol **11:** 181–187.
9. ALLEN, T.M., P. WILLIAMSON & R.A. SCHLEGEL. 1988. Phosphatidylserine as a determinant of reticuloendothelial recognition of liposome models of the erythrocyte surface. Proc. Natl. Acad. Sci. USA **85:** 8067–8071.
10. VERHOVEN, B., S. KRAHLING, R.A. SCHLEGEL & P. WILLIAMSON. 1999. Regulation of phosphatidylserine exposure and phagocytosis of apoptotic T lymphocytes. Cell Death. Differ. **6:** 262–270.
11. VERHOVEN, B., R.A. SCHLEGEL & P. WILLIAMSON. 1995. Mechanisms of phosphatidylserine exposure, a phagocyte recognition signal, on apoptotic T lymphocytes. J. Exp. Med. **182:** 1597–1601.
12. FADOK, V.A., D.R. VOELKER, P.A. CAMPBELL, J.J. COHEN, D.L. BRATTON & P.M. HENSON. 1992. Exposure of phosphatidylserine on the surface of apoptotic lymphocytes triggers specific recognition and removal by macrophages. J. Immunol. **148:** 2207–2216.
13. PRADHAN, D., S. KRAHLING, P. WILLIAMSON & R.A. SCHLEGEL. 1997. Multiple systems for recognition of apoptotic lymphocytes by macrophages. Mol. Biol. Cell. **8:** 767–778.
14. DUVALL, E., A.H. WYLLIE & R.G. MORRIS. 1985. Macrophage recognition of cells undergoing programmed cell death (apoptosis). Immunology **56:** 351–358.
15. FADOK, V.A., J.S. SAVILL, C. HASLETT, D.L. BRATTON, D.E. DOHERTY, P.A. CAMPBELL & P.M. HENSON. 1992. Different populations of macrophages use either the vitronectin receptor or the phosphatidylserine receptor to recognize and remove apoptotic cells. J. Immunol. **149:** 4029-4035.

16. CALLAHAN, M.K., P. WILLIAMSON & R.A. SCHLEGEL. 2000. Surface expression of phosphatidylserine on macrophages is required for phagocytosis of apoptotic thymocytes. Cell Death Differ. **7:** 645–653.
17. KRAHLING, S., M.K. CALLAHAN, P. WILLIAMSON & R.A. SCHLEGEL. 1999. Exposure of phosphatidylserine is a general feature in the phagocytosis of apoptotic lymphocytes by macrophages. Cell Death Differ. **6:** 183–189.
18. DEVITT, A., O.D. MOFFATT, C. RAYKUNDALIA, J.D. CAPRA, D.L. SIMMONS & C.D. GREGORY. 1998. Human CD14 mediates recognition and phagocytosis of apoptotic cells. Nature **392:** 505–509.
19. FLORA, P.K. & C.D. GREGORY. 1994. Recognition of apoptotic cells by human macrophages: Inhibition by a monocyte/macrophage-specific monoclonal antibody. Eur. J. Immunol. **24:** 2625–2632.
20. SCHLEGEL, R.A., S. KRAHLING, M.K. CALLAHAN & P. WILLIAMSON. 1999. CD14 is a component of multiple recognition systems used by macrophages to phagocytose apoptotic lymphocytes. Cell Death Differ. **6:** 583–592.
21. FADOK, V.A., M.L. WARNER, D.L. BRATTON & P.M. HENSON. 1998. CD36 is required for phagocytosis of apoptotic cells by human macrophages that use either a phosphatidylserine receptor or the vitronectin receptor (Alpha(v)Beta(3)). J. Immunol. **161:** 6250–6257.
22. SCHLEGEL, R.A. & P. WILLIAMSON. 1988. Role of phospholipid asymmetry in cellular membrane fusion. *In* Molecular Mechanisms of Membrane Fusion. S. Ohki, D. Doyle, T.D. Flanagan, S.E. Hui & E. Mayhew, Eds.: 289–301. Plenum. New York.
23. VAN DEN EIJNDE, S.M., J. LIPS, L. BOSHART, C. VERMEIJ-KEERS, E. MARANI, C.P. REUTELINGSPERGER & C.I. DE ZEEUW. 1999. Spatiotemporal distribution of dying neurons during early mouse development. Eur. J. Neurosci. **11:** 712–724.
24. WU, Y.C. & H.R. HORVITZ. 1998. The *C. elegans* cell corpse engulfment gene Ced-7 encodes a protein similar to ABC transporters. Cell **93:** 951–960.
25. LUCIANI, M.F. & G. CHIMINI. 1996. The ATP binding cassette transporter ABC1, is required for the engulfment of corpses generated by apoptotic cell death. EMBO J. **15:** 226–235.
26. MARGUET, D., M.F. LUCIANI, A. MOYNAULT, P. WILLIAMSON & G. CHIMINI. 1999. Engulfment of apoptotic cells involves the redistribution of membrane phosphatidylserine on phagocyte and prey. Nature Cell Biol. **1:** 454–456.
27. SEIGNEURET, M., A. ZACHOWSKI, A. HERMANN & P.F. DEVAUX. 1984. Asymmetric lipid fluidity in human erythrocyte membrane: new spin-label evidence. Biochemistry **23:** 4271–4275.
28. MORROT, G., S. CRIBIER, P.F. DEVAUX, D. GELDWERTH, J. DAVOUST, J.F. BUREAU, P. FELLMANN, P. HERVE & B. FRILLEY. 1986. Asymmetric lateral mobility of phospholipids in the human erythrocyte membrane. Proc. Natl. Acad. Sci. USA **83:** 6863–6867.
29. TANAKA, K.I. & S. OHNISHI. 1976. Heterogeneity in the fluidity of intact erythrocyte membrane and its homogenization upon hemolysis. Biochim. Biophys. Acta **426:** 218–231.
30. WILLIAMSON, P., J. BATEMAN, K. KOZARSKY, K. MATTOCKS, N. HERMANOWICZ, H.R. CHOE & R.A. SCHLEGEL. 1982. Involvement of spectrin in the maintenance of phase-state asymmetry in the erythrocyte membrane. Cell **30:** 725–733.
31. DUSTIN, M.L. & T.A. SPRINGER. 1989. T-cell receptor cross-linking transiently stimulates adhesiveness through LFA-1. Nature **341:** 619–624.
32. SMITH, D.M., P.L. WILLIAMSON & R.A. SCHLEGEL. 1993. Plasma membrane lipid packing and leukocyte function-associated antigen-1-dependent aggregation of lymphocytes. J. Cell. Physiol. **156:** 182–188.
33. MOFFATT, O.D., A. DEVITT, E.D. BELL, D.L. SIMMONS & C.D. GREGORY. 1999. Macrophage recognition of ICAM-3 on apoptotic leukocytes. J. Immunol. **162:** 6800–6810.
34. VAN OSS, C.J. 1978. Phagocytosis as a surface phenomenon. Annu. Rev. Microbiol. **32:** 19–39.

Opsonization of Apoptotic Cells

Implications for Uptake and Autoimmunity

DROR MEVORACH[a]

The Laboratory for Cellular and Molecular Immunology, Tel-Aviv Medical Center, Hadassah Medical Center and the Hebrew University, Jerusalem, Israel

ABSTRACT: As a part of innate immunity, soluble host proteins called opsonins, which include complement ligands and immunoglobulins, initially coat microorganisms that penetrate the mammalian sterile milieu. The main purpose of opsonization is to allow subsequent clearance of opsonized particles by specific receptors on the surface of leukocytes. Similarly, several proteins that may act as opsonins and have a role in uptake of apoptotic cells and bodies include thrombospondin I, the complement system, β2GPI, immunoglobulins, CRP, and some unidentified others. The surface changes that lead to opsonization include the appearance of phosphatidylserine that acts as an activator molecule for some known opsonins as the complement system and β2GPI. The consequence of altered opsonization is demonstrated by the development of autoimmunity in C1q deficient mice, and the pro-inflammatory response by macrophages ingesting apoptotic cell opsonized by an autoantibody.

KEYWORDS: Apoptosis; Opsonins; Innate immunity; β2GPI; Autoimmunity

INTRODUCTION

The mechanisms whereby apoptotic cells are identified, removed, and degraded by phagocytes in mammalian cells are not completely understood. However, studies in the nematode *C. elegans*, and in human and murine systems elucidated some of the mechanisms involved. Recent studies showed several receptors and molecules as being important in the uptake of apoptotic cells in mammalians.[1–13] Most of the *in vitro* studies of uptake done so far were done in the absence of serum, and in our recent work[11] we have noticed that serum may have an important effect on uptake, at least in some homologous and heterologous systems.

As a part of innate immunity, soluble host proteins called opsonins, which include complement ligands and immunoglobulins, initially coat microorganisms that penetrate the mammalian sterile milieu. The main purpose of opsonization is to allow subsequent clearance of opsonized particles by specific receptors on the surface of leukocytes. There is evidence that, similarly, proteins allowing more efficient uptake opsonize apoptotic cells and bodies. Several proteins that may act as opsonins and have a role in uptake of apoptotic cells and bodies include thrombospondin I,[14] the

[a]Address for correspondence: Dror Mevorach, M.D., The Laboratory for Cellular and Molecular Immunology, Department of Medicine, Hadassah Medical Center and the Hebrew University, P.O. Box 12000, Jerusalem 91120, Israel. Voice: 972-2-0776295; fax: 972-2-643-3935.
mevdm@netvision.net.il

complement system,[11,15,16] β2GPI,[17] immunoglobulins[18] and possibly serum amyloid P,[19] CRP, and others. In the present paper we look at β2GPI binding to apoptotic cells and the role of phosphatidylserine in complement activation. The consequence of lack of opsonization is further shown by the effect of intravenous load of apoptotic cells or apoptotic cells lysate on autoimmunity in C 1q deficient mice.

METHODS

Apoptotic Cells

Balb/c or C3H/SnJ mouse thymocytes were used as the source of apoptotic cells in all the experiments. Apoptosis of murine thymocytes was induced by serum deprivation. The optimal conditions for thymocyte apoptosis without necrosis were 2 hours of incubation at 37°C / 5%CO_2 in RPMI medium without serum. In these conditions, > 60% of the cells bound annexin-FITC (R & D) and > 93% and > 98% of them, excluded propidium iodide and trypan blue, respectively. Apoptosis was induced in some experiments by irradiation or dexamethasone as described elsewhere.[18]

Liposome Preparations

Liposomes were purified as previously described[3] with slight modifications. Briefly, pure phospholipids (Avanti Polar Lipids, Pelham, AL), phosphatidylcholine, and phosphatidylserine were suspended in chloroform/methanol (90/10) and added to disposable glass tubes in mixtures of either 100% phosphatidylcholine, or 30/70 phosphatidylserine /phosphatidylcholine. The solvent was evaporated under N_2 gas, and lipids were resuspended in PBS by vortexing. The mixture was then sonicated for 5 minutes, and the total lipids contents were determined by lipid chromatography.

Purification of β2-glycoprotein-1 (β2GPI)

1.25 ml of 70% perchloric acid was added to 50 ml human AB serum (GIMINI) at 4°C, mixed 15 minutes, and centrifuged at 5000 × *g*. The supernatant was neutralized with base and dialyzed against 20 mM Tris-HCL, O.O2% NaN_3, pH 8.0, and passed over a heparin-Sepharose affinity column (Pharmacia). Absorbed proteins were eluted stepwise with 20 mM Tris-HCL, O.O2% NaN_3, pH 8.0, containing increasing molarity of NaCl. Fractions between 0.25 to 0.35 M were pooled, concentrated, and shown to have single band of 55 kD on Coomasie SDS-PAGE, which was detected specifically by anti-human β2GPI on a western blot.

Heterologous Phagocytosis Assay

Interaction between human macrophages and apoptotic cells was performed as described elsewhere.[11] Briefly, human macrophages were isolated from peripheral blood monocytes of normal donors and cultured on Chamber-Tek glass slides (Nunc, IL). Six to seven days post-isolation, the macrophages were washed three times with serum-free Iscove's medium after which 2 × 10^6 apoptotic cells were offered to 25,000 macrophages (approximate ratio of 80:1) and incubated for 1 hour at 37°C/5%CO_2. The interaction was terminated by washing with ice-cold PBS. The slides were fixed with methanol and Wright stained, and the numbers of both surface bound and internalized apoptotic cells were scored by light microscopy. Two observers in each assay counted at least 300 macrophages. The phagocytosis index (P.I., the number of apoptotic cells bound to or internalized by 100 macrophages) was calculated.

Inhibition Studies

Prepared micelles and annexin V (Sigma) were added to the media containing the apoptotic cells prior to the interaction with the macrophages. For each receptor blocking experiment, monoclonal antibodies were used at a final concentration of 20 µg/ml, and were added to the macrophages 15 minutes before the macrophages were exposed to apoptotic thymocytes. Monoclonal antibodies were obtained from the following sources: Anti-CR1, 3D9, (R.P. Taylor, VA), anti-CR3 (I domain), MN-41, (V. Vetvicka, KY), anti-CR4 (anti-CD11c, Sigma), anti-CD36, (FA6; R.L. Silverstein, NY, IgM; Sigma), anti-CD14, 63D3 (ATCC) and 61D3 (C.D. Gregory, UK). Each experiment was done in triplicate on at least four separate occasions.

Flow Cytometry

Apoptotic cells were stained with FITC-conjugated annexin V, propidium iodide, and PE-conjugated IgG_{2b} mouse monoclonal anti-human-iC3b (Quidel, CA). Flow cytometry analysis was performed on a FACScan (Becton-Dickinson). Mouse IgG_{2b}k was used as isotype control for anti-C3bi.

Immunization Protocol

C1q deficient mice were kindly provided by P.P. Pandolfi, Sloan-Kettering Institute, New York. Thymocytes were prepared from 6–8 week-old mice as described,[18] and 1×10^7 syngeneic cells injected i.v. into each mouse recipient. The injections were performed every week for a total of four injections. Some mice received apoptotic lysate following lysis by three freeze/thaw cycles or necrotic cells. Necrosis was induced by 20 min at 56°C. Total immunoglobulins, anticardiolipin, and anti-ss DNA were measured at 2 week intervals.

Statistics

The Student's *t*-test was used to compare the results.

RESULTS

Complement Components Are Required for Efficient Uptake of Apoptotic Mice Thymocytes by Human Macrophages

Since several receptors were reported to have a role in uptake of human apoptotic cells,[2,4,5,7,9–13] and in order to increase specificity to complement activation, we used a heterologous system. Heat inactivation of serum decreased the uptake of apo-

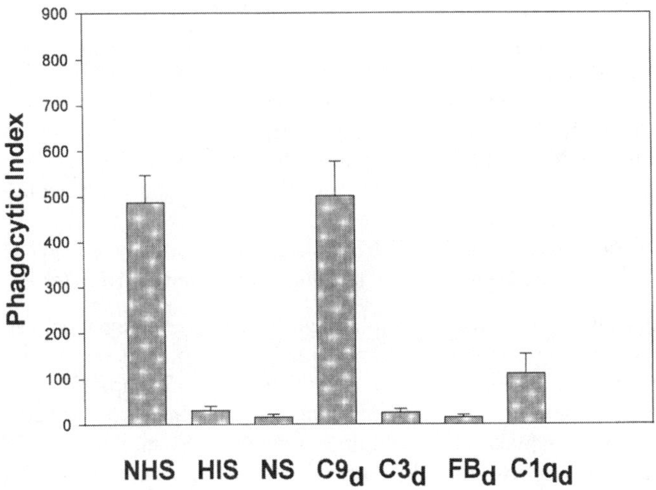

FIGURE 1. Complement needs to be activated for uptake of apoptotic thymocytes by human macrophages. Phagocytosis Index of monocytes-derived human macrophages following one-hour interaction with murine apoptotic thymocytes. Phagocytosis Index is shown in the presence of normal human AB serum (NHS), heat inactivated serum (HIS), in the absence of human serum (NS), and in serum deficient with complement factors, C9 ($C9_d$), C3 ($C3_d$), factor B (FB_d), and C1q ($C1q_d$). The results of three different experiments are shown.

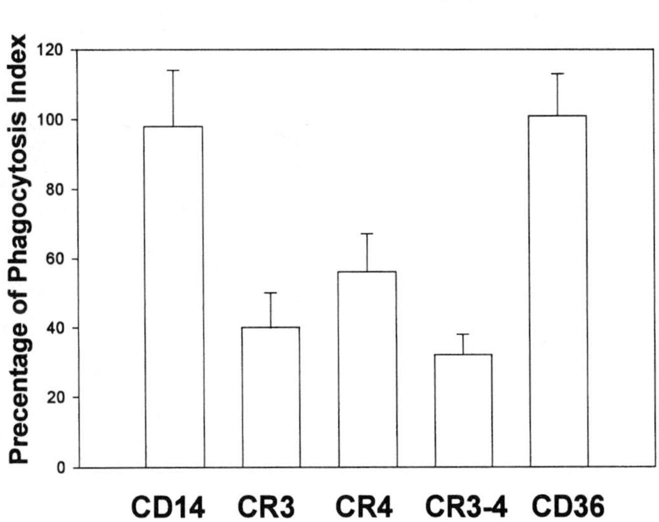

FIGURE 2. Complement receptors mediate binding of apoptotic thymocytes to human monocytes-derived macrophages. Adding monoclonal antibodies to the complement receptors, CR3, and CR4, or both (CR3-4), inhibited uptake. Adding monoclonal antibodies to CD14 (61D3) and CD36 did not.

ptotic cells by over 10-fold and serum deficient in C3, C1q, or B proteins but not deficient in C9 markedly inhibited the P.I. (FIG. 1). Add-back experiments restored uptake (not shown). Flow-cytometry analysis showed more than 90% deposition of C3bi on apoptotic thymocytes (PS-FITC positive, data not shown and ref. 11). Viable thymocytes (PS-FITC negative) showed no deposition of C3bi. Inhibition of CR3 and CR4 reduced the binding of apoptotic cells by 52% and 45%, respectively. Inhibition of receptors such as CD14 (61D3) and CD36 did not decrease uptake (FIG. 2).

C3bi Deposition on Apoptotic Cells Is Decreased if Apoptotic Cells Are Previously Exposed To Annexin-V or Phosphatidylserine Micelles

In order to verify the role of phosphatidylserine in complement activation, apoptotic cells were exposed to of 2mM annexin V for 15 minutes and then exposed to serum. There was a 53% inhibition of C3bi deposition on apoptotic in the presence of annexin V, but no inhibition in the presence of albumin at the same concentration (data not shown). Uptake of apoptotic thymocytes was inhibited by 78% by PS micelles, but not by phosphatidylcholine micelles (FIG. 3).

β2GPI Binds to Apoptotic Thymocytes and Increases Uptake of Apoptotic Thymocytes in the Absence of Serum

Following exposure of apoptotic thymocytes to human serum, PE-conjugated-β2GPI was seen on apoptotic thymocytes but not on viable thymocytes (not shown). Phagocytosis assay was done in the absence of serum and in the presence of

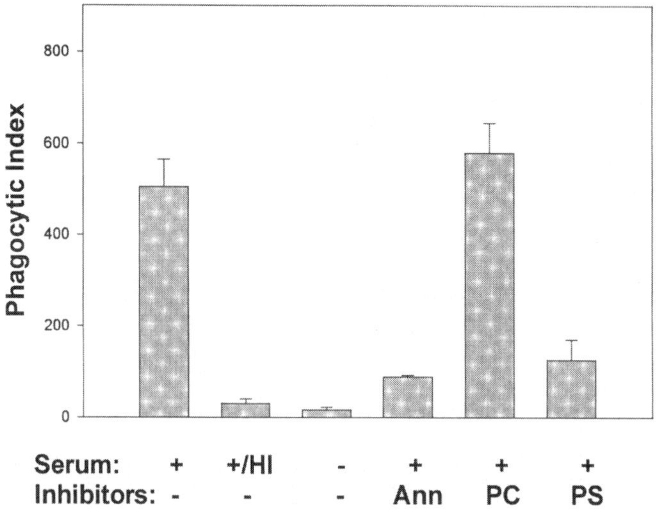

FIGURE 3. Inhibition of uptake of apoptotic thymocytes by heat-inactivated (HI) serum, annexin V (Ann), and phosphatidylserine micelles (PS). Adding Ann or PS, prior to the interaction between apoptotic thymocytes and human macrophages inhibited uptake. Adding phosphatidylcholine micelles (PC) did not.

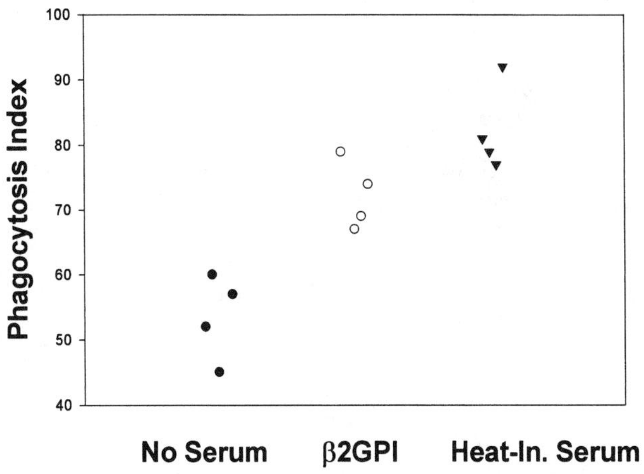

FIGURE 4. β2GPI increases uptake of apoptotic thymocytes by human monocytes-derived macrophages. In the absence of serum, an increase in uptake of apoptotic thymocytes is seen ($p < 0.005$) following 15 minutes incubation with 200 μg/ml purified (see METHODS) human β2GPI. Four experiments comparing assays without serum, addition of β2GPI (β2GPI), or heat inactivated serum (Heat-In. Serum) are shown.

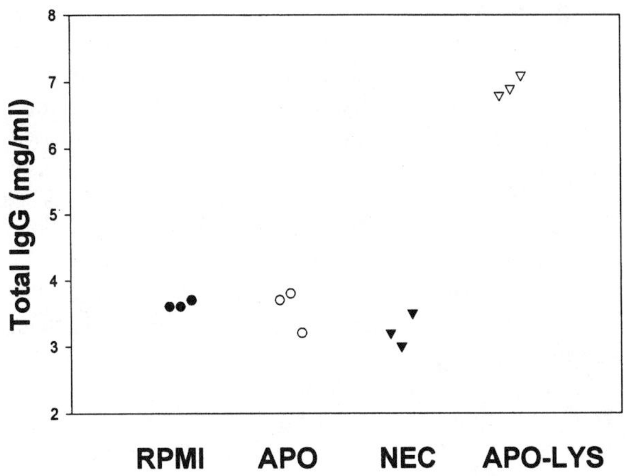

FIGURE 5. Increased total immunoglobulins levels following i.v. injection of apoptotic lysate. Total IgG levels of 4-month-old mice deficient in C1q that were injected with RPMI (RPMI) is compared to mice injected with 10^7 syngeneic apoptotic cells (APO), necrotic cells (NEC), or apoptotic cells-lysate (APO-LYS).

200 μg/ml β2GPI (FIG. 4). Phagocytosis index slightly increased from 53.5 ± 6.5 to 72.2 ± 5.5 ($p < 0.005$) when β2GPI was added. Heat inactivated serum had an additional serum effect on phagocytosis index ($p < 0.06$).

Injection of Syngeneic Apoptotic Material Increases Total Immunoglobulin Levels in C1q Mice

As shown in FIGURE 5, when total IgG was measured as an indication for the development of autoimmunity, no significant increase was seen upon i.v. injection of 10^7 apoptotic syngeneic thymocytes, four consecutive times. Nor was an increase in the levels of immunoglobulins seen when 10^7 necrotic cells were injected. However, when following apoptosis the cells were lysed and the lysate injected, a dramatic increase in total IgG was seen.

DISCUSSION

Macrophages are capable of secreting opsonins such as thrombospondin I[14] and complement.[20] However in the presence of serum, apoptotic cells may be opsonized more rapidly and more efficiently. Opsonization not only may facilitate uptake,[11,17] but may also direct pro- or anti-inflammatory opsonization.[21] Opsonization of apoptotic cells may provide a novel rationale for the acute phase response and proteins such as C reactive protein and serum amyloid P may take an important role in clearance of apoptotic cells and apoptotic debris.[19] Complement activation on the surface of apoptotic cells was shown to occur in a variety of cells: human T-lymphocytes, human cell lines, BL41 and Jurkat,[11,22] human umbilical vein endothelial cells,[23] and human lung cancer[16] as well as in murine thymocytes.[11] β2GPI was shown both in the present study and in a previous study[17] to facilitate uptake of apoptotic cells. Interestingly, β2GPI has some structure similarities to the complement proteins and contains the typical short consensus repeats, but no functional interactions are known so far. Previously, we were also able to show that immunoglobulins, most probably natural autoantibodies, participate in clearance of I.V. load of syngeneic apoptotic cells.[18] Transient rise in anticardiolipin and anti-ssDNA was seen along with mesangial deposits of apoptotic cells-IgG or IgM-and C3, indicating a role in *in vivo* clearance for immunoglobulins and complement. Other serum proteins are probably involved[24] and as summarized in FIGURE 6, different proteins may take part in different situations as specific opsonins facilitating clearance of apoptotic cells. Different situations such as natural turnover, inflammation, and a disease state could each determine a major pattern.

A question that can arise is what on the surface of an apoptotic cell can cause opsonization. In the example of complement, down-regulation of complement regulatory molecules CD55, CD46 (MCP), and CD35 (CR1) had been shown in human apoptotic neutrophils,[25] but enzymatic removal of CD55 and inhibition by CD46 by monoclonal antibody did not induce activation of the complement,[23] suggesting that the most important factor is the expression of an activator molecule that overcomes the inhibition input by CD55 and CD46. We were able to show that phosphatidylserine exposure on apoptotic cells is associated with the appearance of complement breakdown products. Phosphatidylserine or phosphatidylsrerine-β2GPI

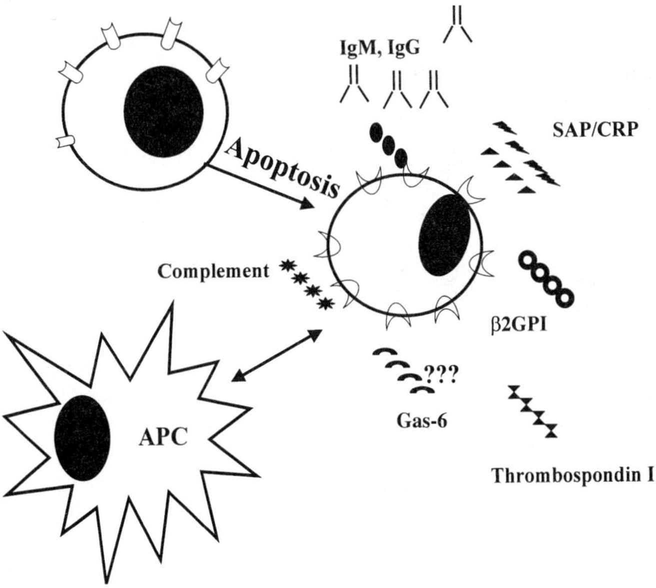

FIGURE 6. Opsonization of apoptotic cells. Following apoptosis surface changes such as the appearance of phosphatidylserine leads to binding of different proteins that may determine phagocytosis pattern and anti- or pro-inflammatory presentation by antigen presenting cells.

complex is likely to mediate also the binding of anticardiolipin. It is not known yet if this is the case for additional opsonins.

The importance of opsonization is well illustrated in the case of systemic lupus erythematosus (SLE). Handling of apoptotic cells was suggested to have an important role in the development of autoimmunity.[26, 27] Non-inflammatory presentation of ingredients of apoptotic cells by antigen presenting cells keeps homeostasis and prevents autoimmune reaction. However, if phagocytosis is inefficient secondary necrosis may occur. As shown recently,[28] phagocytosis of necrotic cells by dendritic cells causes maturation with expression of CD83 and costimulatory markers CD40 and CD80. Otherwise, there is phagocytosis of apoptotic cells that do not mature to dendtritic cells in the absence of conditioned media. Our demonstration and that of others that apoptotic cells activate complement may partially explain the high occurrence of SLE among patients deficient in the classical components of the complement system. In addition, pathological autoantibodies like anticardiolipin were shown to deviate the immune response from a non-inflammatory response to pro-inflammatory response demonstrated by TNF-α secretion. Thus, as illustrated in FIGURE 6, opsonization may determine rapid and efficient uptake as well as immune deviation to a non-inflammatory response. Further studies will demonstrate the specific role of each opsonin as well as novel molecules that may take part at the crosstalk between the phagocytes and the apoptotic cells. Interestingly, apoptotic cells lysate rather than apoptotic cells load aggravated autoimmunity in C1q deficient mice showing that, even in the absence of C1q, some compensation is possible as long as the cell stays apoptotic and does not undergo secondary necrosis.

In summary, opsonization of apoptotic cells may have a critical role in directing different patterns of clearance of apoptotic cells. Initial implications on systemic autoimmunity were already demonstrated and other physiological and disease states are studied.

ACKNOWLEDGMENTS

This research was partially funded by the Chief Scientist of Israel (D.M.), the Israel Cancer Research Funding (ICRF) (D.M.), and the Arthritis Foundation, New York Chapter (D.M.). I. Frolkis (The Lab for Cellular and Molecular Immunology, Israel) helped in most of the experiments. P.P. Pandolfi, (Sloan-Kettering Institute, New York), provided the C1q deficient mice. The experiments with the C1q deficient mice were performed at the KB Elkon Lab (Cornell, NY). Rick Lockshin kindly helped in editing the manuscript.

REFERENCES

1. Duvall, E., A.H. Wyllie & R.G. Morris. 1985. Macrophage recognition of cells undergoing programmed cell death (apoptosis). Immunology. **56**(2)**:** 351–358.
2. Savill, J. *et al.* 1990. Vitronectin receptor-mediated phagocytosis of cells undergoing apoptosis. Nature **343**(6254)**:** 170–173.
3. Fadok, V.A. *et al.* 1992. Exposure of phosphatidylserine on the surface of apoptotic lymphocytes triggers specific recognition and removal by macrophages. J. Immunol. **148**(7)**:** 2207–2216.
4. Ren, Y. *et al.* 1995. CD36 gene transfer confers capacity for phagocytosis of cells undergoing apoptosis. J. Exp. Med. **181**(5)**:** 1857–1862.
5. Sambrano, G.R. & D. Steinberg. 1995. Recognition of oxidatively damaged and apoptotic cells by an oxidized low density lipoprotein receptor on mouse peritoneal macrophages: role of membrane phosphatidylserine. Proc. Natl. Acad. Sci. USA **92**(5)**:** 1396–1400.
6. Fukasawa, M. *et al.* 1996. SRB1, a class B scavenger receptor, recognizes both negatively charged liposomes and apoptotic cells. Exp. Cell Res. **222**(1)**:** 246–250.
7. Devitt, A. *et al.* 1998. Human CD14 mediates recognition and phagocytosis of apoptotic cells [see comments]. Nature **392**(6675)**:** 505–509.
8. Platt, N. *et al.* 1996. Role for the class A macrophage scavenger receptor in the phagocytosis of apoptotic thymocytes in vitro. Proc. Natl. Acad. Sci. USA **93**(22)**:** 12456–12460.
9. Moynault, A., M.F. Luciani & G. Chimini. 1998. ABC1, the mammalian homologue of the engulfment gene ced-7, is required during phagocytosis of both necrotic and apoptotic cells. Biochem. Soc. Trans. **26**(4)**:** 629–635.
10. Albert, M.L. *et al.* 1998. Immature dendritic cells phagocytose apoptotic cells via alphavbeta5 and CD36, and cross-present antigens to cytotoxic T lymphocytes. J. Exp. Med. **188**(7)**:** 1359–1368.
11. Mevorach, D. *et al.* Complement-dependent clearance of apoptotic cells by human macrophages. J. Exp. Med. **188**(12)**:** 2313–2320.
12. Schwartz, B.R. *et al.* 1999. A novel beta 1 integrin-dependent mechanism of leukocyte adherence to apoptotic cells. J. Immunol. **162**(8).
13. Fadok, V.A. *et al.* 2000. A receptor for phosphatidylserine-specific clearance of apoptotic cells. Nature **405**(6782)**:** 85–90.

14. SAVILL, J. *et al.* 1992. Thrombospondin cooperates with CD36 and the vitronectin receptor in macrophage recognition of neutrophils undergoing apoptosis. J. Clin. Invest. **90**(4)**:** 1513–1522.
15. TAKIZAWA, F., S. TSUJI & S. NAGASAWA. 1996. Enhancement of macrophage phagocytosis upon iC3b deposition on apoptotic cells. FEBS Lett. **397**(2–3)**:** 269–272.
16. HARA, T. *et al.* 1996. Homologous complement activation on drug-induced apoptotic cells from a human lung adenocarcinoma cell line. Immunobiology **196**(5)**:** 491–503.
17. BALASUBRAMANIAN, K. & A.J. SCHROIT. 1998. Characterization of phosphatidylserine-dependent beta2-glycoprotein I macrophage interactions. Implications for apoptotic cell clearance by phagocytes. J. Biol. Chem. **273**(44)**:** 29272–29277.
18. MEVORACH, D. *et al.* 1998. Systemic exposure to irradiated apoptotic cells induces autoantibody production. J. Exp. Med. **188**(2)**:** 387–392.
19. PAUL, E. & M.C. CARROLL. 1999. SAP-less chromatin triggers systemic lupus erythematosus. Nat. Med. **5**(6)**:** 607–608.
20. WHALEY, K. 1980. Biosynthesis of the complement components and the regulatory proteins of the alternative complement pathway by human peripheral blood monocytes. J. Exp. Med. **151**(3)**:** 501–516.
21. MANFREDI, A.A. *et al.* 1998. Apoptotic cell clearance in systemic lupus erythematosus. I. Opsonization by antiphospholipid antibodies. Arthritis Rheum. **41**(2)**:** 205–214.
22. MATSUI, H. *et al.* 1994. Activation of the alternative pathway of complement by apoptotic Jurkat cells. FEBS Lett. **351**(3)**:** 419–422.
23. TSUJI, S., K. KAJI & S. NAGASAWA. 1994. Activation of the alternative pathway of human complement by apoptotic human umbilical vein endothelial cells. J. Biochem. (Tokyo) **116**(4)**:** 794-800.
24. LICHT, R. *et al.* 1999. An assay for the quantitative measurement of in vitro phagocytosis of early apoptotic thymocytes by murine resident peritoneal macrophages. J. Immunol. Methods **223**(2): 237–248.
25. JONES, J. & B.P. MORGAN. 1995. Apoptosis is associated with reduced expression of complement regulatory molecules, adhesion molecules and other receptors on polymorphonuclear leucocytes: functional relevance and role in inflammation. Immunology **86**(4)**:** 651–660.
26. MEVORACH, D. 1999. The immune response to apoptotic cells. Ann. N.Y. Acad. Sci. **887:** 191–198.
27. STEINMAN, R.M. *et al.* 2000. The induction of tolerance by dendritic cells that have captured apoptotic cells. J Exp Med. **191**(3): 411-6.
28. SAUTER, B. *et al.* 2000. Consequences of cell death: exposure to necrotic tumor cells, but not primary tissue cells or apoptotic cells, induces the maturation of immunostimulatory dendritic cells. J. Exp. Med. **191**(3)**:** 423–434.

Index of Contributors